UNDERGROUNDS IN INSURGENT, REVOLUTIONARY, AND RESISTANCE WARFARE

Primary Research Responsibility
Andrew R. Molnar

With Research Collaboration of
William A. Lybrand
Lorna Hahn
James L. Kirkman
Peter B. Riddleberger

I0095711

SPECIAL OPERATIONS RESEARCH OFFICE
The American University
Washington, D.C.   20016

November 1963

Reproduction in whole or in part is permitted for any purpose of the United States Government.

Social Science Research on military problems is performed in support of requirements stated by the Department of the Army staff agencies and other Army elements. The research is accomplished at The American University by the Special Operations Research Office, a nongovernmental agency operating under contract with the Department of the Army.

The contents of SORO publications, including any conclusions or recommendations, represent the views of SORO and should not be considered as having official Department of the Army approval, either expressed or implied.

Comments are invited and should be addressed to—
Deputy Chief of Staff for Military Operations/DA
Washington, D.C. 20310
ATTN: OPS SW

TASK UNDERGROUNDS

Research Completed June 1963

Published by Conflict Research Group.

First published by Special Operations Research Office in 1963

ISBN: 978-1-925907-27-8

CONFLICT
RESEARCH
GROUP

PREFACE

The task of one who attempts to write about the topic of undergrounds is formidable. The reasons are immediately apparent. By definition, undergrounds are organized and function in such a way as to minimize knowledge about them—clandestinely and with extreme "security" precautions. Few members know one another, first hand observation and data recording are usually impossible, and few written records are kept (those that are kept frequently are destroyed or become highly classified if the insurgents are successful).

At the time of the preparation of this report, there was no single unclassified document which was focused on underground organizations. There were bits and pieces of information about undergrounds scattered around in many open sources, but a compendium which would synthesize available information into a format useful for a variety of purposes was clearly needed.

For the military user, this report is designed to complement the information presented about undergrounds in existing field manuals. As such, it should be most useful in those military schools offering courses related to the counterinsurgency mission. The report also contains much helpful background information for the formulation of counterinsurgency policy and doctrine.

To avoid misunderstanding, a few words concerning format and style are in order. The report tries to be comprehensively descriptive; in so doing, the information presented ranges from material that can be considered simple, obvious statements of fact to intricate abstractions. Some elements of the report, then, demand the indulgence of the neophyte, while other elements may try the patience of the knowledgeable person. Organization of the report into two parts, and into relatively autonomous chapters within each part, should facilitate selective use of the material presented—and such was its intent. For ease of reading, numbered footnotes appear at end of each chapter. Inevitably there is some redundancy of content for the reader who goes through the report from cover to cover.

With the exception of the last chapter in Part I on government countermeasures, the report reflects an insurgent's viewpoint of undergrounds rather than a counterinsurgent's. By this is meant only that much of the material is presented in terms of the effectiveness of underground operations—and not how to effectively counter them. Design of effective countermeasures depends on first understanding undergrounds.

Finally, treating the information from the viewpoint of the effectiveness of undergrounds has resulted in a prescriptive writing style in some sections of

Part I of the report. Statements phrased in terms such as "must," "should," and "are," imply a greater degree of certitude or a more firm doctrinaire position than the evidence may warrant. On the one hand, this style had the virtue of clarifying propositions and principles. On the other hand, the reader is cautioned not to accept all such statements uncritically, but to view them primarily as hypotheses based on limited evidence.

SORO wishes to express its thanks to a number of consultants whose expertise and advice were invaluable. Mr. Slavko Bjelajac, of the Assistant Deputy Chief of Staff for Military Operations (Special Operations), Director of Special Warfare's Office, provided invaluable guidance throughout the project and special assistance on the case study of Yugoslavia underground movements. Three men reviewed the entire report, and offered specific suggestions in their areas of special competence: Dr. Paul Linebarger, of the School of Advanced International Studies at Johns Hopkins University, who provided useful insights and ideas for carrying out such a project; Dr. George K. Tanham, of the Rand Corporation, who was particularly helpful in clarifying the portions dealing with revolutionary warfare; and Dr. Jan Karski, of Georgetown University, whose personal experiences as an underground worker and knowledge of Soviet methods contributed immeasurably to the sections on Communism.

The descriptions in Part II of the report of the undergrounds in seven countries benefited greatly from the painstaking critique of several area specialists: Dr. Marcell Vigneras, of the Research and Analysis Corporation, who reviewed the study of the French resistance; Mr. David Martin, Administrative Assistant to Senator Thomas Dodd, who reviewed the Yugoslavia study; Dr. Frank Trager, Professor of Politics at the National War College, whose intimate knowledge of Southeast Asia contributed greatly to the Malayan and Philippine studies; Dr. Hans Kohn, Professor Emeritus of the City College of New York, who reviewed the Algerian summary; Dr. William McNeill, of the University of Chicago, whose help was invaluable on the Greek underground; and, Dr. J. C. Hurewitz of Columbia University, who reviewed the Palestine study.

Although the above named persons contributed much of value to this report, final responsibility for its content rests solely with the Special Operations Research Office.

TABLE OF CONTENTS

Table of Contents

Table of Contents

Table of Contents

LIST OF ILLUSTRATIONS

LIST OF CHARTS AND MAPS

SUMMARY

PURPOSE OF STUDY

BACKGROUND

SYNOPSIS

METHODOLOGICAL NOTES

PURPOSE OF STUDY

The objective of this study is to develop a comprehensive introduction to the subject of undergrounds. It is designed to bring together existing information about undergrounds through—
    (1) A generalized description of
        —the strategic roles of undergrounds in insurgency, revolutionary,* and resistance warfare
        —the administrative and operational missions performed by undergrounds for revolutionary and resistance movements
        —typical techniques utilized by undergrounds in accomplishing their missions
        —Communist use of undergrounds
        —countermeasures utilized by incumbent governments to suppress or eliminate undergrounds
    (2) Historical illustrations of the activities of undergrounds in both revolutionary and resistance movements.

BACKGROUND

The overwhelming destructive potential of nuclear weapons has focused attention on internal (revolutionary) wars in the developing nations and on the relationship of such wars to U.S. interests and security. Undergrounds are the clandestine elements of indigenous politico-military revolutionary organizations which attempt to illegally weaken, modify, or replace the governing authority, typically through the use or threat of force. Such movements may or may not be supported by external powers attempting to subvert the established government. Neither are undergrounds uniquely a revolutionary phenomenon. When the politico-military movement is directed against an external occupying power, usually in times of "hot" international conflict such as World War II, undergrounds are the clandestine elements of such resistance efforts. Finally, of course, neither revolutionary nor resistance movements are necessarily limited to the developing nations, although it is the revolutionary ferment in these nations that commands attention on the international scene at the present time.

Whatever the strategic politico-military context, it is vital to U.S. defense efforts, particularly those of the U.S. Army, to have as complete an understand-

*For purposes of this study, the terms insurgency and revolution are considered synonymous and are hereafter used interchangeably.

ing as possible about the nature of undergrounds—their origins, membership, organization, missions, strategies, methods of action, and relationships to other elements of the total revolutionary movement, such as guerrilla units. For many situations, such knowledge is needed to assist established governments in suppressing or eliminating revolutionary movements. Other situations are conceivable in which it may be in the U.S. interest to assist revolutionary undergrounds, as it assisted a number of resistance movements during World War II.

SYNOPSIS

The first chapter of Part I is summarized as an inventory of 14 selected statements of fact, principles, or propositions about the strategic role and importance of undergrounds in resistance and revolutionary warfare. This style was chosen because this chapter has broader relevance to U.S. defense policy and planning than the other chapters. The remaining chapters in Part I are in the form of brief narratives, made up of key statements, designed to highlight aspects of the subject rather than to summarize it.

Part II, the seven historical illustrations, is treated in yet another way. After an initial résumé of important characteristics of each, a series of qualitative and quantitative comparisons based on the seven undergrounds are made. The last part of the summary presents a chart on the numbers of underground members, guerrilla members, total population, and security forces involved in the seven undergrounds reviewed, with interpretative comments.

PART I

Chapter 1: The Role of Undergrounds

1. Undergrounds have been the base of resistance and revolutionary movements throughout recorded history.
2. Clandestine organizations, such as undergrounds, are not the particular product of—
   a. political or religious ideology;
   b. cultural, ethnic, national, or geographical grouping of persons;
   c. structure or form of government;
   d. segment of society or social class;
   e. stage of a society's economic or technological development.
3. Although undergrounds assume different forms, all are characterized by the following:
   a. their *goals* are illegal in terms of the *de facto* governmental system against which they act;
   b. their *activities* are generally both legal and illegal;
   c. their *members* usually play legal roles within the society, with their underground membership concealed.

4

4. Most revolutionary movements (excluding coup d'etat, a revolution "from the top") go through five distinct phases of evolutionary development:
   a. clandestine organization phase;
   b. psychological offensive phase;
   c. expansion phase;
   d. militarization phase;
   e. consolidation phase.

5. Undergrounds typically perform the following functions for a revolutionary movement up to the militarization phase:
   a. organize the revolutionary movement;
   b. control and coordinate all revolutionary activities;
   c. provide internal administrative functions for the revolutionary organization, such as recruitment, training, indoctrination, finances, logistics, communications, security;
   d. undertake subversion of the existing government's personnel and institutions;
   e. conduct psychological operations, both propaganda and actions, among the people and in important foreign nations;
   f. establish "shadow" governments which are to assume power if the revolutionaries win;
   g. collect and disseminate intelligence information for the revolutionary movement;
   h. carry out sabotage;
   i. set up escape and evasion networks for members of the movement.

6. After the militarization phase, undergrounds perform the same functions, proceeding in "nonliberated" areas with the additional critical role of supporting overt guerrilla units of the revolutionary movement by all means—intelligence, recruits, food, clothing, weapons, ammunition, medicines, and other supplies.

7. Undergrounds are as important as guerrilla units, if not more so, to the success or failure of a revolutionary movement.

8. The strength of undergrounds lies in—
   a. the lack of geographical restrictions on their operations;
   b. the varied nature of their membership;
   c. their clandestine nature.

9. Characteristically, guerrilla units emerge only after undergrounds have prepared the way and are almost wholly dependent for their support on an active underground.

10. The differences between undergrounds in resistance and revolutionary movements are their strategic politico-military goals, and not the functions.

11. Resistance movements, and their associated undergrounds, may be important strategically in "hot" international conflict (e.g., affecting length of campaign or nature of a postwar political situation), but

they have not been decisive; undergrounds can properly be considered to be of secondary or ancillary importance in such conflicts.

12. In revolutionary warfare, there is no other, more decisive, military struggle (as in "hot war") on which the outcome of the internal war depends; undergrounds are of primary and vital importance to "cold war" conflict outcomes when such revolutions are tacitly accepted as a major battleground of that conflict.

13. The loyalty of the people is the primary target in revolutionary wars; this political objective is sought by undergrounds through a combination of political and military activities.

14. As a part of a government's counterinsurgency effort, internal security forces must—

   a. protect the people and the government's instruments of political control and influence from the coercive and destructive actions of the undergrounds;

   b. perform a substantial role in influencing the people to support the existing government's claim to legitimacy as a governing authority;

   c. be prepared to perform noncoercive, quasi-military activities, as well as military activities, all with the same ultimate political objective—the loyalty of the people to the governing authority.

Chapter 2: Underground Administrative Techniques

Like other human organizations, undergrounds require essential "housekeeping" administrative functions. Operating in hostile environments, they must balance the need for expansion and aggressive action with the need to limit membership to trustworthy individuals. For this and other reasons, an underground's *organizational structure* will vary through time. Undergrounds generally seek to have agents in every geographic area and among all social and ethnic groups within the country. Although command and responsibility are centralized within the organization, each unit usually operates autonomously under mission-type orders from higher commands.

Undergrounds utilize a number of standard techniques to ensure their survival and growth. For example, they organize on the "fail-safe" principle wherein if one element is compromised the entire system will not be incapacitated. *Parallel units* for every type of activity are maintained. At the base of the organization, workers are organized into *cells* and their contacts limited to members of their own cell. This prevents a captured member from implicating other than fellow cell members. For the same reason, a leader in the chain of command has contact only with his immediate superior and immediate subordinates.

Communications within the underground are subject to the strictest security precautions. Parallel nets are used for important messages, verification of message receipt is usually required, and back-up messages are frequently used. Couriers generally carry compromising documents. Maildrops (places where messages may be left by one courier and picked up by another) are used by the couriers to eliminate the need for them to know each other's address.

Recruitment varies with the stage of underground development. In the early stage, it is highly selective; only individuals of known reliability are accepted. Later, however, more emphasis is placed upon obtaining the support of the general populace. The insurgents move into communities and rural areas and assist the people in any way possible. Once a feeling of indebtedness is created, the insurgents ask help in return and ultimately recruit or "draft" young men for the movement.

Undergrounds may tap external sources such as foreign governments or fraternal societies for *financing*. To raise money within their own country, secret loans from wealthy individuals or concerns sympathetic to the movement may be secured. Various items may be sold by door-to-door canvasses or through "front" stores. Robbery is another means of obtaining funds; victims are usually restricted to wealthy individuals and "impersonal" business firms, since undergrounds avoid arousing the hostility of the general populace. Persons with more modest incomes may be coerced to make "contributions" under threat of reprisal. In areas controlled by the underground, taxes may be levied.

The underground usually is the internal *supply* arm of the guerrillas. It buys supplies on the black market or on the legal market through front organizations. It may steal goods from warehouses, or conduct open raids to get supplies. Sometimes undergrounds manufacture weapons and ammunition. Rural workshops may be small and mobile so that they can be moved to avoid enemy forays; those in cities are ordinarily disguised since they are stationary. Urban manufacturing usually is done in established or front shops. Goods and materials may also be systematically collected from the population, though this requires a high degree of underground influence and freedom of action. In making such collections, undergrounds generally try to avoid the label of "bandits" by making at least nominal payments, or giving IOU's for the goods. Sources outside the country also may be tapped. Firms engaged in foreign trade may import equipment under noncontraband labels. Foreign governments may supply undergrounds through such means as parachute drops.

To protect the personal anonymity of their members, many jobs may be filled with persons who can perform their duties while engaging in their legal occupations; such members appear "normal" and escape the suspicion of neighbors or security personnel. Those who must live under false identities are given the necessary documents. These may be forged in underground printing shops; or authentic papers of dead or missing persons may be slightly altered and used. In choosing meeting places, neighborhoods of persons known for antiregime activities and likely to be under surveillance are avoided.

Security of communications may be protected by using old men and women, or children as couriers, for their movements are least likely to arouse suspicion. The missions of couriers may be made part of routine trips, to escape the suspicion of observant persons and provide a good excuse for travel. Couriers are usually expected to arrive exactly on time for a rendezvous, as an individual who must wait for his contact risks being arrested for loitering. Members meeting for the first time often establish their identity by recognition signals which appear innocent to an observer.

Other security measures are designed to prevent betrayal from within. Loyalty checks are conducted on prospective recruits and on members suspected of collaboration with the enemy. Loyalty oaths may be given to new members to impress them with the seriousness of the cause and the need for secrecy. The general practice is to hold records to a minimum, on the principle that only information which cannot be memorized or which is needed for future reference should be put in writing. When it is necessary to record the names and addresses of underground workers, only cover names are generally used.

Chapter 3: Underground Operational Techniques

In *subverting organizations*, an underground member seeking a leadership post represents himself as dedicated and loyal. He takes the initiative in planning activities and volunteers for any job, no matter how time-consuming or unpleasant, while avoiding any contacts or activity which might jeopardize his position in the organization. At the same time, his candidacy is supported by underground members who belong to the rank and file. Although they may be few in number, they can thus influence the decisions of the whole organization. Its membership may be controlled through a system of rewards and sanctions, once the underground is master.

Front groups are frequently used. If it is unable to infiltrate existing organizations, an underground often creates organizations to do its work behind an innocent façade. These usually espouse some worthy cause which will enlist the support of respectable members of the community, while the leadership is kept firmly in the hands of the underground members. When a subversive group is small, it often seeks to draw legitimate groups into a *united front*, and gain the prestige of speaking for a larger number of people. After a united front has been formed, the subversive group will try to discredit its leaders and take over.

Paper groups are also used. Organizations with one or two members obtain charters from some national or international body and then send delegates to its confederation or congress. The paper group often gets equal representation with larger, legitimate groups, and if delegations vote as a bloc, it can give the impression of wide popular support of a particular individual or policy. Such groups created by an underground can disproportionately influence larger organizations.

Standard techniques are used in the conduct of *psychological operations*. The means of communicating the propaganda message approximate those used by governments or legal organizations: radio broadcasts, where the populace possesses radios; newspapers and pamphlets; word-of-mouth; slogans or other symbols displayed upon walls. Overt mass responses are sometimes evoked for their psychological effect; mobs are formed and demonstrations created to show dissatisfaction with the government. Passive resistance is encouraged, not only for the material effect it may have upon production, but also for its psychological effect upon the government and the people. When persuasion alone does not get the desired response from an audience, an underground sometimes resorts to coercion, threats, and terror.

In preparing for the assumption of power, a *provisional government* is frequently set up to lend an air of legitimacy to the underground movement, and to get financial help from those foreign governments who recognize it. In areas which it controls, an underground may create *shadow governments* which operate courts and schools and supply police, public health services, etc. Underground *intelligence* operations assist guerrilla units by providing them with scene-of-battle information. They usually gather tactical rather than strategic intelligence and rely upon quantity rather than quality. Transportation and communication facilities are reconnoitered by undergrounds in preparing for sabotage attacks, which may be supervised by specialists from military units. Secret scientific and military information as well as political intelligence is also obtained by undergrounds.

Techniques of underground *sabotage* can be viewed in two general categories: selective sabotage and general sabotage. In the first, the underground tries to incapacitate installations which cannot easily be replaced or repaired in time to meet the government's crucial needs. Tactical targets, such as a bridge essential for transporting troops and supplies to a battle area, are concentrated on since strategic targets, such as factories, must be incapacitated for a much longer period. An underground may undertake sabotage, not only to hamper the government's military effort, but to encourage the populace to engage in general acts of destruction. Such acts serve as a form of propaganda and commit people more firmly to the cause.

Undergrounds generally use simple *explosives*, since their members usually are not trained demolition experts. Plastic explosives are ideal, since they are easily stored and simple to use. Members receive training through manuals, directives, clandestine newspapers, leaflets, radio broadcasts, or personal instruction from military units. To foster general sabotage the underground often instructs the population in the use of such simple devices as Molotov cocktails, tin-can grenades, and miscellaneous devices for causing fire or damage to small equipment.

In conducting *escape and evasion* operations, an underground may remove persons from an area, in which case egress routes are usually directed away from lines of battle, or it can hide fugitives in secret lodgings within the area. If they are to be hidden for only a short time, it is only necessary to provide some food, a place of rest, and directions for travel. If, however, a fugitive must live in hiding for an extended period of time, extra ration cards may have to be obtained and cover stories formulated to explain his presence in case of discovery. A third alternative is to put fugitives in remote camps in wooded or mountainous areas, or with roving guerrilla units. To guard against infiltration of the escape and evasion network by government security personnel, strangers seeking the assistance of the underground may be subjected to tests. They may be questioned repeatedly to uncover any discrepancies in their accounts of places of residence, jobs, friends, and reasons for asking help.

Summary

Chapter 4: Communist Use of Undergrounds

The international Communist movement has participated in a wide variety of resistance and revolutionary movements. In some instances, Communist wartime resistance movements became the vehicles by which Communist parties launched or completed revolutions. The case of Yugoslavia is well-known but it is sometimes overlooked that anti-Japanese activities helped the postwar Communist movements in China, Indochina, Malaya, and the Philippines.

Regardless of their locale or particular method of operations, Communist parties have followed a general pattern in their attempts to gain power through revolution. Leaders trained in Moscow return home to form small elite groups and prepare to operate both legally and illegally. Two apparatuses are usually formed, one more or less overt, the other tightly secret and open only to the more elite. The open party rarely attempts to gain support on ideological grounds, but appeals to local, regional, and national interests. It joins in united front activities to improve its image and credibility. In the meantime, the secret elite are *infiltrating* legal organizations and gaining key positions, without revealing their Communist affiliation. Once in power, they may persuade such organizations to join united fronts with the open Communist Party. When the revolutionary situation is considered ripe, two organizations are formed by the open party—a guerrilla unit and a civilian support unit. Both assume the names of popular causes (e.g., liberation, independence) and recruit membership in that name. Recruits need not give allegiance to communism, and frequently do not understand communism, or know that the movement is Communist-led.

Key leadership positions are maintained by Communists in the revolutionary movement, and any opposition—people or parties—is eliminated, making the Communist Party the only alternative for the people. Throughout, membership in the core elite is kept very selective and small; should the "open" party be defeated, the cadre is likely to remain intact for another attempt at a later date.

Chapter 5: Government Countermeasures

It is difficult, if not impossible, to eliminate most undergrounds in their initial stages of development, as to do so would require a very large and comprehensive intelligence system. A better course of action is to contain their influence by measures that will command the respect and loyalty of the people. Once an underground movement has become militant, however, the government can take steps to cut off foreign sanctuary and support, and to separate the guerrillas from the underground and the local populace from which it must obtain supplies. By offering rewards for information, by regrouping people away from the guerrillas, and by civic action designed to win the confidence and active cooperation of the people, the government may force the underground into a defensive position.

Since underground activities are directed toward gaining the control and loyalty of the people, government forces can meet this threat through population control measures and pacification. In administering a pacification program it is important to understand the nature of human behavior under stress. Civic action, information campaigns, relocation, and retraining are some of the techniques which have been used in reducing stress and winning active support.

To penetrate the underground, intelligence sources can be established, potential targets can be identified, the *modus operandi* of underground agents determined, and informers acquired. Once members have been identified it is important to keep them under surveillance and identify their contacts within the underground. The most successful way of destroying underground initiative and obtaining information seems to be infiltration.

In combating undergrounds, it is preferable to centralize military and civilian commands under one authority so that concerted political-military action can be taken against the underground while security and assistance can be provided to the populace. Although military units should follow the overall direction and policy of a central authority, they also should be allowed to function locally in such a manner as to take advantage of "live" intelligence and changes in the tactical situation.

PART II

In France during World War II, the underground carried out sabotage and small raids while building a secret army for the purpose of assisting in the Allied invasion. Some of the complexities of coordination of underground activities with external forces, in this case the Allies, are illustrated.

In Yugoslavia, Milhailovic operated on a similar strategy. So as not to put the populace in jeopardy of Nazi retaliation, he carried out small guerrilla attacks which would not implicate civilians and organized a secret army anticipating an Allied invasion. Tito's partisans were less concerned about the character of the attacks and engaged the occupiers whenever it was to their advantage. Foreseeing the day when the resistance would end, the Communists sought to gain control of the country by forming both open and clandestine governments to provide the needed civil administration. Yugoslavia was the scene not only of resistance against the occupier but also competition and fighting between the Communist and non-Communist factions for control of the resistance movement and of the country.

Algeria is an example of a successful non-Communist revolution fought to win independence from the French. The Algerians struggled against a major world power for 7½ years and won primarily through political rather than military means. Noteworthy in this case is the importance of extraterritorial support obtained by the Algerians via Tunisia and Morocco.

In Malaya the Communist Party (MCP) organized a military force (MRLA) and a civilian underground (the Min Yuen) from remnants of the anti-Japanese resistance movement. The MCP was comprised almost entirely of one ethnic group in Malaya, the Chinese. This factor, and the collapse of their underground supply organization through British relocation of the rural population, contributed heavily to their defeat.

In Greece, another Communist movement was defeated after the loss of a sanctuary and base of supply in Yugoslavia. Failing to develop an effective

Summary

internal underground supply organization, the guerrillas were forced to retreat to the Albanian borderland and fight the Greek army in a conventional type battle which they ultimately lost.

In the Philippines, an underground Communist movement developed from the anti-Japanese resistance and carried on a revolutionary campaign to overthrow the government. Through social, political, and military reform measures and the able leadership of Ramon Magsaysay, the cooperation of the people was obtained and the Communist insurrection suppressed.

The unusual factors of large-scale, illegal immigration and the special political status of the mandate in Palestine created a unique situation for an underground to operate in. Drawing on former members of World War II undergrounds, the Haganah developed an effective intelligence network. It also transcended national boundaries and utilized the sympathy of other nations for the refugees to aid the movement.

Some Qualitative Comparisons

A characteristic of the underground organizations reviewed was the development of cadres. The non-Communist underground groups grew up from prewar labor, political, military, and religious organizations. The leaders of these groups were instrumental in organizing the underground groups and in setting up civilian administrations throughout occupied territories. On the other hand, the Communist cadres were composed of professional organizers who had been involved in clandestine work for many years before the war. All of the undergrounds reviewed were organized on a cellular basis.

The undergrounds primarily relied on couriers and radios for communications, although they often improvised when conventional means were lacking. They were frequently dependent upon external support for finances and supplies. In most cases, undergrounds supplied guerrillas with food and safe housing, and manufactured crude weapons and ammunition. In order to maintain the security of the organization, counterintelligence units were established to infiltrate government military posts and civil organizations with civilian workers. Terror squads were characteristic of all the movements studied; they provided necessary security for clandestine operations by enforcing threats and punishing traitors and collaborators.

Psychological operations were carried out through clandestine newspapers but primarily through word-of-mouth communications. Radio broadcasts from outside the country were used for propaganda purposes in Yugoslavia, Palestine, Algeria, Malaya, and Greece. Ethnic and religious differences as well as group prejudices played an important role in the propaganda campaigns in these five countries. In the revolutionary movements reviewed, psychological operations were perhaps the most important single activity.

The undergrounds generally were responsible for tactical rather than strategic intelligence and tended to rely on quantity rather than quality of information. In World War II the Allies, although working with the underground, sent agents from the United States Office of Strategic Services (OSS) and

British Special Operations Executive (SOE) on special assignments to gather strategic information. Sabotage was used not only to destroy the enemy's lines of communication and supply depots, but also to take the pressure off guerrillas. One of the most useful and successful functions of an underground was the development of escape and evasion networks.

Countermeasures against undergrounds usually involved a two-pronged attack. The first was aimed at obtaining the cooperation of the population and the second at attempting to destroy the movement. In order to protect and separate the people from the guerrillas and underground, relocation to safe areas was used in Malaya, Greece, the Philippines, and Algeria. The German device of deportation was another variation of relocation. Relocation provided a means by which the populace could be protected from threats of the insurgents, and acted as a method to separate the underground and their local supporters. By the use of cordon and search, and population control measures, many members of undergrounds were captured. Amnesty and rewards were made to induce defection from the underground. These measures, coupled with rehabilitation programs, often placed the underground on the defensive, making it more concerned with defections, informers, and infiltration than with expansion and aggressive activities.

Some Quantitative Comparisons

In comparing selected quantitative information on the seven undergrounds studied (see fig. 1), some interesting observations can be made.

In each of the movements, the underground was larger than the guerrilla force. The size of the ratio of underground members to guerrillas varies from two-to-one to twenty-seven to one, with an average of nine underground members to every guerrilla. In addition, the guerrillas were organized into (1) roving guerrilla units, (2) regional troops, and (3) militia. The latter two usually remained close to the village or community from which they came and operated in a quasi-clandestine manner. The fact that, even in the peak of the militarization phase, there are more underground members than guerrillas emphasizes the political nature of resistance movements and internal wars.

The highest ratios of underground to guerrilla strength occurred in Greece, Malaya, and the Philippines. These three countries have several factors in common which may suggest the reason for this. The terrain and cutoff of external supply required that food, ammunition, and arms be furnished through an internal underground organization. All three movements were Communist directed, and the Communists characteristically use large underground organizations.

And yet these are precisely the three countries in which a successful counterinsurgency effort occurred and in two of them (Greece and the Philippines) the insurgents greatly outnumbered the security forces. Clearly, the number of combatants is less critical than the measures taken by either side. Although accomplished by different means, the end-result of all three counterinsurgency campaigns was the cessation of underground support of the guerrillas which directly or indirectly broke the back of the revolutionary movement.

Summary

Country	Population (millions)	Insurgents			Security forces Military, militia, police	Ratios and percentages		
		Underground	Guerrillas	Total		$\frac{\text{Underground}}{\text{Guerrilla}}$	$\frac{\text{Combatants}}{\text{Population}}$	$\frac{\text{Insurgents}}{\text{Security forces}}$
France	41.0 (1946)	300,000	100,000	400,000 (1944)	500,000 (1940–44)	$\frac{3}{1}$	$\frac{1}{46}$ (2.2%)	$\frac{1}{1.25}$
Yugoslavia	16.0 (1940)	Partisans ----	----	200,000 (1944)	200,000 (1940–44)	$\frac{3}{1}$	$\frac{1}{34}$ (2.9%)	$\frac{1.3}{1}$
		Royal Army 50,000	15,000	65,000 (1944)				
Algeria	10.3 (1956)	FLN 21,000	8,000	29,000 (1956)	800,000 (1958)	$\frac{3}{1}$	$\frac{1}{12}$ (8.3%)	$\frac{1}{13}$
		OAS ----	----	60,000 (1960)				
		----	----	25,000 (1962)				
Malaya	4.9 (1950)	90,000	5,000	95,000 (1950)	300,000 (1951–54)	$\frac{18}{1}$	$\frac{1}{12}$ (8.1%)	$\frac{1}{3}$

14

Greece_ _ _ _ _ _ _	7. 9 (1946)	675, 000	25, 000	700, 000 (1948)	182, 000 (1948)	$\frac{27}{1}$	$\frac{1}{9}$ (11. 2%)	$\frac{4}{1}$
Philippines_ _ _ _	19. 2 (1946)	100, 000	12, 000	112, 000 (1952)	37, 000 (1950)	$\frac{8}{1}$	$\frac{1}{129}$ (0. 7%)	$\frac{3}{1}$
Palestine_ _ _ _ _	2. 0 (1948)	30, 000	15, 000	45, 000 (1948)	95, 000 (1948)	$\frac{2}{1}$	$\frac{1}{14}$ (7. 0%)	$\frac{1}{2}$

Underground/Guerrilla
X Median 3/1
X Mean 9/1
Range 2/1 to 27/1

Combatants/Population
X Median 1/14
X Mean 1/36
Range 1/129 to 1/9

Insurgents/Security Forces
X Median 1/1.25
X Mean 1.4/1
Range 1/13 to 4/1

Figure 1. Insurgents, Security Forces, and the Populace.

The sources for the figures are referenced in each of the case descriptions. The figures are estimates of the insurgent and government strengths at the peak of the movement.

15

Summary

Another interesting aspect is the percentage of combatants (insurgent and government security forces) engaged in the conflict, full time or occasional, at the peak of the movement. These range from 0.7 to 11 percent with an average of 6 percent of the population being involved. This leaves a large proportion of the populace who, for personal or other reasons, were not directly involved, and raises some interesting questions about the meaningfulness of referring to the "people" in an undifferentiated manner. It may be to the government's advantage to make finer distinctions than those between "the guerrillas" and "the people" and plan countermeasures that do not alienate the uncommitted, or passive supporters, and focus them on the underground, as well as the guerrillas.

METHODOLOGICAL NOTES

The development of a single source book of information on undergrounds, which would reflect the current state of knowledge about undergrounds, was undertaken as an initial study. Sources used in the study were all open (unclassified). In this literature, very few sources treat undergrounds systematically and nonpersonally as the major subject. Rather, undergrounds are discussed as an ancillary topic in historical accounts of larger campaigns as in *Rearming the French* by Marcel Vigneras, analytic studies of particular situations such as *Guerrilla Communism in Malaya* by Lucien W. Pye, and foreign sources, as *Guerrilla Warfare* by Ernesto Guevara. Information from these sources was supplemented by cautious and selective use of more detailed and focused accounts of underground actions found in autobiographical memoirs and other personal documents.

A direct methodology was used in research. First, a cross-section of the available literature was reviewed; an outline for categorizing and describing the collected information was developed for the first part of the study. From this framework, the chapters of Part I evolved inductively, by generalizing from specific actions, and deductively, through a functional analysis of performance requirements as derived from general goals and objectives of varying strategic politico-military situations. Thereafter, an iterative process, consisting essentially of collecting additional information, evaluating it, and categorizing it, was used. The framework was modified and refined to accord with the additional information. The whole process was then repeated. When the framework was finally established near the end of the data collection and analysis effort, subsequent information rarely modified it. Instead, the information primarily clarified the general propositions of the outline, (e.g., basic security techniques) or illustrated variations in application of a technique (e.g., secret message delivery).

For the second objective of the study, preparation of historical illustrations, seven relatively recent (during or after World War II) instances of underground movements were selected. They were chosen on the basis of the extent to which the experience emphasized an important characteristic of the nature of undergrounds (see introduction to Part II), and on availability of information about the underground. Sources describing the resistance movement involved were reviewed, information pertinent to the study abstracted, and the underground's "story" reconstructed. However, the cases in Part II are not histories of the undergrounds; they represent only selected aspects of the underground's role and functions.

As a final methodological note, the draft manuscript of this report, after intensive internal review and revision, was submitted to the expert consultants identified in the Preface. Revisions and additional literature review were made at their suggestion. It will be noted by the reader that treatment of topics within the study is uneven; this reflects the necessary time and effort limitations under which the study was conducted; there may be available unclassified information which could fill some gaps in our knowledge as presented in this study. Some of these gaps might be considered as candidates for additional research, not only of a descriptive nature, but for analytic and predictive studies utilizing recent advances in military social science research. The study will be used as one foundation for future SORO research programing aimed at increasing our knowledge of all aspects of unconventional warfare and its relationship to the revolutionary process.

PART I

UNDERGROUNDS: STRATEGY, TACTICS, AND COUNTERMEASURES

CHAPTER 1

THE ROLE OF UNDERGROUNDS IN RESISTANCE AND REVOLUTIONARY WARFARE

HISTORICAL ANTECEDENTS OF TODAY'S UNDERGROUNDS

Today's revolutionary movements in Asia, the Middle East, Africa, and Latin America and World War II resistance movements against German and Japanese occupying forces are but the latest examples of major world events in which undergrounds have played a critical role. Throughout history undergrounds have been at the base of resistance and revolutionary movements, frequently directing political activities, performing many different kinds of organizational and operational functions, and in most cases supporting overt guerrilla warfare.

During the Peloponnesian War in the fifth century B.C., underground agents in Athens and Sparta tried to subvert the governments of each other's city states. Spartan agents working within democratic institutions of Athens were particularly effective in their propaganda and agitational activities which were designed to create internal dissension and develop popular distrust of the government.

In China it early became common strategy for underground movements to form guerrilla bands and organize support groups among citizens in order to overthrow a local warlord or mandarin. During the T'ang dynasty, in the ninth century A.D., a popular leader named Huang Ch'ao set himself up as a sort of Robin Hood, taking goods from the wealthy and from government officials, distributing them to starving peasants, and organizing the peasants behind him. The underground and guerrilla tactics of this era provided lessons which were absorbed by subsequent revolutionary groups and were ultimately reflected in the actions and writings of Mao Tse-tung.[1]

In Europe during the Middle Ages there were underground movements among underprivileged or persecuted groups. Among these were the *Jacquerie*, a loosely knit organization of beggars and thieves in 13th-century France; the peasant groups in southern Germany; and the clandestine heretical religious sects, such as the Albigensians in southern France.

Underground movements also played a role in numerous dynastic conflicts, such as the "War of the Sicilian Vespers." Around 1280, Charles of Anjou, the French King of the Two Sicilies,* was preparing to invade the Byzantine Empire. The elderly Byzantine Emperor, Michael VIII (Paleologus), anxiously sought by all possible means to prevent Charles from carrying out his plans. Michael was aware of considerable unrest in Sicily which had resulted from Charles' harsh rule and was also aware of the frustrations of Spain's King of Aragon, Pedro III, who asserted claims to Sicily. After a promise of aid from Pedro, Michael's agents aggravated the unrest

---

*The Kingdom of the Two Sicilies included southern Italy as well as the island of Sicily.

23

among the Sicilian people, and as a result of their underground activities a popular uprising in Sicily resulted in the massacre of eight thousand French troops in 1282. Soon after this, Pedro, pretending to lead a crusade to the Holy Land, changed course to land on the shores of Sicily and help the insurgents against Charles. The War of the Sicilian Vespers* lasted for 20 years and eventually culminated in the Aragonese conquest of Sicily from the Angevins.

The Sicilian spirit of independence, which first manifested itself during the War of the Sicilian Vespers, was to lead to further revolts against a variety of rulers in subsequent years. Though not a political independence movement, the Mafia developed from a group of outlaws who, for pay, protected landed estates, to an extralegal organization which strongly influenced the political activity of the island through economic means. To the present day the Mafia claims to be the spearhead for the true interests of the Sicilian people.

The 14th century also provides some interesting cases of people resisting an outside invader and regaining national independence via movements involving undergrounds. For instance, after the second conquest of Scotland by Edward I of England, a Scottish resistance movement was begun in 1306 by Robert Bruce, a former vassal and ally of Edward. Bruce combined ambush and surprise attacks against English garrisons in Scotland with raids extending into the border regions of northern England. Using these tactics, Bruce had driven the English from most of Scotland by 1314 and was crowned Robert I, King of Scotland. In the same year he defeated the invading army of Edward II, son of Edward I, at the Battle of Bannockburn and thus restored the independence of Scotland.

The Renaissance and Reformation were marked by the development of many groups with the primary goal of plotting political change. Perhaps the most notable of these were the numerous small clandestine groups planning coups throughout the major cities of central and northern Italy during the 15th and 16th centuries. Intrigue and assassination were common weapons in these struggles for wealth and power, aptly described by Machiavelli. The Counter Reformation saw the growth of international movements organized in the name of the Catholic and Protestant faiths, but dedicated to achieving political as well as religious goals. Elaborate intelligence and counterintelligence systems, propaganda campaigns and other forms of political warfare aimed at influencing the rulers of Europe characterized these movements.

New concepts in the 17th century stressing the importance of the individual led to the formation of reform movements which brought about major changes in many countries. The overthrow of Charles I of England, for example, was effected primarily by religious groups which had been forced to meet clandestinely and which turned to political and military action in order to achieve a wide range of political as well as religious reforms. Later, clandestine groups favoring a restoration of the monarchy obtained funds from persons friendly

---

*The name "Sicilian Vespers" is given to the war because the massacre began at the hour of prayer.

to Crown Prince Charles and infiltrated offices of the English Government. In 1660, after years of civil war and interregnum, they brought Charles to the throne.

In the 18th century are two outstanding examples of underground movements which were well-organized and active on a wide variety of fronts. In America the Secret Committee for Correspondence and affiliated groups evolved into an intercolonial movement engaging in large-scale propaganda and other psychological warfare campaigns against the British colonial authorities. They trained militia groups, established friendly contacts abroad, and eventually played a leading role in gaining independence for the 13 colonies.

In France the Revolution of 1789 had its genesis in underground groups. Once in control, the Jacobins organized networks of sympathizers in other countries dedicated to subversion and revolution. The subsequent overthrow of the radical Jacobins was followed by the establishment of the 5-man Directory, which was soon in difficulties both at home and abroad. Disorder was rampant, the financial situation was desperate, discontent was everywhere, and the Republican armies suffered reverses. The government, not knowing which way to turn, swayed one day to the Jacobins, the next to the moderates. Under these circumstances, Gen. Napoleon Bonaparte, with the aid of several political leaders, was able to come to power through a coup d'etat.

Napoleon's conquest gave rise to a series of resistance movements throughout Europe and Latin America. In such notable cases as Spain, Mexico, and Venezuela, armed guerrilla uprisings occurred in the name of restoring or gaining national independence.

The "Spanish Ulcer," as Napoleon termed the war in Spain, was one of the principle factors leading to the defeat of the French Emperor. France invaded Spain, defeated the regular army, and captured the main cities and towns. However, it was not able to control the people living in the rugged Spanish countryside. They organized into small irregular bands and made surprise attacks upon units of the French Army. They were estimated to have taken a total of 100 soldiers a day. This guerrilla war* diverted French military resources badly needed for operations elsewhere. With the aid of English forces in Portugal and Spain, the Spanish guerrillas were able to drive the French from Spain and assist Wellington's forces in defeating Napoleon.

After the Congress of Vienna in 1815 and with the subsequent spread of nationalist aspirations, underground resistance and revolutionary movements became particularly important in effecting political change. Especially noteworthy are the *Carbonari* and Young Italy movements on the Italian peninsula, which were led by such men as Mazzini and Garibaldi. These men sought not only independence for the Italian states under foreign rule but also the ultimate unification of Italy into one nation.

The *Carbonari* began during the Napoleonic era as an open resistance movement with cells throughout most of Italy. In 1813, after an abortive insurrection, they were driven underground. Though operating clandestinely, the

---

Guerrilla is a Spanish word meaning "little war," or "body of partisans," and it was here that the term *guerrilla* was first used.

movement managed to enroll nobles, army officers, small landlords, government officials, peasants, and priests. It was not until 1861 that the goals of the movement were achieved.

The Irish nationalist movement, though unsuccessful for many years, attracted considerable attention throughout the 19th century by its constant use of political warfare and terrorist techniques. The Irish movement for independence was certainly one of the most prolonged of its kind, spanning nearly four centuries.

Underground movements throughout the Russian Empire during the late 19th century, many of which employed extensive terrorism, were almost as influential as Marxism itself in shaping Lenin's concepts of party organization and tactics. This was particularly true with regard to the use of terror for obtaining popular compliance with party demands. The Bolshevik revolution was brought about by well-organized underground activity.

World War I was precipitated by the assassination of the Austrian Archduke Francis Ferdinand by a Bosnian student working with the Black Hand Society. This clandestine group, organized in Serbia, financed by the Russian-led Pan-Slav movement, and including among its members several high Serbian officials, was dedicated to liberating Slavs in the Austro-Hungarian Empire and organizing them into a Greater Serbia.

From this brief historical overview, it is clear that clandestine organizations are not the product of a particular: political or religious ideology; cultural, ethnic, national, or geographical grouping of persons; structure or form of government; segment of society or social class; or stage of a society's economic or technological development. Also it is clear that undergrounds are not new nor unique to the contemporary world scene, although such an impression is easily created by the pressure of current problems. Underground movements directed toward changes in governing authority have appeared in societal life throughout recorded history.

WHAT IS AN UNDERGROUND?

DEFINITION

Before turning to a closer examination of undergrounds in resistance and revolutions, the term *underground* needs to be defined as it has been used in the past in a wide variety of ways, and as it will be used in this study.

(1) *Underground* has been used synonomously and interchangeably with *resistance*. The various European resistance movements of World War II have been called undergrounds, with no distinction being made between organizations performing guerrilla fighting and those performing clandestine support activities.

(2) *Underground* is frequently used to describe a revolutionary movement in its early stages, as in "Castro's Cuban Underground."

(3) *Underground* has been used to refer to civilian organizations which support military or guerrilla forces of a resistance or revolutionary movement. In the Malayan insurgency, the revolutionary movement was made up of a military organization (the Malayan Races Liberation Army—MRLA) and a civilian organization (the *Min Yuen*) of which the latter was considered the underground.

(4) *Underground* is used to mean espionage or spy rings. Agents, foreign or indigenous, working for a foreign government in order to provide "intelligence" information about an existing government are called members of an underground.

The word *underground* is also used in the phrase "to go underground," which means to disappear from view, to escape and evade security forces, to work undercover or subversively, or to hide for months or years in order to conceal one's subversive ties so that one may be useful at some later date. *Underground* is also used to describe a secret, concealed, or illegal object such as an "underground press."

Ideally a definition of the term *underground* would include not only what an underground is, but what functions it performs and how it goes about doing its functions. In a real sense, then, this entire study is aimed at completing a full definition of an underground, because the emphasis in the study is on describing the functions and techniques of undergrounds. As a starting point, however, undergrounds are defined for this study as: *Clandestine organizational elements of politico-military movements attempting to illegally weaken, modify, or replace an existing governing authority.*

DISCUSSION OF DEFINITION

Although undergrounds assume different forms, all are characterized by the following: (1) their goals are illegal under the system within which they exist; (2) their activities may be either legal or illegal, but they use primarily illegal means to achieve their goals; (3) they attempt to conceal the identity of their members and organization from the governing authority.[2]

A political party which exists legally and openly, and whose goals and activities are legal, falls outside of the definition even though it may be working to modify or replace an existing governing authority. Likewise, a secret society which conceals its goals and activities from nonmembers does not constitute an underground so long as both goals and activities are legal. On the other hand, a crime syndicate, although both its goal and activities are illegal and it conceals its identity, is not considered an underground since it typically does not aim at changing existing governing authority. A formerly legal political party which is declared illegal by the governing authority is not an underground even though it continues to exist, unless it then attempts to overthrow the government through clandestine, illegal means.

Clandestine may mean that the existence of the underground is unknown. More often, however, it means that responsibility for actions such as sabotage cannot be determined precisely. In this instance, it is known that an underground exists, but its members are unknown to the government. Most underground members conceal their identity by playing normal, legal roles within the society while simultaneously working for a resistance or revolutionary movement. They live at known addresses, work, or appear to work, at legal occupations, and may participate in some legal, nonoccupational activities. This is in contrast to a guerrilla member, who ordinarily leaves his normal societal pursuits to take up arms in quasi-military guerrilla units and to live outside the community.

Members of undergrounds are of several different types. Leadership cadres constitute the hard core of the organization; they devote most, if not all, of their effort to the movement and plan their lives around the movement and its needs. They frequently do not play a role in the society, but rather work full time for the movement with the cooperation of other underground members. Active workers are those underground members who, while maintaining their role in society, knowingly and voluntarily perform assigned tasks on a regular basis and attend meetings regularly. Auxiliary, or part-time workers, are those members of the movement who will perform a particular task when called upon, but who are not "regular" for one reason or another. Finally, unorganized sympathizers, who do not join the movement for one reason or another, may help the movement by participating in activities such as passive resistance, or mass demonstrations, or by withholding information from the governing authority.

Politico-military movements are those resistance or revolutionary organizations whose primary strategy of change is centered on disrupting the political and physical (primarily military) mechanisms of the existing government's influence on or control of the people, and on establishing new mechanisms of influence and control by their movement. Excluded by this definition are those organizations who work for societal change, sudden or gradual, in political or physical mechanisms through nonviolent, legal means, such as by changing legislation, long-term education, or even nonviolent persuasion of so-called "pressure groups."

The true goals of politico-military movements may or may not be known or suspected by the government. The goals of the non-Communist anti-Nazi resistance movements were well known. The aims of Communist-led partisans and many Communist front groups, on the other hand, have usually been hidden. The true goals of a movement may even be concealed from some of its members, as was the case with such groups as the Vietminh and Malayan Races Peoples Union, who thought they were struggling for "national liberation" *only*. Undergrounds may also employ the services of individuals who know little or nothing of either announced or real goals, but who participate for adventure, mercenary motives, or out of fear or expediency.

The primary means for implementing a movement's strategy is through a combination of political and military actions. The functions and techniques used by undergrounds are developed in the remaining sections of this Chapter and in Chapters 2 and 3. Generally, however, the actions of an underground may be violent or nonviolent. Some undergrounds restrict themselves to basically nonviolent activities such as the formation of front groups. Most undergrounds plan to work eventually with armed groups and to conduct violent activities themselves. Undergrounds will use whatever political or military means necessary to accomplish their strategic goals of obtaining human and physical resources for the movement and withholding human and physical resources from the government.

A useful distinction can be made between urban and rural undergrounds. The former ordinarily carry the brunt of a politico-military movement's activities in the cities. The latter are concerned more often with support of guerrilla units. Urban undergrounds can be used to relieve governmental pressure on guerrillas (and vice versa). Through mass demonstrations, raids, and sabotage, urban undergrounds can draw security forces to cities. On the other hand, increased guerrilla activities can draw security forces back to rural areas.

Finally, it should be noted that "governing authority" is broadly conceived to include: the representatives of government who control the symbols of government and the instruments of force, the established bureaucracy, and the legally sanctioned institutions and procedures (however they came to be established). It is recognized that attempts to change these may only be efforts to change fundamental ideologies, values, beliefs, and other basic concepts of a "way of life." However, in this study emphasis is on changes in governing authority as defined, whether these changes are sought as ends in themselves or only as means to other ends. The emphasis here is on the mechanisms of influence and control—the symbols and instruments of political control, and the symbols and instruments of military power, whether the government be *de facto* or *de jure*.

UNDERGROUNDS IN RESISTANCE

The distinction between resistance and revolution is somewhat artificial; some resistance movements, in fact, evolved into revolutionary movements, as was the case in Yugoslavia, Indochina, Greece, Malaya, and the Philippines after World War II. Nevertheless, this classification does provide a convenient structure within which to describe organizational and operational functions of undergrounds, because these functions do vary as a result of basic differences in aims, in the nature of the enemy faced, and resulting different patterns of development. For example, resistance movements being usually dependent upon some outside power or coalition for aid, or being an auxiliary fighting force to the outside power, need not themselves plan self-contained major actions that will be decisive to the outcome of the total struggle. Revolutionary movements, however, must plan to conduct their own conclusive actions.

RESISTANCE

Resistance has been defined by one authority as "operations directed against an enemy, behind his lines, by discontented elements among the enemy or enemy-occupied population." [3]   Such operations are of increasing importance, he states, because the ideological aspects of modern conflicts transcend national boundaries, because modern means of communication and transport make it possible to contact and supply people behind enemy lines, and because total war makes every part of the community a legitimate target for attack.   A resistance movement, therefore, is characterized by the fact that in addition to drawing support from the local population, it is likely to receive assistance from some outside source in order to defeat a common enemy.   Also, a resistance movement is organized only after a foreign power has taken over the government through a sponsored revolution or invasion.

The goals of resistance movements are directly related to the military power requirements of an occupier.   The occupier of an area may generally be said to want the territory for its strategic value, its natural resources, or for the prestige of territorial expansion.   More specifically: the acquisition of territory may be required for reasons of position, communication, or transportation; the territory may be on the occupier's frontiers and may be useful in absorbing the shock of a surprise attack, or it may protect the flank of a strategic position; the conquest of territory, or territorial expansion, may add to the prestige and morale of peoples at home and abroad.   In all of these, the occupier is necessarily concerned with the resources of the occupied country—human, industrial, and natural.   Therefore, his combat power cannot be measured only by the number of troops and amount of firepower in the field; the communication and transport network and the supplies en route and at the source of supply must also be considered.   It is these latter components of combat power—the communication, transportation, and supply networks—which are the main targets of the resistance movement.   Actions against these targets cause the enemy to suffer in three ways: he has increased difficulties in coordination; he receives less combat materiel for his troops; and he has to use troops needed for combat to protect his resources and lines of supply.   In short, the resistance attempts to extract a higher price for occupation than the enemy is willing to pay.

Resistance, of course, can take a variety of forms.   It may be characterized primarily by partisan warfare, as in Russia.   Or, due to difficulties imposed by terrain, supply, or concentration of enemy troops, the underground may constitute the entire movement, as in Denmark, Holland, and Belgium.   In still other instances, both guerrillas and the underground may form the basis of resistance, as in Yugoslavia.   In all forms, undergrounds can and usually do play a critical role.

UNDERGROUNDS IN RESISTANCE DURING WORLD WAR II

After the German occupation of a large part of Europe and the Japanese occupation of China and Southeast Asia, resistance movements became a way of life for many people in those areas. In the mountains, forests, and swamps of European countries such as Yugoslavia, Greece, Albania, and western Russia, partisan bands fought guerrilla wars.[4] In the jungles of Indochina, Burma, Malaya, and the Philippines, small groups of guerrillas harassed the Japanese supply lines and communication centers. In both areas people organized into clandestine groups to provide food and clothing for the guerrillas and to gain information about enemy movements.

(U.S. Army Photograph)

Disbandment ceremony for one group of the Malay Peoples' Anti-Japanese Army (December 1945).

In Norway, where the climate was cold and bitter and the mountains lacked cover and the snow provided German security forces with clues for tracking, the resistance took the form of urban underground organizations which carried on sabotage, intelligence work, and passive resistance against the occupier. The flat terrain of Denmark, the Netherlands, Luxembourg, and Belgium made it relatively easy for the Germans to concentrate their forces rapidly and destroy guerrilla bands; therefore, guerrilla warfare was difficult. There were a few scattered groups in the wooded and mountainous areas of Belgium and Czechoslovakia which harassed the Germans.

Although some of the terrain of Poland is suitable for guerrilla warfare, the Poles did not engage in large-scale guerrilla warfare, mainly because of the difficulty of obtaining supplies. Instead, they performed sabotage and diversionary actions and maintained an intelligence network. They also organized a secret "state" with officials who, among other things, ran secret schools and courts. Throughout the war they organized and maintained a secret army

31

whose main function was to rise at the defeat of the German Army or collapse of the Nazi regime.[5] In Germany, where security measures were severe and there was popular support for the government, guerrilla warfare was extremely difficult.[6]

In all of these countries, throughout the population centers and the rural areas "resistance" predominantly meant underground resistance. These underground groups in Europe provided the Allies with intelligence, aided special agents dropped into these countries for special assignments, disrupted supply and communication lines, and helped downed Allied fliers to escape. They provided information to the Allies about the disposition of German forces, their movements, capabilities, attitudes, and morale; and also located, described, and mapped targets for Allied and guerrilla attacks. Intelligence was transmitted by radio from occupied Europe to Allied forces in London and Cairo. Undergrounds also performed acts of sabotage, carried on strikes and slowdowns to hinder war production, and engaged in passive resistance to lower the morale of the German soldiers. In order to maintain national morale and identity, they published clandestine newspapers and distributed them throughout the country.

Undergrounds throughout Europe recruited secret armies which emerged during the Allied invasion, took up arms against the Germans and prevented the retreating Germans from demolishing bridges, harbors, and other strategic points. Working behind the German lines, they coordinated their activities with the advancing Allied forces, and also organized provisional governments to maintain law and order until Allied troops arrived.

These resistance groups were composed of all types of people from all walks of life—from political parties, church groups, unions, students, or Boy Scouts. Some groups sprang up spontaneously in various localities after the occupation and eventually organized into large national undergrounds whose leaders attempted to coordinate their activities. Other groups were organized by local Communist parties which had been well organized before the war and in some cases were already carrying out subversion. In several countries, such as France and Yugoslavia, Communist-led groups carried a large share of the resistance effort after the Allied invasion. In order to unify disjointed groups into larger, more effective organizations, the Allies dropped liaison officers into these countries to coordinate underground efforts with those of the Allied military forces. In return, the Allies provided the underground with supplies and other forms of aid.

SUMMARY OF UNDERGROUND FUNCTIONS IN RESISTANCE

It is clear that in many cases the underground is the total resistance movement and the functions of the two are identical. In these cases, the underground *clandestinely* performs the following functions, usually with the aid of external supporting powers:

(1) Organizes and controls all resistance activities, coordinating them with external supporting powers;

(2) Performs internal administrative "housekeeping" functions for the resistance movement;

(3) Maintains intelligence and escape and evasion networks for external supporting powers, as well as for the movement itself;

(4) Conducts sabotage, strikes, passive resistance against the transportation, communication, and supply sources and mechanisms of the occupying power;

(5) Conducts psychological operations and other activities among the occupied peoples to win or maintain influence and control over them (e.g., "shadow governments," "governments-in-exile," recruiting a secret army which will "rise up" at a given time).

In resistance cases in which organized guerrilla units can operate, the underground performs the following additional functions:

(6) Coordinates activities with guerrilla units;

(7) Provides intelligence information for guerrilla units;

(8) Provides recruits, shelter, food, clothing, and other supplies for guerrilla units.

UNDERGROUNDS IN REVOLUTION

REVOLUTION

Revolution is frequently used interchangeably with other terms such as rebellion, coup d'etat, insurgency, or insurrection. *Webster's Dictionary*, the *Encyclopedia of the Social Sciences*, and various writers on the subject do not agree on a common meaning of the word. Revolution is used usually to refer to any sudden change with far-reaching consequences. Occasionally revolution is used to indicate gradual change that is suddenly recognized as having had far-reaching consequences. Without qualifying adjectives—such as cultural, scientific, economic, industrial, or technological—revolution most frequently means political revolution and it will be so used in this study. More precisely, revolution as used in this study means *the illegal modification, or attempted illegal modification, of an existing governing authority at least partially through the use, or threat of use, of force, by a group of the indigenous people of a nation.*[7]

A revolution may be undertaken for a variety of reasons. A small group of people whose only objective is to be in political power may feel they cannot obtain it through legal means and may therefore seek to gain control of the government through illegal means. A nationalist group within a colony or protectorate may seek to obtain independence, or autonomy as a step toward independence. A foreign power may encourage individuals with whom it is sympathetic to form a group, then assist them in taking over a government.

Whatever the motivation, the revolutionary objective is illegal transfer of internal political power.

A number of terms are used to designate different forms of revolutionary activity. A revolution in which a small group of men within the government or its armed forces seizes control of government is generally viewed as a *coup d'etat*. The use of open, organized, and armed resistance in which the inhabitants of an area seek to obtain automony or independence but make no attempt to alter or overthrow the central government is a *rebellion*. *Revolutionary warfare* refers to all aspects of revolutionary efforts to displace an existing government by force, and also to the efforts of a *de facto* government to defend itself against such displacement. An *insurrection* is the initial stage of a revolution which is still localized and limited to seeking modifications of government policy and not yet a serious threat to the state or the government in power.

In analyzing political power and the process of revolution, one observer describes two types of illegal transfer of power—revolution from the top and revolution from the bottom.[8] The discussion which follows is based largely on that author's analyses.

In a revolution from the top, or coup d'etat, a small group tries to obtain control of, or to neutralize, the armed forces and other instruments of power, usually with little or no violence. It usually does not seek public support until after the coup has been initiated or has succeeded. Here men in top echelons of government seize the instruments of power, such as mass-communication media, military materiel, transportation facilities, power stations, and finally the symbols of power such as administrative and legislative buildings in the government capital city. On the other hand, revolution from the bottom, which usually involves more violent seizure of power, is largely the result of social disorganization and unrest. A spontaneous mass movement encouraged or directed by an underground group develops slowly in a long revolutionary process that explodes suddenly through a precipitant event. There may be a mixture of these two forms of revolution, as there was in Russia, where a movement from below was combined with a Communist coup from the top.

REVOLUTION FROM THE TOP

In effecting a coup, insurgents typically must cope with three major conditions: (1) the sympathies of the nation's armed forces, (2) the state of public opinion, and (3) the international situation.[9] If any one of these factors is unfavorable and not properly assessed, the coup can fail. If the revolutionaries do not alienate the support of the armed forces from the government, civil war may develop; without public dissatisfaction with the existing government, it is difficult to implement control (and without the support of public opinion, it is difficult for the revolutionaries to maintain control once it is gained); and if international conditions are unfavorable, other nations may step in to protect their own interests.

The steps usually involved in carrying out a revolution from the top are—

(1) *Organization.* A small clandestine group attempts to recruit people who can secretly influence or actually gain control of top commands in the military.

(2) *Infiltration.* The next task is to infiltrate trusted agents into the army, police, and centers of power such as mass-communications media and power stations.

(3) *Seizure.* A decisive blow is struck to seize the instruments and the symbols of power and finally the bureaucratic apparatus.

(4) *Consolidation.* Consolidation of power and elimination of the opposition are the final steps.

Neutralization of government leaders is critical, for the control of the state does not rest in any impersonal system: "It [control] is the function of men. When men are disposed of, their control also vanishes. There are two ways of neutralizing personalities of the existing regime—they must either be killed or captured." [10] Once the seizure is effected, the group directs propaganda at those segments of the population whose support is needed for them to stay in power. Plans are made to destroy or neutralize opposition elements and to explain the coup to the outside world.

The Egyptian coup d'etat of 1952 followed this pattern very closely.[11] A clandestine group was established within the army and a junta was formed. Strategic posts within the army and the government were infiltrated, and at the proper moment came the following decisive moves:

(1) *Capture of the government leaders.* Cairo was divided into four sectors, and teams of four officers and two enlisted men, each with a jeep and a van, arrested all officials in their sector.

(2) *Separation of government leaders from the instruments of violence.* Small bands surrounded the barracks in Cairo in order to prevent loyal officers from reaching the troops.

(3) *Capture of the means of mobilization.* The Egyptian state broadcasting system, the central telephone exchange, rail stations, airfields, and the army general headquarters were taken; tanks were deployed at strategic points in the city.

(4) *Capture of the symbols of power.* Key government buildings were taken.

(5) *Prevention of foreign intervention.* Officers were sent to the British and United States Embassies to explain that this was an internal affair and that foreign intervention would not be tolerated.

In order to carry out a successful coup d'etat, the instruments of power must be in the hands of small elite groups. Since the number of men who control and operate the government is usually small, their power rests in the willingness of the people to be ruled by them and in their ability to rapidly mobilize sufficient instruments of violence to destroy any organized resistance.

REVOLUTION FROM THE BOTTOM

In a revolution from the bottom, the ultimate objectives are the same, but the strategy is quite different. As the rebels do not have access to key governmental or military posts, they must aim to (1) eliminate or neutralize the government's instruments of violence and (2) gain control and the support of the people. In laying their plans, the revolutionaries typically consider several basic factors which can influence the outcome of the struggle.

Environmental Conditions

The size of the country, its terrain and climate, and its geographical position with respect to its neighbors are important. A revolution from the bottom is usually a long, drawn-out war. A large country with extensive isolated areas, such as China, provides an ideal situation for a revolutionary war. (In fighting the Communists, Chiang-Kai-shek's government was forced to spread its troops over huge areas, thereby reducing concentrations in any one place.) The long communication lines necessitated by wide dispersal of troops are vulnerable to attack and this in turn limits the government's ability to mobilize a counterforce rapidly. Rugged terrain, with mountains, forests, or swamps, usually handicaps the government forces, which are generally organized in large units with heavy equipment for larger, external wars; the same terrain is advantageous to the smaller forces of the revolutionaries, usually lightly armed and specially trained for small surprise raids. On the other hand, readily accessible sources of food are crucial for a small force which lives off the land. The better the climate and the more food there is available, the easier it is for a guerrilla force to operate.

Sanctuary

A sanctuary, or safe-area, in the territory of an external power or located in some remote rural area is needed where supplies can be gathered and stored and men trained. The safe-area also provides a place for escape and recuperation beyond the reach of government forces. In countries with large territorial expanses, sanctuary is found in remote, isolated areas. If the country is small, a neutral country, preferably with a common frontier, may provide such refuge.

The sanctuary is used primarily for organizational rather than strategic purposes, and fulfills the same function as the rear for a regular army. It also serves as a political training ground for testing mass appeals and molding mass organizations.[12] Postwar Greece provides an illustration of the importance of the sanctuary: the Communist insurgency collapsed shortly after Yugoslavia closed its borders to the rebels. In Indochina, where the Communists were successful, most of the initial organizing and staging took place in neighboring China, and this is considered to have contributed significantly to the ultimate outcome of the campaign.

Foreign Assistance

A revolutionary movement usually must have outside support in the form of arms, money, and supplies. In seeking this support, the revolutionaries must consider the international situation. Rebels in countries whose locations are of strategic importance to larger, more powerful nations often receive assistance from them.

Having launched a revolution, the leaders typically pay a good deal of attention to foreign reactions, as the incumbent regime usually has allies who may have a vested interest in its maintenance of power. Usually they must decide, for example, whether to announce their aims and affiliations or whether to conceal them in order to neutralize possible opposition from other countries or groups in other countries who have different values.

EVOLUTIONARY DYNAMICS OF REVOLUTIONARY MOVEMENTS

In revolutions from the bottom, most revolutionary movements seem to evolve through five distinct phases. The phases can be identified by the predominant activity being carried on by the movement. It should be recognized that a phase is not completely separate from the others: there is some overlap of its predominant activity into other phases. Neither do the phases have minimum or maximum time-lengths; they will vary from revolution to revolution depending on the successes or failures experienced within phases. Finally, evolution can be uneven within a very large revolutionary movement: in some areas of the nation, the movement may be in one phase, while in other areas it is in a later or earlier phase.

Clandestine Organization Phase

In this phase, clandestine cells typically are established in heavily populated areas, organizations and industries are infiltrated, and cadres are recruited, trained, and tested. Clandestine escape and evasion nets usually are established to provide means of escape in the event that the group's subversive activities are discovered. External support may be secretly solicited in the form of money and often weapons. A base is usually established in a safe-area or sanctuary in order to develop the necessary training schools, create a supply and logistics system, and try out various political and organizational appeals upon the local inhabitants. Once the hard-coreworkers have been recruited and trained, and the skeletal framework of an organization established, the insurgents are ready for more overt actions.

Psychological Offensive Phase

The object of this phase is to bring about the loss of effectiveness of the social and governmental institutions of the country and to create unrest and disorder. Both real and conjured sources of tension and discontent usually are exploited such as scarcity of the basic economic necessities of life; social, political, and economic discrimination; lack of adequate housing or education; economic or political dependence on a foreign power; etc. Agitators may lead demonstrations, start riots, and organize strikes. Typically propaganda campaigns are organized and newspapers and slogans are circulated to emphasize, and frequently to create the theme of discontent. Agitators may spread rumors which cast doubt on the government's good faith and administration of justice. They usually spread stories of graft or corruption within the government and armed forces, and of oppression of citizens. They frequently make exorbitant demands for reform which they do not expect will be met. They capitalize on the fact that much of the support for a government is based on the people's confidence in the trustworthiness of its administrators. In most cases, the mere fact that honesty of the officials is questioned is enough to make many people lose confidence in the government. Selective terrorism—threats, intimidation. and assassination—is directed against officials at all levels of government.

Expansion Phase

During the unrest and uncertainty created by its disruptive activities, the movement typically attempts to crystallize public support for a strong organization which can reestablish order. The objective of the expansion phase is to recruit people through mass organizations and to win popular support. United front groups usually are organized. Clandestine cells within various groupings in the country, such as laborers or students, attempt to enlist the support of these groups. Revolutionary movements actively seek malcontents, victims of ill fortune or of social or economic discrimination; adventurers, men without property or family ties, the unemployed, and people with political aspirations. The movement also attempts in this phase to eliminate all opposition; thereby it destroys other alternatives and leaves only one solution to the chaos—popular support of the revolutionary movement.

The revolutionary group must demonstrate continually that it exists and that it is determined to gain control of the government in the name of the people or a large majority of the people. Terrorism is one way of drawing attention to the movement; it usually exaggerates the movement's strength. When the existing government imposes restrictions and controls, the insurgents can use the hardships created by these measures as further ammunition against the government. To prove their determination to rule the country, the insurgents often seek as soon as possible to establish a provisional government. Formal "committees" are set up in other countries to lobby on behalf of the revolutionary movement. At that point in time, when the existing government begins to lose popular support and the revolutionary ranks begin to fill, the movement enters the militarization phase.

Militarization Phase

In order to overcome the government's military forces, a guerrilla army is usually formed.[13] Guerrilla fighting is characterized by absence of fixed fronts, infiltration, quick concentration of guerrilla forces for action, and their immediate disengagement and dispersal after fighting.

Much guerrilla strategy generally follows the three-stage campaign outlined by Mao Tse-tung.[14] The first stage is termed strategic defense. Since the government forces are usually superior, the guerrillas concentrate on harassment, surprise raids, ambushes, and assassinations; they try to force the government troops to extend their supply lines. Since their primary aim is control of people rather than of territory, they readily trade territory to preserve the guerrilla force.

The second stage is said to begin when the government forces stop their advance and concentrate on holding territory in order to stop guerrilla action. As men, arms, and supplies are acquired, the guerrillas attack larger government forces and installations. Government forces are usually organized in large units, armed with heavy weapons, and prepared to fight conventional wars. The guerrillas are dispersed in small units through widely separated areas. They seek rough terrain, where the mobility of larger government units is minimized, and capitalize on the speed and mobility of their own smaller units. Thus, harassment is used to wear down the government troops while the guerrillas are organizing and building their army. As Mao says, "Our strategy is one against ten and our tactics are ten against one." [15]

The third stage referred to by Mao is the counteroffensive, which begins when the guerrilla army becomes sufficiently well-trained and well-equipped to meet the government forces. The guerrillas seek to create liberated areas and within these areas of control they build up additional military forces. Local militia units are organized and armed. Members work at civilian jobs and fight only when the government forces approach their village. Roving territorial units are created and carry on sabotage, ambush, and intelligence work. These units act as a screen and security force for the regular guerrilla army. They attack government troops, disperse, and reassemble at some preplanned point. Guerrilla members infiltrate government lines and concentrate to attack large government installations. Civilians in villages are organized into groups which form food networks and intelligence nets, manufacture or procure weapons and ammunition, and prepare the battle area with caches of supplies for the guerrilla force.

Consolidation Phase

Throughout the militarization phase the revolutionary movement does not cease its political and economic actions. If the guerrilla force cannot decisively defeat the government's military forces, the drain on the economy and the effect of political pressure by the insurgents are often enough to ensure success. Indochina and Algeria are two cases in which heavy economic costs and political conditions were significant factors in the outcome of the revolutions. Once in control, revolutionaries frequently use instruments of force to eliminate opposition, create new mass organizations, and establish surveillance systems to prevent a counterrevolution.

SUMMARY OF UNDERGROUND FUNCTIONS IN REVOLUTION

In terms of the revolutionary dynamics of a revolutionary movement just described, all revolutionary functions up to the militarization phase usually are clandestinely performed *within* or by the underground, which—

(1) organizes the revolutionary movement;

(2) controls and coordinates all revolutionary activities;

(3) provides internal administrative functions for the revolutionary movement, such as recruitment, training, indoctrination, finances, logistics, communications, security;

(4) conducts subversion of the existing government's personnel and institutions;

(5) conducts psychological operations, both propaganda and actions (e.g., popular fronts) among the uncommitted people of the nation, and among important foreign nations;

(6) establishes "shadow" governments which are to assume power when the revolutionaries win;

(7) provides intelligence information for the revolutionary movement;

(8) conducts sabotage of sources of government control over the people of the nation, including assassination of political leaders; disruption of economic institutions and processes, communication, transportation, armed forces, police, and militia, and "all those things which make the lot of the people a happier one."

(9) provides escape and evasion networks for members of the movement.

During the militarization phase and thereafter, the underground functions in all of the above ways in "nonliberated" areas—areas in which the existing government has a continuing "presence" in one form or another (e.g., military forces political officials, militia, police) and the underground element continually needs to conceal the identity of its members. In this situation, *roving guerrilla units may conduct their own internal command, organization, and training function, but the underground usually performs all other essential functions for those guerrilla elements.* Only in areas where the government "presence" may be sporadic at best—either because it was driven out or because it never had enough troops to effectively control the area—can the revolutionary movement operate overtly, forming provisional governments and controlling the people. The concept of an underground is inappropriate in such a situation, because the people can openly provide food, clothing, shelter, and supplies.

In the consolidation phase, when the underground emerges into overt identification with the new revolutionary government, the government may continue to maintain underground elements to perform counterrevolutionary functions.

THE IMPORTANCE OF UNDERGROUNDS

It is apparent from the preceding two sections that undergrounds are as important, if not more so, as guerrilla elements to the success or failure of resistance or revolutionary movements—if on no other basis than the sheer number of functions performed. But such an assessment of the significance of undergrounds is supported by historical accounts as well. Resistance and revolutionary movements have sometimes succeeded with little decisive assistance from guerrilla troops (e.g., Cuba) and have sometimes failed not so much from defeat of the guerrillas as from successful operations against the underground itself (e.g., Malaya, Philippines). On the basis of the functional analysis and the historical accounts, a number of observations can be made which may account for the strength and significance of undergrounds, whether in a resistance or revolutionary situation.

(1) Because the underground is part of the community and sustains itself from normal sources, there are no geographic restrictions on its operations comparable to those which limit guerrilla operations. Both resistance and revolutionary underground operations can be successfully conducted under a wide variety of rural and urban conditions.

(2) The variety of personnel in an underground movement is the source of a second significance. Because it chooses members from all parts of the population, an underground has direct access to more varied support than the government which it opposes. This direct access to the people is less important in resistance than in revolutionary situations, because the resistance underground is operating against an occupying power whose claim to govern is based on military conquest and whose governing is accomplished chiefly by the use, actual or threatened, of military coercion. A revolutionary underground, on the other hand, is operating against an indigenous government which claims to govern on its legitimacy, which is based ultimately on the cooperation and loyalty of a significant part of the people governed. With this legitimacy as a primary target, the revolutionary underground has, obviously, a critical advantage in its broadly based membership. And given "real" and widespread sources of discontent in a population or an indigenous government whose governing relies more on physical coercion than on legitimacy, broadly based membership facilitates efficient, effective organization and operations.

(3) The clandestine nature of an underground, whether resistance or revolutionary, contributes to its significance. For the individual, membership in an underground does not always require an open, irreversible commitment, while it does give the individual the satisfaction of participating in something of general importance. For the underground, then, the clandestine character of its organization enlarges and to some extent facilitates recruitment.

(4) In considering underground functions vis-à-vis guerrilla operations one should not be misled by the more dramatic nature of guerrilla action. Characteristically guerrilla elements emerge only after undergrounds have prepared the way and, when formed, guerrilla troops are almost wholly dependent for their support on an active underground. The number of guerrillas is typically much smaller than the number of underground members. When a government has destroyed a few guerrilla bands, it has by no means destroyed the underground movement whose clandestine cells in every community can carry on the struggle.

Strategy, Tactics, and Countermeasures

It is also critical, when assessing the significance of undergrounds, to examine resistance and revolutionary movements in terms of the international context in which they occur.

A resistance underground operates in a time when there is an overt international military conflict and in the situation in which one country has already occupied another militarily. The major decisive struggles, on which the success or failure of the resistance movement ultimately depends, is being waged primarily by military means of parties other than the members of the resistance movement.

The usefulness of resistance movements as an aid to conventional forces in World War II was attested to by Generals Eisenhower and Montgomery who referred to the large number of enemy divisions "tied down" by European resistance movements.[16] And support or nonsupport of a particular resistance movement can significantly affect the length of a military campaign and the nature of a postwar political situation. Thus, while resistance movements may be important strategically in "hot" international conflict, they may not be decisive, and the contending parties can treat resistance movements (and their undergrounds) as of secondary or ancillary importance.*

A revolutionary underground, on the other hand, usually operates in a time when there is no "hot" international conflict and in situations in which the contending parties are indigenous—in short, in an internal war. There is no other more decisive *military* struggle on which the outcome of the internal war depends. Rather, the revolution itself is the major and ultimate arena of conflict for the people directly involved. When other nations become indirectly involved, such as in a "cold war" international conflict, by identifying their own interests with the success or failure of various revolutionary movements and heavily supporting "their" side with political, social, economic, or military resources, the outcome of each revolutionary situation is primary and decisive for the international conflicts. Provided such internal conflicts do not escalate to overt international military conflict, the cumulative successes or failures of revolutionary movements represent the ultimate yardstick of victory in a cold war international conflict, and the contending parties *cannot* treat the revolutionary movements (particularly their undergrounds) as of secondary or ancillary importance.

The implications of this assessment of the strategic importance of undergrounds are manifold in terms of the problems faced by internal security forces attempting to counter undergrounds in a revolutionary situation.

Indiscriminate use of internal security measures can alienate the uncommitted, or even those who support the government; for every underground worker eliminated, two more may be created. On the other hand governmental moves increasing the freedom of the people also increase the opportunity for underground members to perform their functions. Major military weapons

---

*This conclusion applies to conventional forms of war, perhaps even wars in which tactical atomic weapons might be used if such wars could occur without escalating to all-out nuclear warfare. In all-out nuclear war the conclusion might not be applicable. In such a situation, it has been argued, resistance movements might be the key to survival, if not victory.[17]

systems—missiles, artillery, bombers, fighter planes—are relatively ineffective against an unidentified enemy subverting persons within one's own camp or influencing people within the town one is committed to help to survive.

Internal security forces are forced to develop and use countermeasures which strike some required balance between physical coercion and political persuasion. The countermeasures must at once (1) protect the people and the instruments of political control against the coercive and destructive actions of the undergrounds and (2) perform a substantial role in positively influencing the people to support the existing government's claim to legitimacy as a governing authority.

FOOTNOTES

1. See S. Y. Teng, *The Nien Army and Their Guerrilla Warfare 1851–1868* (Paris, La Haye: Mouton and Co., 1961), p. 193. The tactics of the Chinese Communists in guerrilla warfare are compared with the tactics of Nien, and the author concludes that Mao Tse-tung's strategy is based on previous Chinese guerrilla movements, not on Marxist-Leninist doctrine.

2. For a discussion of the legal status of unconventional forces see Phillip M. Thienel, *The Legal Status of Participants in Unconventional Warfare* (Washington: Special Operations Research Office, December 1961). The 1949 Geneva Conventions on the "Treatment of Prisoners of War and the Protection of Civilian Persons in Time of War" extends prisoner-of-war status to guerrillas in an organized resistance and in a civil war if they are declared belligerents and if they (1) are commanded by a person responsible for his subordinates, (2) have a fixed distinctive sign recognizable at a distance, (3) carry arms openly, (4) conduct operations within the laws and customs of war. Underground members would not meet these requirements. Hence the occupier could impose the death penalty on a civilian if the person is guilty of espionage, or performs serious acts of sabotage against the occupier, or intentionally causes the death of one or more persons.

3. Julian Amery, "Of Resistance," *In the Nineteenth Century and After*, LXLV (1949), 138–149.

4. Hugh Seton-Watson, *The East European Revolution* (New York: Frederick A. Praeger, 1951). The author discusses World War II resistance movements in Eastern Europe. See also Ronald Seth, *The Undaunted: The Story of the Resistance in Western Europe* (New York: Philosophical Library, 1956), for a discussion of resistance movements in Western Europe. See also Arnold and Veronica Toynbee (eds.), *Hitler's Europe (Survey of International Affairs, 1939–1946)*, (London: Oxford University Press, 1954), for a discussion of the German occupation policy and European resistance efforts.

5. T. Bor-Komorowski, *The Secret Army* (London: Victor Gollancz, Ltd., 1943), Passim.

6. See Jon B. Jansen and Stefan Weyl, *The Silent War* (New York: J. B. Lippincott, 1943), for a discussion of the anti-Nazi underground.

7. Paul A. Jureidini, et al., *Casebook on Insurgency and Revolutionary Warfare* (Washington: Special Operations Research Office, 1962). This study discusses several definitions of revolutions and insurgency and offers some new definitions. The report presents a systematic summary of 23 revolutions from World War I to the present and describes the historical background, environmental conditions, leaders, organizations, techniques, and outcomes of the various revolutions.

8. See Feliks Gross, *The Seizure of Political Power in a Century of Revolutions* (New York: Philosophical Library, 1958).

9. Donald J. Goodspeed, *The Conspirators* (New York: Viking Press, 1961), p. 209.

10. Ibid., p. 225.

Strategy, Tactics, and Countermeasures

11. See Gross, *The Seizure*, for an analysis of the coup, and K. Wheelock, *Nasser's New Egypt* (New York: Frederick A. Praeger, 1960), for a more detailed account.

12. Otto Heilbrunn, *Partisan Warfare* (New York: Frederick A. Praeger, 1962), p. 44.

13. There are many sources for discussion on the principles of guerrilla warfare. See Hope Miller and William Lybrand, *A Selected Bibliography on Unconventional Warfare* (Washington: Special Operations Research Office, 1961). Standard Communist sources are *The Selected Works of Mao Tse-tung* (London: Lawrence and Wishart, Ltd., 1954), a discussion of the Chinese Communist strategy and tactics; Vo Nguyen Giap, *The War of Liberation and the Popular Army* (Hanoi: Foreign Languages Publishing House, 1950), a discussion of guerrilla activities in Indochina; and Ernesto Guevara, *Guerrilla Warfare* (Washington: Command Publications, 1961). A useful source is Heilbrunn, *Partisan Warfare*, for a discussion of guerrilla and counterguerrilla operations. Franklin Mark Osanka (ed.), *Modern Guerrilla Warfare* (New York: The Free Press of Glencoe, 1962), presents a collection of papers on guerrilla warfare and counterguerrilla procedures and policies.

14. Mao Tse-tung, "On the Protracted War," *Selected Works*, Volume II.

15. Mao Tse-tung, "Problems of China's Revolutionary War," *Selected Works*, Volume I.

16. David Lampe, *Danish Resistance* (New York: Ballantine Books, 1957), pp. 28–31; also F. O. Miksche, *Secret Forces* (London: Faber and Faber, Ltd., 1950), p. 141.

17. F. A. Gleason, "Unconventional Forces—The Commanders' Untapped Resources," *Military Review*, 39 (1959), 25–33. The author describes how resistance forces can be used and the kinds of activities which they can perform which are of value to a military command. See also G. T. Metcalf, "Offensive Partisan Warfare," *Military Review*, 32 (1952), 53–61, for a discussion of the use of partisans in future wars.

CHAPTER 2

UNDERGROUND ADMINISTRATIVE FUNCTIONS AND TECHNIQUES

INTRODUCTION

The underground must develop an effective organization in order to carry out its operational missions. An effective organization in turn requires the underground to perform certain essential "housekeeping" administrative functions. Over the years many and varied techniques have been devised for performing both administrative and operational functions. In discussing these functions in this chapter and the next, an attempt has been made to establish the major problems likely to be faced by most undergrounds and to describe alternate ways in which the problems have been handled.

It should be recognized that usually there is no one best way to accomplish a given function. The choice and effectiveness of a given technique depends largely upon the countermeasures used by the security force. As undergrounds adopt certain practices, the security forces invariably develop countermeasures that destroy the effectiveness of these practices. Consequently, both governments and undergrounds are constantly changing techniques and developing new ones.

The use of techniques is also governed by the resources available to the underground. Although sophisticated communications techniques such as "spurt" radio transmission (messages are electronically condensed and transmitted in a very short time interval) and microphotography are sometimes used by espionage rings, undergrounds generally use simpler techniques. In Angola, for example, the nationalist underground has relied upon tribal drums for communications. Diplomatic and intelligence services use complicated, mechanical ciphers to conceal the meaning of messages; undergrounds, on the other hand, rely upon a courier's memory or the use of special jargon to hide message content. Whether a simple or complicated device is used, the ultimate criterion is whether or not it works. Since undergrounds usually have limited means, in this chapter—and the next—the simpler, easy-to-use techniques are presented for each function and, where possible, alternate ways of accomplishing the function are described.

ORGANIZATION

There is no single organizational structure which is the best for an underground under all conditions. There are a number of major factors which must be considered by those responsible for organizing the underground, and on the basis of this consideration, an organizational pattern is used which will best provide for successful accomplishment of underground functions. The four major factors which must be considered are discussed next, after which, patterns of organization that reflect critical organizational principles are described. (The case histories in Part II illustrate various specific organizational structures.)

FACTORS DETERMINING ORGANIZATIONAL STRUCTURE

Predominant Strategy

As indicated in the previous chapter, resistance and revolutionary movements operate on both political and military fronts. The choice of which front an underground will emphasize depends on the likelihood of success in one area rather than another and on unexpected events which compel the underground to revise its predominant strategy. It may be that underground infiltration of key positions in both the government and the military is blocked, as is usually the case in resistance situations and often the case in revolutionary ones. With seizure of the instruments of control impossible, the underground chooses the predominantly political strategy of trying to weaken the government's effectiveness by sabotage and by subverting the people's support of the government.

An underground may shift from a predominantly political strategy to one predominantly military because it is losing popular political support. The leaders may feel that unless they act immediately and aggressively they may lose their last opportunity to seize power. In the Philippines, the Communists were restrained from continuing their activities on the political front when they were denied access to seats in the legislature. To offset the drop in popular support created by this curtailment of their activities, they consequently turned to military action. Also, popular sentiment against the existing government may be so strong that the leadership may feel that they can use this popular discontent so quickly seize power, as the Communists in Greece believed after World War II.

Once having decided to shift primary emphasis from either the political or military front to the other, the entire underground may require major reorganization. General Vo Nguyen Giap reports that in Indochina the transition from political to armed struggle caused a great change in the principles of organization and work.[1] In Malaya, the strategy shifted from political efforts, through the organization of united fronts and infiltration of unions, to the militarization phase of terrorism and guerrilla warfare. After this failed, they turned back to political activity. With each shift in strategy, various organizational units had to be either organized, disbanded, or reorganized, all at no small cost to the movement. These included: escape and evasion nets for both guerrillas and underground; intelligence nets in urban areas, as well as throughout the countryside; secret supply depots and supply routes; and recruiting teams for both underground and military units.

The expected duration of the underground also has a very important effect upon its organization and its activities. Movements which are organized for a very short period of time usually do not have a complex organization, elaborate security procedures, or selective membership (excluding the coup d'etat type revolution). Underground movements which expect to exist for several years usually develop more complicated organizations and undertake a wider range of activities.

The important point is that different predominant strategies will give different emphases to different underground functions, and performance of these different functions, in turn, requires different organizational structures.

Origins and Leadership

In periods of political unrest, or after a country has been invaded, it is extremely risky to attempt to form a completely new clandestine organization. Consequently, undergrounds usually develop within existing social, political, or military organizations. The underground organizer usually must rely upon past friendship so that he will not be betrayed and upon the informal communication channels of organizations with which he is familiar. As a result, many undergrounds have emerged from outlawed political parties or disbanded military organizations. In these cases, they are usually constructed along previous organizational lines.

Historically, the organizational character of the underground movement has differed according to whether the movement has its beginning in political or military organizations and whether it was led by a political or military leader.[2] Organizations controlled by politicians are frequently found in revolutionary warfare. Also, many of the European underground organizations of World War II developed from prewar political parties and were directed by political leaders. Consequently, the organization reflected the influence of the political leader.

On the other hand, although many resistance groups may organize spontaneously, national groups are usually sponsored and supported by external conventional military forces. Since it is usually through this military assistance that the movement gains its stature, the choice of underground leaders is influenced by their knowledge of strategy, tactics, and military affairs, and their activities are oriented toward providing aid to conventional forces. In Yugoslavia, for example, the Allied Forces withdrew support from Mihailovic and recognized and supported Tito because they felt he was carrying on aggressive action against the Germans and would thereby be of greater assistance to the military mission. This support, both political and military, greatly affected the growth and expansion of the partisan organization.

Some movements have both political and military leaders who function independently of each other. In this form of organization, the underground operates politically in the populated areas, while the guerrilla forces are led by a military man in the rural areas. This occurred in the anti-Fascist resistance in Italy during World War II and in Morocco during the fight for independence (1953–55).

Strategy, Tactics, and Countermeasures

Types of Organization

Organizational structure varies with the organizational theories of the resistance or the revolutionary leaders.

Mass Organization

The assumption is made that a large number of people are necessary to overcome the power of the governing authority and its instruments of force. This type of organization was advocated by the Mensheviks in the Russian revolution. Membership is open to anyone who wishes to join and the objective is to recruit as many people as possible. One disadvantage of this organizational structure is the loose security measures associated with it. The members are usually not practiced in security precautions and the identity of underground members is easily obtained through loose talk and careless, overt actions. However, organizations of this type have managed to minimize the threat of informers primarily through the public sympathy for the movement and through the use of terrorism.[3] In the early days of the OAS in Algeria, Gen. Raoul Salan moved about openly, even while the French Government was seeking him and had posted a reward for him; the populace was terrorized into remaining silent. Another disadvantage of this type of organizational structure is in its command and control structure. It is difficult to obtain concerted action against the governing authority, usually due to the lack of training and discipline of members.

Elite Organization

The theory here is that a small elite organization can make up in skill and discipline what it lacks in size and that at the proper moment, a small militant group can accomplish more in one blow than a large mass organization can accomplish over a prolonged period of time. The membership in a movement such as this is small and each individual is carefully screened and tested before he is permitted to join. Once a member, he is subjected to intensive training and discipline to develop the skills necessary for clandestine work. This type of organization usually works toward a coup d'etat, or a revolution from the top. In a police state, where the mechanisms of internal security are extensive, this is the most common form of underground. An elite organization generally must have some condition of internal confusion, such as the assassination of the head of state or rivalry between several major political factions, before it can strike with any hope of winning. The disadvantage of this type of organization is that it must remain relatively inactive while waiting for the proper moment, and inactivity usually works against a movement because its members may lose their enthusiasm.

Elite-Front Organization

The Communists usually work from this type of organizational theory. Recruitment is very selective and the party itself does not expand rapidly. Instead, a "front" organization is created or an existing organization is infiltrated which claims to seek some popular objective such as liberation or independence. Within the front movement, military and civilian organizations are established. Since the Communists organize these groups, they are usually in leadership positions. If the movement fails, the Communist underground is not damaged either organizationally or by reputation since it is the front group and not the Communists who lose the insurgency. On the other hand, if they are successful, the Communists are in firm control of the revoluntionary organization.

Conflicting Needs of Security and Expansion

An underground operates in a hostile environment in which government forces attempt to seek it out and destroy it. In order to survive, and at the same time achieve its ultimate objectives, an underground must adapt to two major factors, government countermeasures and its own successes and failures.

Changes in the organization and its size are dependent upon the effectiveness and size of the government security forces. As the government becomes more effective, the underground must emphasize security—and this usually means smaller and fewer organizational units.

An underground may, after a few spectacular successes, find itself deluged with new recruits. If it fails to expand its organization quickly by relaxing security measures, it may pass up an opportunity to seize power and attain its objectives. On the other hand, if it recruits unreliable persons, a serious security leak may enable the security forces to destroy the entire organization. Also, rapid expansion may provide government security forces with the opportunity to infiltrate the underground.

Thus, undergrounds face a dilemma of conflicting goals. In order to achieve their objectives, they must be expansive and aggressive. In order to survive, they must take precautions and prize security. Organizational patterns and size must be juggled so as to achieve an optimum balance between the need to expand and the need to maintain security.

Summary

In summary, the following major factors affect an underground's organizational structure:
(1) Predominant strategy of the resistance or revolutionary movement, which in turn is determined by—
    (a) Access to positions in the government and the military;
    (b) Popular support;
    (c) Expected duration of movement;
(2) Organizational antecedents and types of leaders;
(3) Size and membership of movement according to prevailing revolutionary theory or doctrine;
(4) The conflicting needs of security and expansion.

Strategy, Tactics, and Countermeasures

PATTERNS OF ORGANIZATION

Command and Control

To be effective an underground needs concerted action. Unless it can establish a centralized command, an underground's activities occur in a haphazard manner and may lose much of their cumulative effect (e.g., units may attack the same targets and draw government attention to each other). However, for security reasons, decentralization of activities is most desirable. Therefore, a compromise must be achieved.[4]

Since a command echelon cannot direct each subordinate unit, it must rely on mission orders; that is, the central command issues orders describing the tactical objective and recommends activities that it believes can best accomplish the objective. Each of the subordinate units, which must place a premium on survival, can devise its own plan for carrying out the orders. Consequently, the subordinate units usually have the authority to make independent decisions on local issues, and to operate autonomously with only general direction and guidance from the centralized command. When special assignments are given, the central command may send a special representative to the subordinate underground unit to supervise the task directly. In this case, the special representative, being responsible to the central command, is usually placed in charge of the unit.

The central command may not know precisely how many members belong to the subordinate units of the organization. It may test the strength of the organization by calling demonstrations, strikes, or other trial actions. This provides the central command with some estimate of underground strength and ability to react without jeopardizing the identity of its members. It can also determine the length of time necessary to mobilize its units through the clandestine communications net, the approximate number who can be reached, and finally the ultimate number who participate.

The underground usually specifies in great detail the transfer of authority as well as the procedure for reestablishing the chain of command in case a leader is captured. This is to ensure that the capture of an important leader will not necessarily lead to the collapse of the organization or an interruption of its activities.

Centralization of Administrative Functions

Undergrounds which function for a prolonged period of time generally centralize many activities in the central command so that subordinate units may receive services which they could not ordinarily provide for themselves. Such activities as the production of false documents, the collection of funds, the purchase of supplies, analysis of intelligence information, and security checks on new recruits may be better performed by a central agency. These centralized activities can best be performed outside of the country, in areas in which government security measures are lax, or in a place of sanctuary. Here members can meet openly and discuss plans and procedures without fear of being captured or of having records fall into the hands of the security forces. In Algeria, the internal command was buttressed by an external political-military command which was usually in Tunisia. In the Philippines, the Communists had not only an internal underground command, the "politburo-in" (in Manila), but a "politburo-out," safely located in guerrilla-held territory where much of the collating of intelligence and planning took place. During World War II, much of this centralized activity was conducted for European governments by the governments-in-exile which were located in England.

Decentralization of Units

The basic underground unit is the cell. The cell usually has from three to seven members, one of whom is appointed cell leader and is responsible for making assignments and checking to see that they are carried out. As the underground recruits more and more people, the cells are not expanded—rather, new cells are created.[5]

The cell may be composed of persons who live in a particular vicinity or who work in the same occupation. Often, however, the individual members do not know the place of residence or the real names of their fellow members, and they meet only at prearranged times.[6] If the cell operates as an intelligence unit, its members may never come in contact with each other. The agent usually gathers information and transmits it to the cell leader through a courier or maildrop.* The cell leader may have several agents, but the agents never contact each other and only contact the cell leader through intermediaries.[7] Lateral communications and coordination with other cells or with guerrilla forces are also carried out in this manner. In this way, if one unit is compromised, its members cannot inform on their superiors or other lateral units.

To reduce the possibility of discovery of its members, the underground disperses its cells over widely separated geographic areas and groups. This extends the government security forces so that they cannot concentrate on any single area or group. For example, the underground attempts to gain as wide a representation as possible among various ethnic and interest groups. The Malayan Communist Party was easy prey for security forces partly because it was almost entirely composed of indigenous Chinese. The OAS in Algeria was composed of Europeans who numbered only one out of every 10 inhabitants. Even more crucial for the OAS, it was centered in three cities, thus allowing the French to concentrate their forces. The FLN, on the other hand, was made up of Muslims, who constituted 90 percent of the population and were distributed throughout the country.

An underground is usually organized into territorial units. The size of the units depends on the density of the population and the number in the underground. Each territorial unit is subdivided into districts and finally into cells. Within each of these geographical districts, and on each level of organi-

---

*A maildrop is a place where a message may be left by one person to be picked up later by another.

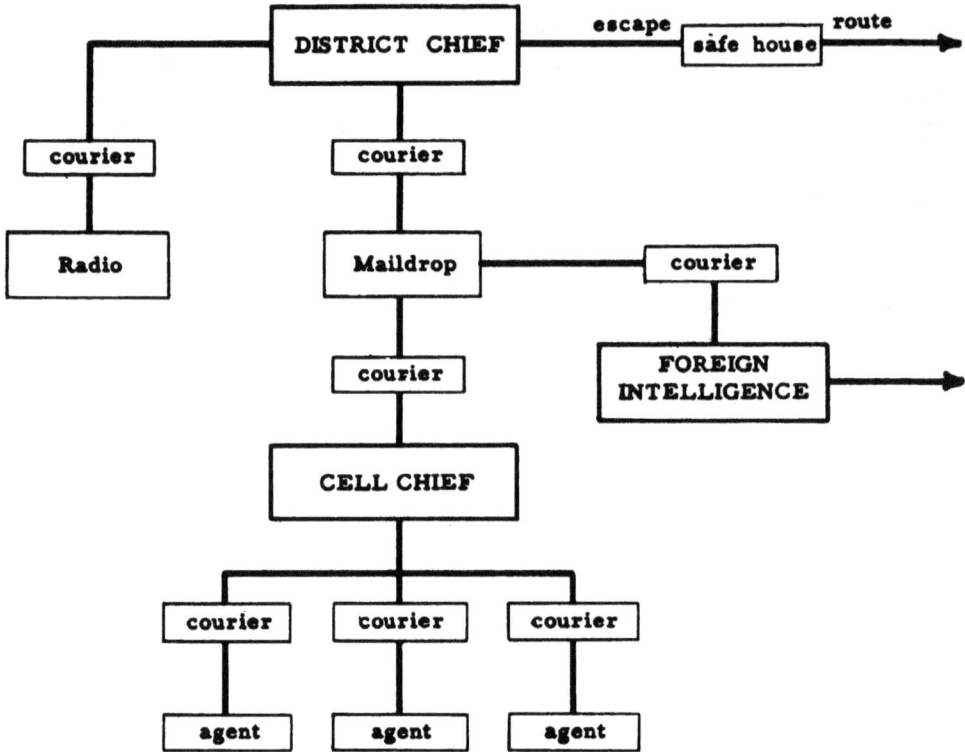

Figure 2. Intelligence Network.

zation, different functions and groups responsible for those functions are represented. In addition, undergrounds have found it convenient to organize units within existing occupational activities such as railroad workers unions. Thus the underground is organized by territorial units, such as state, county, city, and cell, as well as by professional and occupational groups which transcend traditional boundaries.

The underground organization and many of its activities are based upon a "fail-safe" principle, that is, it is organized so that if one element fails, the consequences on the total organization will be minimal. Almost all clandestine organizations which are susceptible to compromise by security forces have parallel organizational units and networks of units. In every case the underground attempts to have a back-up unit which can perform the same duties as the primary unit if the latter is compromised. It usually takes a long time to establish a unit or net, and the underground must plan for contingencies such as the compromise of the primary unit or increased government security measures. Thus, the organizational expansion of undergrounds is usually in a lateral direction by duplicating units and functions. The decentralization extends to all functions. The underground usually does not jeopardize intelligence units by demanding that they perform sabotage as well, for sabotage operations may draw attention to individuals and compromise their usefulness as intelligence agents.

COMMUNICATIONS

Critical messages, such as emergency warnings to other units, must get through with speed and certainty. Therefore, undergrounds use parallel communication nets—that is, an important message is sent by two routes in order to ensure delivery. For less vital messages, the underground uses a back-up system: the message is sent, and without immediate confirmation that the message arrived, another message is sent through another channel. In all clandestine operations, acknowledgment of the receipt of a message is crucial. If positive confirmation is not received, or if there is considerable delay, the underground must assume for security reasons that the message has been intercepted and that the network may be compromised. Here again the fail-safe principle is applied. If a message is believed to have been intercepted, members who are most susceptible to compromise either prepare cover activities or proceed to some safe-house and await events to determine whether the ring has been compromised and whether it is necessary for them to use an escape and evasion net.

Parallel communications are important because it takes considerable time to set up a communication net in a hostile environment. Every effort is made to have alternative means of communication available in case there should be a breakdown in any one channel. In Malaya, the Communist Party was forced to rely entirely on runners to deliver messages and this constituted one of its major handicaps. A message may go to one address, be picked up by a courier and delivered to another place, and thence be delivered by another courier to the recipient. In this way the message moves through a chain-like system of maildrops. Obviously, for routine messages, it is advisable to use maildrops or couriers since they ensure the security of the various units. Emergencies may require that the message be sent directly by courier or by radio.

RECRUITMENT

Techniques of recruitment vary with the stage of underground development. In the clandestine organization stage, recruiting is on a highly selective basis with an emphasis on security. After leaders and cadres are trained, however, recruitment takes on the character of mass enrollment. In the early stages, a great deal of emphasis is placed upon the abilities and reliability of the recruit, since these men will eventually be the leaders of the movement and will possess important organizational information. During the mass recruitment stages, the underground enlists the support of the entire populace in order to obtain sufficient numbers to overcome the security forces. Mass recruitment techniques place more emphasis on active support than on security.

SELECTIVE RECRUITMENT

For many tasks, no qualifications beyond a certain degree of intelligence, emotional stability, and reliability are required of recruits. Some activities, however, call for special qualifications.

Leadership Tasks

In recruiting for leadership positions, an underground seeks persons who possess what one former underground leader termed "ability and talent in handling people." [8] Of course, this is especially important in an underground since units are usually operationally autonomous and depend heavily upon their leaders to give tactical direction, enforce discipline, and sustain morale. Sometimes these persons are recruited from the ranks of national or local leaders in various occupations, since they already have proved leadership abilities as well as influence. An example of this type of recruitment was seen in Algeria, where the Arab *Organisation Secrète* directed part of its first recruitment effort at Arab tribal chieftains. Similarly, the underground might recruit politicians, labor organizers, priests, etc., as leaders.

Intelligence Tasks

Some intelligence tasks call for the recruitment of persons with access to important information. Gathering intelligence is a major activity of an underground, and to do it effectively an underground contacts and develops sources in all areas of interest. The range of persons so utilized may be wide, varying from a peasant woman living near an enemy convoy route to a person in a sensitive governmental bureau.

Immediate access to valuable information is not the only criterion for selection, however. Sometimes the underground recruits individuals to provide information which is valueless in order to keep them on a string, in the hope that they may advance in their profession and be in a position to supply useful information someday. If time permits, it may be better to wait for someone to work his way up into a position of trust rather than to seek to recruit someone in a high position with the attendant risk of failure or compromise. [9]

Special Tasks

A variety of technical skills is frequently required. Doctors may be recruited on a standby basis to treat wounded personnel. Skilled carpenters are enlisted to construct false partitions in rooms to hide equipment or refugees. Metalworkers are used to construct small arms and bombs. Also, bookbinders, who can bind messages into the covers of books, may be brought into membership. Printers are often contacted to assist in publishing propaganda tracts. In addition, locksmiths might be recruited to make special keys, and chemists to mix explosives. One underground leader had a close assistant whose sole, permanent assignment was to secure the services of persons with such technical competencies.[10] · To obtain food for refugees or guerrillas, an underground may establish contact with some food distributor. Recruits from the police force aid in the escape of imprisoned underground personnel, and provide needed passes and documents. Also, trucks and drivers are sometimes needed to carry underground personnel on raids and to transport food and arms to friendly military forces.

MECHANICS OF THE APPROACH

Selecting Reliable Recruits

The Use of Friends
Naturally, only persons whose apparent attitudes indicate they will be loyal to the movement are contacted for recruitment. This leads to the frequent use of members' friends as a source of recruits, since much is already known about their views. Such individuals are not likely to inform on friends in the organization who try to recruit them. Friends are particularly useful when quick recruitment is needed and there is not enough time to appraise and check other prospects. British citizens who had spent time in France were often sent to that country by the Special Operations Executive (SOE) to establish resistance cells among former acquaintances.[11]

Many friends, however, would not be willing to join, perhaps because of family commitments. A member may attempt to determine whether his friends are available before revealing his subversive interests, thus concealing his underground connections from as many outsiders as possible. The tactic used by an underground organizer in Nazi Germany illustrates this point. Werner Michaelis had been expelled in 1933 from his job as leader of a mining union local, for being politically unreliable. In 1934 he began systematically to look up all his old friends in the district. He would casually drop in on them, and if he noticed that a former acquaintance seemed nervous, he would explain that he was looking for a job, drink a cup of coffee, and go. But if he got the feeling that his host had remained loyal to the movement, he would stay and ask the person to join a political discussion group. In this way, Michaelis established a network of likely underground workers without prematurely revealing his illegal motives.[12]

Not all friends willing to join are suitable for recruitment. Those who are politicians, labor leaders, newspaper editors and reporters, professors, student leaders, or religious leaders in known opposition to the government are sometimes avoided, since they may be under police surveillance. In Germany, immediately after the Nazi takeover, an underground organizer who was a member of the Communist Party avoided contacts with party leaders. He knew that the Gestapo was making frequent arrests of central committee members, and the only party people that he approached were functionaries and members of the party with lesser-known backgrounds.[13]

Strategy, Tactics, and Countermeasures

Development of Recruits

To expand its membership an underground may have to develop recruits. An example of this is the recruitment activities in front groups in postwar Malaya. Many unions were Communist led, and it was not uncommon for the leaders to give party-line lectures to union members several times a week. Party members had a standing assignment to notice the workers' reactions. Those who seemed interested and receptive to this educational prodding and who possessed leadership ability were marked as prospects. It should be noted that the Communists did not always stress doctrinal themes, but rather based their appeals on noncontroversial generalities such as Chinese solidarity, general social welfare, trade unionism, and anticolonialism, and transferred the widespread loyalty evoked by these themes to the Communist Party.[14]

Surveillance of prospective recruits by the Communists in Malaya was a long process. Interviews with one group of 60 ex-party members revealed that those who had been party leaders were scrutinized for an average period of 18 months in unions and other front groups before being admitted to probationary membership, while those who had been rank-and-file members were under observation for an average of 2 years. Party members engaged in surveillance were often forbidden to make recruitment propositions to the prospects. Instead, they were instructed to pass on the names to outside recruiters for further action. In this way, the members doing surveillance concealed their underground connections and escaped the danger of being denounced by prospects whose attitude they had misjudged.[15]

Recruitment

After being cleared in security investigations, prospects are recommended for recruitment by the members with whom they have had the closest contacts. To prompt members to make thoughtful appraisals before sponsoring a prospect, undergrounds sometimes hold members personally responsible if the recruits they recommend should provide unreliable. The Malayan Communist Party (MCP) Constitution even provided for the expulsion of members who recommended such individuals.[16] In practice, however, this step was seldom taken, because recommendations had become mere formalities and it was considered unfair to expel members under these circumstances.[17]

In the first direct proposition to a prospect, care is taken to give him only the bare minimum of information. This precaution is taken not only because he may refuse to join—and thereafter would be considered a potential informer—but because recruitment at this stage may be only tentative, with checks and double checks still to be made.[18]

Probationary Periods

Sometimes it is not possible to run full security checks immediately. In such cases the prospects may be recruited but not permitted to meet other underground members until full checks have been conducted. These recruits deal with only one person from the cell, and at some location other than the cell's regular rendezvous. One anti-Nazi cell in Germany made it a practice to have one of its cell members meet unchecked recruits in a crowded coffeeshop or in woods. Only after the background of the recruits had been checked were they permitted to come into contact with the other cell members.[19]

Recruits often undergo a period of trial membership before being accepted into regular membership. They are assigned menial tasks until they learn proper security procedures and demonstrate their abilities. In the MCP, the periods were supposedly set by the party Constitution: workers, peasants, and soldiers were to undergo a 2-month probation; skilled workers, tradesmen, members of the intelligentsia, students, and government workers were to be tested for 3 months; members of the "petty bourgeoisie" were to be subjected to 4-month periods; and former party members were to be put on trial for 1 year before being granted full status.[20] To fill depleted ranks, though, these periods often were greatly reduced.[21]

MASS RECRUITMENT AND POPULAR SUPPORT

The underground must rely on mass support and not merely on conspiracy if a revolutionary movement is to succeed.[22] Furthermore, public sympathy is vital since it provides a shield for the underground activists against the security forces. Once the underground leaders and cadres have been recruited and trained, the next objective is to obtain support from a large segment of the populace and to expand the base of the underground.

Urban Areas

Underground workers organize front groups with humanitarian or other legitimate goals and infiltrate large organizations to obtain positions of leadership. When they have thus developed mass followers, the underground workers introduce the objectives of the underground and divert the membership to support their own cause. In Indochina, Vo Nguyen Giap reports that in order to prepare for insurrection, they had first to develop and consolidate organizations, then expand other organizations in cities, mines, plantations, and provinces, for only on the basis of strong political organization could semiarmed organizations be set up.[23]

Cells are organized in cities throughout the country in a unified effort to win the sympathy and ultimately the support of the populace. Underground workers are instructed to look for and to help families who have lost everything, those who have not received needed assistance from the government, the unemployed, and the sick. In this manner, social indebtedness is created, enabling the worker to gain entree to a family or neighborhood. Once the underground has gained the attention and loyalty of a large number of people, pressure is applied to make them repay the indebtedness by joining the movement. A favorable word from a mother, father, or relative, or close

Strategy, Tactics, and Countermeasures

friend can be a more powerful persuader than any impersonal propaganda message.[24]

Rural Areas

In rural areas and small villages it is difficult to secretly organize underground cells because of the close personal contacts among the villagers. In this case a different technique is used. A rebel force marches in and takes over the village. They "elect" a local government. They work in the fields, help in production work, assist farmers during floods and droughts.[25] They assist the villagers by performing civic and public functions which the central government is unable to handle. After providing many useful services which create a villagewide social obligation, it becomes relatively easy to begin the steps of indoctrination, exacting taxes, and recruiting members for the guerrilla bands as porters or fighters and others for clandestine underground cells. These cells are organized to provide the guerrilla force with food, shelter, and intelligence concerning the tactical movements of the government forces. If the guerrillas are driven from the area, it is extremely difficult for the reentering government force to determine which villagers cooperated with the rebels out of expediency or fear for their lives and which cooperated out of sympathy for the movement.

SPECIAL TECHNIQUES

Bribery

An underground may desire the services of persons who cannot be recruited in the usual manner because of their apathetic attitude toward the movement. In such a situation, the underground may resort to bribery. This needs little explanation, except to say that bribery need not be blatant. For example, an underground worker may know of someone who is suffering financial troubles and give or loan that person "just a little to help out." As one author describes it:

> The transaction thereafter can be managed with a smoothness born of centuries of Old World practice. "Obligate the fellow" is the venerable Foreign Office formula. And perhaps without ever being aware of taking bribes the victim will do his bit in a slippery underground enterprise . . . .[26]

Blackmail

In recruiting individuals the underground may force cooperation by using blackmail, whether it be intimidation or threat of public exposure of private misconduct. Also, any persons who have assisted the underground in any way may be forced to join by a threat to expose their illegal acts. For example, one Malayan plantation worker was enlisted to supply tobacco to the *Min Yuen*.

One day he was told that one of his contacts had surrendered and was going to reveal everything. In this way the Communists tricked the worker into leaving his job and joining the *Min Yuen*.[27]

Other Means

If the underground has succeeded in demonstrating its capacity to destroy the security of a community, perhaps through bombing or assassination, an atmosphere of terror may prevail throughout the area so that the inhabitants are afraid not to comply with underground requests. In such a situation, the threat of physical harm need not be explicit. Peasants may know that unless they assist the local underground in feeding guerrilla troops, they will suffer reprisals. If an underground cannot develop an atmosphere of general insecurity, it may have to threaten directly the individuals whose services it requires. The threats may be conveyed through letters, phone calls, or personal confrontations.

FINANCES

THE USE OF FINANCES

Payment of Underground Expenses

Depending upon their activities, undergrounds may need money to meet the following expenses: the salaries of full-time workers in the organization; advances of money to persons traversing an underground escape route who need money to pay contacts or buy food; the purchase of paper, ink, and equipment for propaganda publications; the purchase of explosives and other materials for sabotage; and the purchase of such equipment as typewriters and radios. An underground may also extend aid to families who shelter refugees, to enable them to buy extra food. This happened in Belgium after the Nazis eliminated many resistance collaborators from the bureaucracy. Previously, these sympathizers supplied fugitives with documents enabling them to switch identities and hold jobs. When this source of papers no longer existed, it was necessary for many evaders to go into hiding. Money to care for them was supplied by the treasury of the *Armée de Belgique*.[28]

Financial aid may be extended to the families of underground workers who have been captured or forced to flee. Typical of this was the support given by the Luxembourg resistance to the dependents of 4,200 persons who were deported and nearly 4,000 who were sent to prisons and concentration camps during the Nazi occupation. *L'Oeuvre Nationale de Secours Grande-Duchesse Charlotte* not only provided immediate care for orphans, but also gave each a 30,000-franc trust fund.[29] At the same time in Belgium *Fonds de Soutien* (Funds for Support) was begun by the *Mouvement National Belge* for the families of workers in hiding.[30]

Strategy, Tactics, and Countermeasures

Money is also needed for bribery. The Japanese students who were paid to take part in the attack against White House Press Secretary James C. Hagerty received 1,000 yen each, or approximately $2.78. For participating in other demonstrations, they were given from 350 to 500 yen. Japanese security personnel estimated that the 5 weeks of demonstrations cost the Communists as much as $1.4 million.[31]

Support of Military Units

An underground also may channel funds to military units to pay salaries and buy supplies. In the Philippines it was a prime responsibility of the Communist Politburo in Manila to obtain money for the Hukbalahap movement;[32] and in Malaya, the *Min Yuen* was the major supplier of money to the rebels, obtaining many funds by extortion from large landowners and transportation companies and by appropriating cash from Communist-dominated unions.[33] Manufactures available only in urban markets outside the control of the military units are often procured by the underground.

EXTERNAL MEANS OF FINANCING THE UNDERGROUNDS

Sources of Money

Foreign Governments
Often an underground is aided by an outside sponsor, usually a government. Much money for the anti-Nazi Belgian resistance, for example, came from franc reserves in London released by the British Government. At one time, 10 million francs a month were forthcoming.[34] Similarly, much of the funds used by the French resistance were remitted from the Bank of England or sent from the Bank of Algiers (after the Allied landing in North Africa).

Outside support is extended to undergrounds for several reasons. The most important is that the activities of an underground often contribute to the defeat of a common enemy. Such aid also enables the sponsor to demand some reciprocity on the part of the underground.

An outside government may give financial assistance to an underground even if there is no common enemy. According to one report, such a case occurred in 1940 when the Japanese Government—not yet allied formally with Germany and Italy—provided some Polish underground persons with financial aid as well as technical equipment and Japanese passports in exchange for intelligence data on the German and Soviet occupying forces.[35]

Friendship Societies
In addition to governmental support, funds may be channeled to an underground by friendship societies or quasi-official aid groups. Perhaps the best known of the latter was the Jewish Agency in the Palestine revolution which had offices or representatives in every part of the Western world. Open appeals for money were made in newspapers and lectures and at charity balls and other social events.

Types of Funds

Cash in the Local Currency

Aid is often given in the form of cash in the local currency, which has the advantage of being easily exchanged for goods or services. The main problem is the physical transfer of the money. Usually this is handled by a front business organization, through diplomatic channels, through clandestine couriers, or by airplane-dropped agents.

Substitute Currency

Hard currency, such as U.S. dollars or British pounds, is sometimes given to an underground when the sponsoring government lacks adequate reserves of the local currency. This makes a good substitute because it is easily exchanged on the black market for local currency or goods. Hard currency is useful also when the local currency is confiscated by the authorities and replaced by scrip, a frequent government countermeasure.[36] This was used by the Castro regime soon after the Cuban revolution.

Dollars were used extensively in financing World War II undergrounds. One British agent in Yugoslavia reported that it was no trouble to use dollars (or gold pieces) since "there was invariably a market for 'good' money in the towns."[37] In France, the organization, *France d'Abord*, was able to exchange dollars for francs by utilizing diplomatic channels. In one case in April 1943, this organization received $45,000 in U.S. money. The money was turned over to an attaché of the Hungarian Legation in Vichy, who took the dollars in a diplomatic pouch to Switzerland, exchanged them into francs at the black-market rate, and brought the money back to France.[38]

Governments-in-Exile

An exile government may raise money for an underground by floating bonds. Since these are often supplied to an underground for sale in the country of operations, this method will be discussed in greater detail under internal sources of financing.

Counterfeit Money

One other way to finance an underground movement is through the use of counterfeit money. Although production of such money is not exclusively the province of a sponsoring government, undergrounds usually lack the necessary facilities and technical competency; therefore, the main effort is generally undertaken by friendly governments. Of course, the use of counterfeit money adds to the dangers already facing underground members. During World War II, this factor reportedly prompted the Polish state underground to reject an offer of counterfeit money from London.

Strategy, Tactics, and Countermeasures

INTERNAL MEANS OF FINANCING THE UNDERGROUNDS

Noncoercive Means

Gifts

Voluntary gifts from wealthy individuals and, occasionally, from commercial enterprises have constituted a good source of income for many undergrounds. A few wealthy Chinese businessmen in Manila made large gifts to the Hukbalahap;[39] the Malayan *Min Yuen* received substantial aid from several Chinese millionaires in Singapore. Many industrialists and bankers provided funds for the anti-Fascist underground in Italy.[40] Donor firms in France during the resistance encountered difficulties in hiding their donations from the Germans, and this hampered the exploitation of this source of revenue. Donations from individuals were more easily covered up. Financial gifts to the underground also come from friends and relatives of underground workers. Given the manpower and opportunity, an underground can make door-to-door canvasses for contributions. Dues levied on underground members also provide needed funds.

Loans

The underground may also borrow funds. The Yugoslav Partisans, for example, floated a 20-million-lira loan which was marketed among the Slovene populace as "Liberty Loans;"[41] and the *Service Socrates* organization of the Belgian banker, Raymond Scheyven, managed to borrow in the name of the government-in-exile over 200 million francs for the anti-Nazi Belgian underground from the end of 1943 to liberation.[42]

A problem that sometimes confronts an underground worker in soliciting funds from strangers is that of convincing them of the agent's good faith. The underground may provide him with an official-looking document authorizing him to collect funds and sign notes. The *Service Socrates* used a more complicated system, however. This organization invited prospective lenders to suggest a phrase to be mentioned on the BBC on a given night. The underground passed the requests on to the London authorities, the phrase was broadcast at the designated time, and the individuals knew that they were dealing with bona fide agents of the underground. To safeguard the *Service Socrates* and the Belgian Government in London against future false claims, lenders were given certificates stating the amount of the loans and bearing a number. Raymond Scheyven, using his pseudonym, "Socrates," signed these certificates, and a copy of this signature was on file in London for comparison at the time of repayment after the war.[43]

If the underground can borrow in the name of some constituted authority such as a government-in-exile, it is more likely to receive a favorable response than if funds are sought in the name of an aspiring underground whose trustworthiness as a debtor organization may be in doubt. As one writer expressed it, governments-in-exile provide necessary "symbols of legalism."[44]

Embezzled Funds

An underground may obtain funds embezzled from government agencies, trade unions, and businesses. An example is the secret appropriations that the Danish resistance received from the Royal Treasury to support the publication, *Information*.[45] Also, misappropriated Grand Duchy revenues constituted perhaps half of the money raised for the anti-Nazi resistance in Luxembourg.[46] Trade union funds were embezzled on a fairly large scale by Communist leaders of Malayan trade unions in the years 1945–47, and provided a major source of income for the MCP until the British replaced the Communists with unionists loyal to the government.

Sales

The sale of various items by a door-to-door canvass or through "front" stores may provide money. Yugoslav Communists once sold fraudulent lottery tickets.[47] The Luxembourg resistance sold lottery tickets as well as photographs of the Grand Duchess.[48] In post-World War II Malaya, the MCP treasury was supplemented by funds obtained from party-owned bookstores, coffeeshops, and even small general stores.[49] Similarly, the Yugoslav Communists raised money through sales made by party-owned clothing stores.[50]

Coercive Means

Robberies

To bring in money, undergrounds frequently resort to holdups. The Hukbalahap in the Philippines, for instance, was able to collect funds by staging train robberies.[51] Likewise, the OAS in Algeria conducted a series of bank robberies. In Malaya, the Communists formed a "Blood and Steel Corps" to engage in payroll robberies and raids on business establishments.[52] Business firms, rather than individuals, are usually the targets of such robberies.

Undergrounds generally avoid outright confiscations from the general populace for several reasons. In the first place, widespread robberies would tend to brand an underground as an outlaw band and destroy its public image as a potential legitimate authority. Secondly, simple confiscations of money would not make the victims compliant servants of the underground, as other forms of coercion can do. Finally, robberies preclude the possibility of exacting continued support under the threat of exposing the affected persons' assistance to the underground.

Forced "Contribution"

Although undergrounds do not rob the general populace, they sometimes coerce individuals into making donations under the tacit threat of reprisals. Aggressive application of this technique is usually reserved for wealthier persons. Typical was the practice of the OAS, which fixed the amounts of contributions to be exacted from persons in the professional occupations, but allowed people of modest means to give what they wanted.[53] A person received a typewritten note in the mail informing him that a *"percepteur"* of the OAS would call in the near future to collect his contribution. The *percepteur* was well dressed, curt, but polite. If his credentials were questioned—some crooks tried to extort money in the name of the OAS—he could show a photostat message signed by the Commander in Chief of the OAS, Gen. Raoul Salan. If the person refused to pay, he would not be threatened, but a week later his car or home would probably be bombed by a charge of plastic explosives. The OAS would then increase its assessment to cover the cost of the reprisal. After a few such object lessons, most of the people approached were willing to make a "contribution," and many agreed to make regular payments.[54]

The Yugoslav Partisans were able to utilize this coercive technique to their political as well as financial advantage. By exacting large amounts from landowners the Partisans were able to weaken their political opponents, and going one step further, they eliminated some of these political competitors by denouncing them to the Germans as helpers of the underground. A check of landowners' financial records sometimes revealed unaccountable deficits, which led to their arrest and the confiscation of their properties.

An underground may suffer a setback, however, if a popular person refuses to contribute and the underground does not dare to make the usual reprisal out of fear of public indignation. A case in point was the widely publicized refusal of the French actress, Brigitte Bardot, to aid the OAS. Such a response serves to weaken the image of infallibility and complete control that the underground tries to cultivate.

Taxes

Taxes may be levied against the general public in areas where enemy forays are not frequent or serious enough to prevent underground municipal administrators from collecting taxes, with the backing of nearby military units. The tax may be levied on a per capita basis, as was done in Philippine areas under Hukbalahap control,[55] or it may be levied on a more selective basis, affecting only persons with regular incomes above a certain level, as was apparently the practice in the Slovene area of Yugoslav Partisan control.[56]

LOGISTICS

Logistical operations are required to meet the materiel demands of both the underground and the guerrilla forces. As underground workers are generally engaged in civilian occupations, they are usually able to provide their own basic supplies of food, clothing, and medicines. What they need are operational supplies—printing equipment, paper, ink, radios, and sabotage implements. Guerrilla logistical needs, including food, clothing, medicines, arms, and munitions, are both basic and operational, and these forces have usually relied in part upon underground logistical operations to provide such supplies.

PROCUREMENT

Purchases

Black Market

Undergrounds sometimes purchase supplies on the black market—from persons who own or have access to certain goods and who are willing to sell or trade those goods in spite of legal restrictions. For instance, some workers in an Italian anti-Fascist underground had the specific assignment of bartering with a black market sponsored by some young Fascists. This market flourished during a period when the demand for staple goods was very high. Reputedly, 220 pounds of salt could be exchanged for an excellent machinegun.[57]

Legal Market

Semifinished items for manufacturing may be purchased from legal firms. In most cases, this is done through a front organization which has a valid need for these items. In World War II Poland the Home Army bought large quantities of artificial fertilizer from two German-controlled factories at Chorzow and Moscice, through agricultural cooperatives and individual farmers. From this fertilizer, the underground extracted saltpeter for use in explosives.[58]

Thefts

Secret Confiscations

Supplies may be removed secretly from plants and warehouses by workers. Italian workers were able to supply the above-mentioned underground with some radios pilfered from stock in factories. The risk in this method was great, however, since inventories were made regularly; further, such confiscations could not be counted upon to produce a steady supply of goods.[59] The problem of inventory checks can be avoided if office clerks are able to account for losses by forging orders and invoices, altering bookkeeping records, etc. This was done by Polish workers in two large pharmaceutical plants in Warsaw to cover the transfer of 5,000 kilograms of urotropine to the Home Army for use in explosives.[60]

Raids

Raids are often made on warehouses or other storage centers. In France during World War II the manager of one warehouse was awakened by 12 masked resistance members who forced him to hand over his keys. There were trucks in the courtyard and 200 men ready to load them. A total of 38 tons of coats, sweaters, shoes, radios, and typewriters were taken.[61] Many such raids were carried out in France after a previous understanding with sympathetic employees.[62]

Strategy, Tactics, and Countermeasures

Manufacturing

Types of Manufactures

Undergrounds frequently engage in the manufacture of such items as mines, flamethrowers, hand grenades, incendiaries, explosives and detonators, boots, mosquito nets, waterproof ponchos, and hammocks. Rarely, however, are they able to turn out heavy equipment because of concealment problems. One exception occurred in France during the Nazi occupation, when workers in a steel mill of Clermont-Ferrand succeeded in constructing four crude tanks out of farm tractors and sheets of steel from the factory. The components were hidden separately inside the plant until they could be welded together and armed with 37-mm. cannons and heavy machineguns.[63]

Rural Manufacturing

The Vietminh achieved a degree of safety in conducting their manufacturing in rural areas under nominal French control by using small, mobile workshops which could be moved from place to place to avoid French forays. The small size and simplicity of these shops aided their mobility—10 to 15 workers generally were involved and frequently manpower was the only source of energy. In spite of their crudeness, these shops were a major source of such items as mines and explosives.[64]

Urban Manufacturing

An underground engaged in urban manufacturing has to use other devices to avoid the enemy. The Polish Army enlisted the services of workers in legally licensed shops, especially metal shops, to manufacture small arms. Production was thus conducted more or less in the open, avoiding the difficulty of completely hiding its noise and bustle. For camouflage, arms were sometimes produced in shops that turned out similar looking items. Hand grenades, commonly known as *"Sidelowski"* since they closely resembled the round cans of Sidel polish, were produced in the same place as the actual cans for the polish, and flamethrowers were made in a factory engaged in the manufacture of fire extinguishers.[65]

In Palestine, the Haganah used the same basic technique, with variations. They established their own shops in industrial sections to avoid attracting attention. These places were devoted primarily to illegal production, although legitimate items were often manufactured at the same time so that production could be switched to "civilian" orders in case of inspections. Posted lookouts were used to warn of the approach of inspectors. Each shop was restricted to the manufacture of parts, which were more easily concealed than the finished products. By bringing the components together only at a well-hidden assembly plant, the underground also avoided the possibility of a raid on a shop in which all of the skilled workers and important machines might be captured. A natural look was also maintained by having open offices, reception desks, and office books which were subjected to inspection by auditors and tax assessors. To further ensure secrecy, only a few men in the underground—those coordinating production—knew the locations and operational features of the shops. Shop workers were selected only after extensive security checks on their backgrounds; they were also encouraged to form their own social milieu, to limit contacts with outsiders and hence lessen opportunities for security leaks.[66]

Urban manufacturing is not always restricted to shops with legal covers, however. The Polish Home Army underground had some small shops that were completely hidden: false walls partitioned rooms and cellars and concealed the quarters of the shops. To conceal the noise of the machine, these shops had to be constructed near places where legal goods were being manufactured. Thus, one was built near a mechanical mangle and another just above a welding shop. Work that involved use of chemicals often had to be done at night so that no one would notice the special colors of smoke rising from the chimneys.[67]

Collections From the Populace

Goods may be systematically collected from the population, although this requires a high degree of underground influence and freedom of action. In rural areas, food is often collected for guerrilla troops. This was done in Greece during World War II. The EAM, through its "Guerrilla Commissariat," supported guerrillas by the levy of regular tithes of foodstuffs from the peasants whom it effectively controlled. In addition to these tithes, for which no payment was made, other foodstuffs were purchased at a scale of prices set by the underground.[68]

To avoid being considered "bandit" organizations, undergrounds often make it a practice to give at least nominal payments or IOU's for goods requisitioned from peasants or other persons of modest means. Ernesto "Che" Guevara of the Cuban *Movimiento 26 de Julio* stated that the fundamental rule is always to pay for any goods taken from a friend. He also stated that when it is impossible to pay simply because of lack of money, one should always give a requisition or an IOU—something that certifies the debt.[69]

Such consideration is not always shown, however—particularly in collecting goods from wealthy manufacturers. For instance, an Italian underground approached industrialists with the attitude that it was their duty to furnish whatever was needed in the field of manufactures. Because they cooperated many of these industrialists were not punished after the liberation.[70]

External Means

Import Firms

An underground may use businesses engaged in foreign trade to import equipment, under noncontraband labels. This occurred in Haganah activities. A textile firm, for example, might order textile machinery, and delivery would be in arms-producing machinery or arms parts. Payment to the firm would be made for goods or services supposedly received, thus keeping all financial records in good order.[71]

(Courtesy of the Norwegian Information Service)

Norwegian resistance member receives a parachute drop from Great Britain during World War II.

Parachute Drops

Supplies may also be obtained from a sponsoring government through parachute drops. Probably the most familiar instance of this type of operation is the drops which the French resistance received from the RAF. Sophisticated radio liaison was necessary in order to work out the details of the drops. Such matters as agreement on drop-zone locations, the exact times of the drops, and ground-to-air recognition signals had to be worked out in advance. Following the drops, which usually took place at night, resistance persons stored the goods in caches near the drop zone so that they might leave the scene immediately and without incriminating evidence. Special liaison agents from abroad were often used to help execute these complex arrangements.[72]

Wartime Equipment

Wartime stores of equipment sometimes provide a postwar source of supplies. For example, the MCP was able to provide guerrillas after World War II with many arms cached during the war. These were arms originally received in air drops from the British, for use against the Japanese. By claiming that many drops were lost, the Communists received extra drops, and only these extra arms were returned to the British authorities after the war. The rest remained in caches and were finally used during the "Emergency."[73]

TRANSPORTATION

By Vehicles

It is often necessary to ship contraband by trucks, in which case a number of devices may be used to hide the cargoes and avoid arousing suspicion. Arms destined by the Haganah for caches in agricultural regions were often hidden in farming implements that were being taken to these places, while consignments to urban areas were frequently put in compressors, gas cylinders, asphalt sprayers, and other industrial pieces. During the orange season, truck cargoes were sometimes covered with layers of oranges which would roll into any hole made in inspecting a cargo. Illegal cargoes were also concealed by tarpaulins covered with fertilizer, preferably with a disagreeable odor. The chances were that policemen, well-dressed and polished, would not insist on a full inspection of such cargo. Another device was the use of trucks of well-known firms such as breweries, whose products were shipped everywhere and in great quantity. These usually escaped suspicion. Underground members dressed as policemen and driving motorcycles sometimes escorted heavy truckloads under the very "auspices of the law." Trucks even succeeded in joining British military convoys, often traveling hundreds of miles and passing many roadblocks with no check at all. It was necessary, of course, to make telephone calls and inform commanders of roadblocks that two or three lorries from another unit had been added.[74]

Strategy, Tactics, and Countermeasures

By Foot and on Animals

Because guerrilla bases are usually in remote areas of difficult accessibility, the transport of supplies to guerrillas has usually not been mechanized. In German-occupied Greece, for instance, the rural "Guerrilla Commissariat" used pack animals as far as they could negotiate the mountain trails, and mountain dwellers carried the supplies the rest of the way.[75] In Vietnam in the early 1950's coolies were used extensively. One Vietminh division required about 40,000 porters to supply its minimum needs. These coolies were local inhabitants organized into what was called the "auxiliary service." On level terrain, the coolies were expected to cover 15.5 miles per day (12.4 at night) carrying 55 pounds of rice or from 33 to 44 pounds of arms. In mountainous areas the day's march was shortened to about 9 miles (7.5 miles at night), and the load was reduced to 28.6 pounds of rice and 22 to 33 pounds of arms.[76]

If several days or nights of travel are required, stopover facilities will be needed. Che Guevara recommends that "way stations" be established for this purpose in the houses of persons affiliated with the movement. According to Guevara, these houses should be known only to those directly in charge of supplies, and the inhabitants should be told as little as possible about the organization, even though they are trusted people.[77]

STORAGE

Supplies are sometimes stored in individuals' houses. More often they are stored in centralized locations, so that fewer persons are subject to capture in the event of searches. Caches are frequently located in remote areas. French resistance people, for example, dug and camouflaged pits at the sites of parachute drops to store equipment until it could be moved to more convenient hiding places.

Remote areas are also utilized as hiding places for the benefit of guerrillas. The Malayan *Min Yuen* collected food in rural areas and delivered it to caches hidden in the jungle, where it was picked up by the guerrillas. In Vietnam, local inhabitants helped "prepare the battlefield" for the guerrillas by storing food near the scene of an impending Vietminh attack. These stores enabled the guerrillas to travel lightly and quickly. Where supplies must be stored for longer than a couple of days, the caches have to be ventilated and insulated against dampness. Of course, the ventilators must be camouflaged. Pipes from Vietminh caches beneath the ground were sometimes covered at the surface by bushes.[78]

SECURITY

SECURITY MEASURES TO CONCEAL OR DISGUISE ACTIVITIES

Personal Anonymity

Conventional Living

If a member of an underground does not appear to be following a normal routine, he may attract the attention of neighbors and enemy security personnel. This is most likely to happen in locales where block wardens provide the authorities with detailed information on the activities of residents.[79] For these reasons, a former leader in an anti-Nazi underground in Germany suggests that members strive to live as "conventionally" as possible:

> You can't hide from the scientific surveillance of a modern police state, but you can mislead the police. And the best way to mislead them is to live as conventionally and as openly as possible. The more you resemble a normal everyday citizen in every respect, the less apt you are to be suspected.[80]

To promote this, undergrounds fill many jobs with persons who can perform their duties to the movement while engaged in legal occupations. Postmen, taxi drivers, traveling vendors, railroad inspectors, or others who travel regularly in their work are frequently enlisted for courier duty. One Vietminh manual covers this point:

> In normal times as well as during periods of operation the cadres and guerrillas must take the occupations of the people into consideration when requesting them to do liaison work. They can ask the merchants, carriers, and hawkers to hide documents in their packs in order to carry them to their address.[81]

Also, an underground may indoctrinate its members with the need for protecting their nonconspiratorial demeanor by avoiding drunkenness, guarding against accidental admission of underground affiliation, and refraining from making boasts of underground exploits and expressions of undue familiarity with underground plans.[82]

Documents

In many countries proper documents are essential; without them a person cannot travel, obtain a job, buy food, or rent a room. Therefore, members who must abandon their normal lives and assume new identities must be supplied with documents such as birth certificates, identity papers, social security cards, employment permits, travel stamps, etc. One technique is to forge the documents: forms are procured, a biography is created and put on the documents, and falsified signatures and stamps are affixed.

The initial problem is procurement of the forms. One underground group in Norway obtained authentic ones from a contact who worked in the police department.[83] Another alternative is to print official-looking blanks in underground printing shops. The Polish Home Army had teams of specialists scattered throughout the country working solely on duplicating official identity cards, labor certificates, movement orders, and other forms used by the authorities.[84]

The fictitious biography that appears on the documents is then created. Of course, it must be memorized so that the user can accurately answer any questions about it. Steps may be taken to make it difficult for the police to check the authenticity of this biography. A frequent practice is to list a birthplace where police cannot easily check birth registration records. The papers of SOE agents in France often carried as birthplaces the names of towns that had been bombed out, and hence whose records had been destroyed. Locales in French colonies were also listed.[85] Another safeguard is to avoid designating the person as a salaried employee or worker. Such men have to be at their jobs, and a check at their stated place of work would reveal the falsity of the documents. Instead, such occupations as peddler, freelance writer, or artist may be used, since these persons are self-employed and need not be at a certain place.[86]

Attention is given to other details to avoid obvious irregularities which would arouse suspicion: signatures and stamps are checked against genuine ones to ensure proper duplications; care is sometimes taken to date the documents when the official who supposedly signed them was not on vacation; the exact color of ink used by the official is used in the forgeries. In preparing birth certificates investigations have been made to determine the style of writing used at the time of the putative birth as well as the stamps and seals used by the issuing authorities at that time.[87]

Many of these problems of authenticity can be avoided if the official documents of another person are used: the legend on these documents is entirely genuine; there is no danger of an imperfection appearing in the signature, etc. It only remains for the user to memorize the details, and for the picture and overlapping stamp to be replaced. Naturally, it is not easy to obtain such documents. Sometimes, however, the underground comes into possession of the documents of persons who have died or disappeared: a Polish Home Army member adopted the identity and used the documents of a fellow Pole of about the same age who had left the country at the time of the German invasion.[88] Another alternative is to borrow documents, although these can only be used for short periods of time. Polish couriers traveling across Europe used the documents of French workers in Poland who were slated to return to France on vacation. Of course, it was necessary to hide and provide for the Frenchmen while their documents were being used.

Meeting Secrecy

Choice of a Site

The preferred meeting place is one where the arrival of a number of persons about the same time will not attract attention nor arouse suspicion. Such a place might be found in a secluded area such as a woods. If this is not con-

(Courtesy of the Natural Rubber Bureau)

Inspection of documents in Malaya during the "Emergency."

venient it may be necessary to assemble in a member's house or apartment. In this case an underground may try to avoid neighborhoods where persons noted for antiregime activities reside, for they are likely to be under government surveillance and the presence of nonresidents in that locale might arouse suspicion. The same caution applies to the homes of persons connected with other underground cells, since they too may be under observation. Also, places near the homes of block wardens may be avoided.[89]

75

Strategy, Tactics, and Countermeasures

Change of Meeting Places

Undergrounds generally change meeting places frequently. If a meeting pattern is fixed, the chances are greater that some outsider will notice the meetings, become suspicious, and report the activities to the enemy authorities. Changing their meeting place may enable underground members to evade a raid by the enemy security forces. There seems to be no fixed rule as to how often meeting places should be changed; however, in general "the greater the number of meetings, the greater the number of changes." One anti-Nazi underground in Germany made it a practice never to hold more than one meeting a week in any one place.[90]

Cover Stories

An innocent explanation for convening a group of people may be announced before a meeting or as the first item of the agenda so that a cover story will be ready in case of inquiries. If possible, the meeting may be arranged to coincide with some genuine, legal occasion for being together. This makes the cover story as plausible as possible. For example, the German underground mentioned previously used birthdays, anniversaries, weddings, and other such occasions as pretexts for assembling.[91]

Miscellaneous Procedures

There are other precautions that an underground may take to protect its meetings. The Communist Party in the United States reportedly employs the following procedures for secret gatherings: arrivals and departures are staggered, since group movements are likely to arouse attention; when meetings are held in homes, members of the family are present to answer door knocks; as few documents as possible are used, in anticipation of searches; and after members have departed, rear guards check for incriminating items that may have been left behind.[92]

Communications Secrecy

Traveling

Unless they travel at night, couriers cannot conceal themselves. Therefore, their activities are generally disguised. Aged men and women and children may be used because it has been discovered that their movements are less likely to arouse suspicion than those of men in the active years of life. An anti-Nazi organization in Italy, for example, found such persons to be best for liaison purposes.[93] Similarly, the couriers of the Polish Home Army were almost exclusively women.[94]

Couriers also disguise their missions by combining them with routine trips, thereby escaping the suspicion of observant persons while at the same time being able to provide a good excuse for their travel in case of inquiry. This is most easily done if travel is a part of a courier's daily legal life; such persons as postmen, taxi drivers, and traveling vendors, therefore, are often used for this work.

Other underground members, as well as couriers, may use the latter technique. For example, the leader of a network in France undertook missions while working as a railroad inspector. Equipped with papers which permitted him to ignore the curfew and move freely about the country, he was well protected while making his frequent underground trips.[95]

Special occasions such as birthdays, anniversaries, and weddings provide credible excuses for travel, as well as for convening a group for a cell meeting. One couple belonging to an anti-Nazi underground in Germany postponed their wedding until a time when two important underground persons could arrive for a rendezvous for which the wedding festivities provided the screen.[96]

Regardless of the cover utilized, the essential point, according to a former underground leader, is that no journey should be made without providing a convincing legal excuse.[97] If he lacks the excuse of the routine trip or special occasion, a member just formulates a very plausible explanation for his travel.

Routine interceptions and searches by enemy personnel cannot always be avoided but they need not result in arrests if the messages can be hidden or disguised. One obvious way to conceal messages is by memorizing them. Messages that must be written can be concealed by the messenger on his person or they can be either coded or integrated into innocuous documents like letters.[98] Disguised messages are less liable to discovery than concealed ones.[99]

Rendezvous in Public

Promptness is important because if an individual is forced to wait for his contact he may risk being picked up for loitering. Resistance people in France, for example, had to be careful of this because loiterers were often arrested on suspicion of being black-marketeers.[100] When one member is late the procedure may call for the other to leave the scene and return later at a previously arranged time. Caution is exercised in returning since the failure of the absent member to keep the initial appointment may mean that he has been arrested.

If underground members are meeting for the first time, they usually establish their identity by recognition signals. These may be visual identification marks and passwords. Generally, passwords are innocent sounding in case the wrong person is approached or the exchange is overheard by bystanders— e.g., a request for directions and an agreed upon reply.

If an extended conversation is anticipated, the members may first settle on the explanation they will give of their personal relationship and their reason for meeting if they should be questioned.[101] This may be omitted if they are meeting only to transmit documents and no conversation beyond the exchange of passwords is required. In this situation, any familiarity between the liaison persons is concealed if they treat each other as strangers and hide the transmission of documents. A common technique is to wrap the documents in a newspaper and lay them where they may be inconspicuously picked up by another person.[102]

Strategy, Tactics, and Countermeasures

Transmittal of Written Messages

Although written messages are sometimes delivered in person, often they are sent through maildrops. One underground used the lavatory in a dentist's office: a courier would leave material under the lid of the tank, and an hour later another would arrive and pick up the papers.[103] Sometimes, individuals are used as maildrops. In this case, people who make numerous public contacts in their daily lives are often selected, because their meetings with underground members would likely go unnoticed. Thus the Polish Home Army frequently used merchants.[104] Likewise, a French underground unit used such persons as butchers, bakers, and tobacconists.[105]

Practice of Random Behavior

Undergrounds generally try to minimize repetitious behavior to avoid an observable *modus operandi*. Activities that may not appear unusual at first may incur suspicion if repeated several times in noticeable succession. Moreover, if security forces detect a routine, they are able to anticipate underground moves and concentrate countermeasures. Courier routes, rendezvous sites, meeting places, codes, ciphers, and perhaps sleeping quarters are therefore changed frequently in most undergrounds.

SECURITY MEASURES TO PREVENT BETRAYAL

Loyalty Checks

Check of Prospective Recruits

One method by which enemy personnel try to penetrate an underground is by infiltrating counteragents. Recruits, therefore, are usually not accepted finally until their past and present records of family life, jobs, political activities, and close associates have been investigated and found satisfactory. Further, most undergrounds require a probationary membership period. If someone is urgently needed before the investigation can be completed, he is sometimes brought into affiliation with a cell and assigned limited tasks, but he is not permitted to come into close contact with the cell members until the full check has been conducted. The usual practice is to restrict his contacts to one member of the cell and to places other than the cell's regular meeting places.[106]

In addition to these background investigations, loyalty tests may be administered to prospective recruits. For example, in Palestine the Shai made it a practice to test some prospective recruits by subjecting them to capture and interrogation by underground persons posing as British security personnel.

Check of Suspected Members

Members suspected of collaboration may also be subjected to loyalty checks. The above-mentioned Shai technique may be used, in addition to others. A suspected person may suddenly be summoned to meet with underground security personnel. If he is indeed a collaborator he may sense pending exposure and try to postpone the confrontation or to desert.[107]   Or the underground may keep a suspected person in ignorance of an important revision, such as a change in the meeting place of top leaders.   Then an observer is posted near the original meeting site.   If enemy security personnel appear, the underground knows that the suspect is an informer since he alone was not informed of the change in plans.

Loyalty Oaths

An underground may administer an oath of loyalty to each new member, mainly to impress him with the seriousness of the job and the necessity for secrecy.   The oath of a World War II Belgian underground is illustrative: members were required to swear that they would "never abandon the fight . . . accept any mission . . . obey all orders . . . never betray the country or organization . . . and serve until death." [108]   Underground members are sometimes warned that a betrayal of confidence is punishable by death.   If the oath is signed, the underground can use it to bring a recalcitrant member into line by threatening to send it to the authorities.

Underground Discipline

Undergrounds require strict adherence to security procedures.   Cell members are generally obliged to report violations, and are subject to punishment if they fail to do so.   To discourage betrayal of fellow workers, traitors are punished severely, often by execution.   Since internal conflicts frequently develop, rules are often established for resolving them by means of courts or higher authorities.   In many cases, hearings and disciplinary actions are taken by the echelon higher than that one in which the infraction reportedly occurred.

MEASURES TO MINIMIZE COMPROMISES

Limited Personal Contacts

Ordinarily a captured member can lead authorities only to members with whom he has had personal contact, since the use of cover names protects the real identities of other persons in the organization about whom he may have heard.   Therefore an underground can minimize the danger of a compromise by minimizing personal contacts among members.   At the base of the organization this is accomplished by organizing workers into cells and by confining their contacts to members of their own cells.   Among leaders in the chain of command this is brought about by limiting one's contacts to his immediate superior and immediate subordinates, and by excluding lateral liaison among administrators in separate branches.[109]

Strategy, Tactics, and Countermeasures

Regulated Liaison

Liaison between echelons is regulated so that in the event a member is captured, he cannot easily lead his captors to the next highest official with whom he regularly conducts liaison. This is done by denying subordinates direct access to their commanders. Contacts with persons in higher echelons are prearranged through intermediaries, with the higher official setting the time and place for the meeting. Since he does not know the superior's address, the subordinate cannot lead police to the superior's place of residence; since he does not control the time of meetings, a compromised subordinate cannot arrange a meeting with his chief before the chief has time to discover the breach.[110] To protect members in lower echelons in case of the capture of a superior official, the usual practice is to conceal from superiors the identities and addresses of those in lower levels, except, perhaps, those with whom the superiors are in direct contact.

Use of Couriers for Liaison

Underground members often face great danger in traveling with illegal documents and false identity papers. To avoid this risk underground leaders frequently use couriers to carry out liaison for them. In Poland, for example, the Polish Home Army used "liaison women" to carry documents and instructions for the leaders. When it was necessary for the leaders to travel, and when they needed material of a compromising nature at the end of their journey, the women would precede their superiors by some distance, carrying the illegal material and assuming the danger in case of police checks.

The problem of avoiding police suspicion was immense, and these couriers were usually uncovered after they had served for a few months. In view of this, these women were not allowed to assume other duties in the underground, so as to limit the information they could be forced to disclose under interrogation. It was also imperative that these women be trained to detect any police who followed them; otherwise they might lead the police to the leaders. A special "observation department" watched their apartments so that if they were arrested or put under surveillance, their contacts could be warned to break off liaison, change their names, and move to new quarters.[111]

Minimization of Records

Official records are frequently held to a minimum, according to the principle that only information which cannot be memorized and which is needed for future reference may be put in writing. When it is necessary to record the names and addresses of underground workers, they are not written "in the clear." Only cover names are used so that if the papers are captured, identities may be protected.[112] This also applies to notetaking at meetings.

Places of Conspiracy

Without some emergency provision it may be difficult for an underground member to reestablish contact with the organization if his superior is captured, removing his only link with the next level of command. To provide a safe method for restoring contact, the underground may use places-of-conspiracy. This arrangement works as follows: each worker is informed of a place where he may go at a certain time of the day, bearing certain identifying marks, to meet a representative of another cell or command. The representative is not recognizable and *he* makes the approach upon noting that the time of arrival is correct and that the worker bears the proper identifying marks. The representative passes details to the organization about the worker's situation, physical appearance, and where he may be reached. After being cleared by a check, the worker is contacted for reassignment. Because of the representative's vulnerable position as a contact for persons in danger, he is limited to this one duty and knows little about other aspects of the underground.[113]

Code Words and Cover Names

In messages, including enciphered texts, underground workers often make maximum use of code words and cover names, which are simply words arbitrarily chosen to designate places, movements, operational plans, and persons. Thus, the code word "Olympus" could represent a rendezvous spot. In assigning cover names, male and female names may be employed without regard to sex. If a message should be intercepted and deciphered, the code words and cover names would still couch the message in a jargon not easily interpreted.[114]

Action in Case of Capture

Reaction of the Underground

If a member fails to keep an appointment or disappears, it is generally assumed that he has been captured and emergency measures are begun. A captured member's family may be hidden or taken from the country so that they cannot be used by the police to intimidate him. This is necessary if the member has maintained his legal identity, for under these circumstances the police can easily check his documents and determine the identity and whereabouts of his family. This precaution, however, does not guarantee that the police will not obtain the information they seek. It merely eliminates one device they can use. For this reason the coworkers of a captured member generally assume that they have been implicated, obtain new documents, and move to new quarters. Also, the underground will probably initiate an investigation to determine the cause of the compromise. One resistance cell in World War II suspected that the owner of a safe-house had caused the capture of a member who had stayed there. The suspect, therefore, was telephoned and told that another member would arrive soon for refuge. The cell then placed the suspect's apartment under surveillance by a young couple who strolled nearby in the role of lovers. Of course, no member arrived, but the Gestapo did. Soon thereafter, the owner was shot and killed by underground executioners.[115]

Behavior of Captured Members

If an underground member is told what to expect in case of capture he will be better able to avoid police tricks or resist their pressures. One method he should expect is police use of an *agent provocateur*, a counterintelligence agent who pretends to be an underground member. He is put into the cell with the prisoner and attempts to win the confidence of his cellmate and obtain information from him. A technique to demoralize a prisoner is to mention a few details about the underground and hint that another underground member has already betrayed the prisoner. Sometimes the police will attempt to obtain information by praising a prisoner's exploits and asking him to explain how he carried out such difficult tasks. A common device is to promise leniency or amnesty in exchange for information. Reinforcing this argument, the police may point out how foolish it is for a prisoner to take all the risks and punishment while the leaders are safe.[116]

Of course, a prisoner may not be able to resist torture. He may try, however, to protect the underground by resisting long enough for his absence to be noticed, which would be a signal to the underground to implement the emergency measures. However, the police may be aware of this delaying tactic. One Danish underground member withstood torture until he noticed by the clock on the wall that the time had passed for a scheduled underground meeting. He then revealed the meeting plans, feeling sure that the members had already departed. His friends were immediately captured because the Gestapo interrogators had expected this tactic and advanced the clock hands by two hours.[117]

Treatment of Released Members

If an underground worker is released, he may not be permitted to immediately reenter the organization. Instead he may be placed under surveillance until it is determined that he is not working with the police or being followed by them. One member of the Polish Home Army was "quarantined" for 6 months after his release.[118] When a member is found to be working with the police, a common practice is to execute him. This is not the only alternative, however, because it may be that he is being forced to cooperate. One released member of a German underground assumed an informant's role after the Gestapo threatened to take action against his family. In view of this man's past loyalty to the organization and the great pressure being exerted on him, the underground decided that the best course was to smuggle him out of the country.[119]

FOOTNOTES

1. Vo Nguyen Giap, *People's War People's Army* (Hanoi: Foreign Languages Publishing House, 1961), p. 76. This book is also available through the U.S. Government Printing Office, Washington.

2. Otto Heilbrunn, *Partisan Warfare* (New York: Frederick A. Praeger, 1962), p. 44.

3. See G. Rivlin, "Some Aspects of Clandestine Arms Production and Arms Smuggling," *Inspection for Disarmament*, ed. S. Melman (New York: Columbia University Press, 1958), pp. 191-202.

4. See J. K. Zawodny, "Guerrilla and Sabotage: Organization, Operations, Motivations, Escalation," *The Annals of the American Academy of Political and Social Science*, 341 (May 1962), 8-18.

5. For descriptions of cell sizes, see Gene Z. Hanrahan, *The Communist Struggle in Malaya* (New York: Institute of Pacific Relations, 1954), p. 90; T. Bor-Komorowski, *The Secret Army* (London: Victor Gollancz, Ltd., 1950), p. 23; and F. B. Barton, *North Korean Propaganda to South Koreans*, Technical Memorandum ORO-T-10, EUSAK (Bethesda, Md.: Operations Research Office, 1951), p. 109.

6. Intracell behavior is described in the following sources: War Department Special Staff, Historical Division (Historical Manuscript File), *French Forces of the Interior—1944* (General Services Administration, Federal Records Center, Military Records Branch), p. 320; Ronald Seth, *The Undaunted: The Story of the Resistance in Western Europe* (New York: Philosophical Library, 1956), p. 259; and Jon B. Jansen and Stefan Weyl, *The Silent War* (New York: J. B. Lippincott, 1943), p. 115.

7. Communication among members of an intelligence cell are discussed in Alexander Foote, *Handbook for Spies* (London: Museum Press, Ltd., 1949), p. 44, and Barton, *North Korean Propaganda*, pp. 151-157. The regulation of vertical liaison is discussed in Rivlin, "Clandestine Arms," pp. 191-202, and Jan Karski, *Story of a Secret State* (London: Hodder and Stoughton, Ltd., 1945), p. 160.

8. Jansen and Weyl, *Silent War*, p. 129.

9. See Foote, *Handbook for Spies*, p. 55.

10. Jansen and Weyl, *Silent War*, pp. 117-118.

11. See Maurice J. Buckmaster, *They Fought Alone* (New York: W. W. Norton, 1958), p. 81. The recruitment of friends is also discussed in Oluf R. Olsen, *Two Eggs On My Plate* (London: George Allen and Unwin, Ltd., 1952), pp. 126-127, and George K. Tanham, "The Belgian Underground Movement 1940-1944" (unpublished Ph. D. Thesis, Stanford University, 1951), pp. 145-146.

12. Jansen and Weyl, *Silent War*, p. 122.

13. Ibid., pp. 91-93.

14. See Lucien W. Pye, *Guerrilla Communism in Malaya* (Princeton: Princeton University Press, 1956), pp. 173-175, 218-247.

15. Ibid., pp. 220-221.

16. Hanrahan, *Communist Struggle*, p. 88.

17. Pye, *Guerrilla Communism*, pp. 243-244.

18. See Foote, *Handbook for Spies*, p. 21.

19. Jansen and Weyl, *Silent War*, pp. 92-93.

20. Hanrahan, *Communist Struggle*, pp. 87-88.

21. Pye, *Guerrilla Communism*, p. 243.

22. Giap, *People's War*, p. 77.

23. Ibid., pp. 77-78.

24. Barton, *North Korean Propaganda*, p. 110.

25. Giap, *People's War*, p. 56, and Alberto Bayo, *Ciento Cinquenta Preguntas a un Guerrillere* (*One Hundred Fifty Questions Asked of a Guerrilla Fighter*) (Havana, 1959), Question 85.

26. R. W. Rowan, *Spies and the Next War* (New York: Robert McBride Co., 1934), pp. 113–114.

27. Pye, *Guerrilla Communism*, pp. 115–116.

28. Tanham, "Belgian Underground," p. 185.

29. Seth, *The Undaunted*, p. 152.

30. Tanham, "Belgian Underground," p. 153.

31. Eugene H. Methvin, "Mob Violence and Communist Strategy," *Orbis*, V (Summer 1961), p. 174.

32. Alvin H. Scaff, *The Philippine Answer To Communism* (Stanford: Stanford University Press, 1955), pp. 34, 146.

33. See Harry Miller, *Menace in Malaya* (London: George Harrap and Co., 1954), pp. 104–105, 113.

34. Tanham, "Belgian Underground," p. 186.

35. Walter Schellenberg, *The Labyrinth* (New York: Harper, 1956), pp. 125–133.

36. Franklin A. Lindsay, "Unconventional Warfare," *Foreign Affairs*, 40 (January 1962), pp. 264–274.

37. Jasper Rootham, *Miss Fire: The Chronicle of a British Mission to Mihailovich 1943–1944* (London: Chatto and Windus, 1946), p. 164.

38. E. Reval, *Sixième Colonne: Un Grand Peuple Lutte Pour sa Libération (The Sixth Column: A Great Nation Fights for Its Liberation)* (Paris: Thonon, S.E.S., 1945), pp. 100–101. Available at the Harvard University Library.

39. Scaff, *Philippine Answer*, p. 34.

40. Charles F. Delzell, *Mussolini's Enemies* (Princeton: Princeton University Press, 1961), p. 297.

41. U.S. Senate, Committee on the Judiciary, *Yugoslav Communism: A Critical Study* (Washington: Government Printing Office, 1961), p. 93.

42. Tanham, "Belgian Underground," pp. 105–106.

43. Ibid. For the use of radio broadcasts to establish identity, see also Maurice J. Buckmaster, *Specially Employed* (London: The Batchworth Press, 1952), pp. 40–41, and Buckmaster, *They Fought Alone*, p. 154.

44. Zawodny, *Guerrilla and Sabotage*, p. 10.

45. David Lampe, *The Danish Resistance* (New York: Ballentine Books, 1957), p. 8.

46. Seth, *The Undaunted*, p. 152.

47. D. A. Tomasic, *National Communism and Soviet Strategy* (Washington: Public Affairs Press, 1957), p. 43.

48. Seth, *The Undaunted*, p. 152.

49. Pye, *Guerrilla Communism*, p. 80.

50. Tomasic, *National Communism*, p. 43.

51. Scaff, *Philippine Answer*, p. 34.

52. Pye, *Guerrilla Communism*, p. 88.

53. Ray Alan, "Brigitte, France and the Secret Army," *The New Leader*, XLIV (1961), pp. 3–5.

54. Ibid.

55. Scaff, *Philippine Answer*, p. 34.

56. U.S. Senate, *Yugoslav Communism*, p. 93.

57. L. Valiani, *Tutte Le Strade Conducono a Roma (All Roads Lead to Rome)* (Firenze: Tipocalcografia Classica, 1947), p. 161.

58. Bor-Komorowski, *Secret Army*, p. 75.

59. Valiani, *Tutte Le Strade*, p. 329.

60. Bor-Komorowski, *Secret Army*, p. 76.

61. French Press and Information Service, "Free France," V (New York, February 15, 1944), p. 140. Available at the Library of Congress, Washington.

62. French Resistance Collection (Hoover Library, Stanford University), Folder 27, No. 7, p. 3.

63. "Free France," Volume VI (October 15, 1944), pp. 264–265.

64. George K. Tanham, *Communist Revolutionary Warfare: The Vietminh in Indochina* (New York: Frederick A. Praeger, 1961), p. 67.

65. Bor-Komorowski, *Secret Army*, p. 77.

66. Rivlin, "Clandestine Arms," pp. 191–202.

67. Bor-Komorowski, *Secret Army*, p. 77.

68. D. M. Condit, *Case Study in Guerrilla War: Greece During World War II* (Washington: Special Operations Research Office, 1961), p. 154.

69. Ernesto Guevara, "La Guerra de Guerrillas" (Guerrilla Warfare), *Army*, 11 (May 1961), p. 63.

70. L. Longo, *Un Popolo alla Macchia* (*People of the Maquis*) (Italy: Arnoldo Mondadori, 1947), p. 305.

71. Rivlin, "Clandestine Arms," pp. 191–202.

72. See War Department Special Staff, *French Forces*, pp. 379–380, 917.

73. Pye, *Guerrilla Communism*, p. 70.

74. Rivlin, "Clandestine Arms," pp. 191–202.

75. Condit, *Case Study: Greece*, p. 155.

76. Tanham, *Communist Warfare*, p. 70.

77. Guevara, "La Guerra," p. 64.

78. Bernard B. Fall, *Street Without Joy: Indochina at War, 1946–1954* (Harrisburg, Pa.: The Stackpole Co., 1961), p. 108.

79. For a discussion of a block warden system, see Joseph Kraft, *The Struggle for Algeria* (Garden City, N.Y.: Doubleday, 1961), p. 105.

80. Jansen and Weyl, *Silent War*, pp. 151–152.

81. Heilbrunn, *Partisan Warfare*, p. 87.

82. See Ladislas A. Farago, *War of Wits* (New York: Funk and Wagnalls Co., 1954), p. 200.

83. Olsen, *Two Eggs*, p. 126.

84. Bor-Komorowski, *Secret Army*, p. 144.

85. Buckmaster, *Specially Employed*, p. 70.

86. David J. Dallin, *Soviet Espionage* (New Haven: Yale University Press, 1955), p. 96, and Buckmaster, *They Fought Alone*, pp. 25, 30–31.

87. Dallin, *Soviet Espionage*, pp. 95–96.

88. Karski, *Secret State*, pp. 122–123.

89. See Jansen and Weyl, *Silent War*, p. 112.

90. Ibid.

91. Ibid., p. 153.

92. J. Edgar Hoover, *Masters of Deceit* (New York: Henry Holt, 1958), p. 282.

93. Valiani, *Tute Le Strade*, p. 158.

94. See Karski, *Secret State*, pp. 229–235, and Bor-Komorowski, *Secret Army*, pp. 59–60. For discussions of the use of these persons as couriers in other undergrounds, see also Buckmaster, *They Fought Alone*, p. 244, and Tanham, "Belgian Underground," pp. 147–148.

95. Philippe de Vomecourt, *An Army of Amateurs* (New York: Doubleday, 1961), pp. 34–36.

96. Jansen and Weyl, *Silent War*, p. 153.

97. Ibid., p. 153.

98. Farago, *War of Wits*, pp. 216–217.

99. For a discussion of simple codes and ciphers, see Fletcher Pratt, *Secret and Urgent* (Garden City, N.Y.: Blue Ribbon Books, 1942).

100. Buckmaster, *They Fought Alone*, p. 27.

101. See Jansen and Weyl, *Silent War*, pp. 152–153.

102. See Farago, *War of Wits*, pp. 212–213.

103. Dallin, *Soviet Espionage*, p. 495.

104. Bor-Komorowski, *Secret Army*, p. 59.

105. de Vomecourt, *Army of Amateurs*, p. 72.

106. See Foote, *Handbook for Spies*, pp. 24–26; and Jansen and Weyl, *Silent War*, pp. 92–93.
107. See Foote, *Handbook for Spies*, p. 59.
108. Tanham, "Belgian Underground," p. 147. Another oath is to be found on page 182.
109. See Ibid., p. 179; Seth, *The Undaunted*, p. 259; and French Resistance Collection, Folder 5, No. 13, pp. 1–4.
110. The regulation of vertical liaison is discussed in Rivlin, "Clandestine Arms," pp. 191–202, and Karski, *Secret State*, p. 160.
111. See Bor-Komorowski, *Secret Army*, pp. 59–60, and Karski, *Secret State*, pp. 229–235.
112. See de Vomecourt, *Army of Amateurs*, pp. 93–94, and Hoover, *Masters of Deceit*, p. 282.
113. See Foote, *Handbook for Spies*, pp. 27–28, 60.
114. See Ibid., p. 57.
115. Lampe, *Danish Resistance*, pp. 66–74.
116. See S. Regalado, "From the Experience of the Underground Work of the Communist Party in Venezuela," *World Marxist Review*, V (January 1962), 57–59.
117. Lampe, *Danish Resistance*, pp. 138–139.
118. See Karski, *Secret State*, pp. 160–161.
119. See Jansen and Weyl, *Silent War*, pp. 177–179.

CHAPTER 3

UNDERGROUND OPERATIONAL FUNCTIONS AND TECHNIQUES

INTRODUCTION

In this chapter an underground's operational missions are discussed—subversion, psychological operations, establishment of shadow governments, intelligence, sabotage, and escape and evasion. As in the preceding chapter, a "standard" technique is presented wherever one has been developed for a function and, when possible, alternative techniques for accomplishing the function in varying environmental conditions. In most instances, examples from past underground experiences are used as illustrations.

SUBVERSION

Undergrounds seek to neutralize or win control of certain individuals or organizations within the society. To accomplish this they use a variety of techniques.

UNDERGROUND LETTER-WRITING CAMPAIGNS

To Divert the Police

In a number of instances, letters have been the device used to tie up the police and channel their energies into work not dangerous to the underground. Such tactics were frequently employed by "Special Action N" cells of the Polish Home Army. According to one source, underground personnel in one instance sent official-looking letters to all German residents in Warsaw instructing each to prepare a parcel of food for wounded German soldiers in the Warsaw hospitals. The letters seemed quite authentic: Nazi Party stationery was used, and every detail was covered—even the number of eggs to be included in the parcels. On the appointed day the German mayor's office and the approaches to it were packed with German civilians. The Gestapo ordered that all present be held for investigation, an action that not only aroused resentment but occupied the police for a full 24 hours.[1] The same source described the sending of anonymous letters to Gestapo headquarters denouncing German officials for acts of disloyalty. Among the accusations were charges that the officials had accepted money in return for favors, that they had secured their future by making deals with the underground, or that they had had intercourse with "racially inferior" Poles. The resultant investigations disproved these charges, but much time and manpower was wasted in surveillance and cross-examination.

Strategy, Tactics, and Countermeasures

To Disrupt Production

"Special Action N" units also forged orders to halt German production in Poland. Communications were sent to factory and workshop managers proclaiming May 1 as "Nazi Labor Day," and stating that all workers were to have a 24-hour leave with pay. This order was received with surprise, because many holidays had already been cancelled in order to step up war production. It was, however, accepted as genuine, since it was couched in the usual Nazi terminology and bore the letterhead of the German Labor Bureau. Because it was sent just before the appointed holiday, there was not time for the Labor Bureau in Berlin to discover and countermand the forgery. As a result, most production in Poland ceased for a day, including that at the important Ursus Tank Works and the gigantic railway repair installations at Pruszkow. Reportedly, the production losses were comparable with those resulting from a minor RAF attack.[2]

To Remove Dangerous Persons

In addition to diverting the police and disrupting production, Polish Home Army members tried to effect the transfer of civilian officials who were particularly active in repressing the underground. Efforts were also made to get rid of *Volksdeutschen* (citizens of German descent and sympathies). A letter bearing the forged signature of one of these persons would be sent to Berlin "volunteering" the individual for service in the German Army. According to the author of many such letters, one might read as follows:

> The Fuehrer has awakened in me the consciousness of the German community. I am at present serving the *Vaterland* as farmer [or merchant, policeman, etc.] I cannot continue any longer to stand by while my German brothers are heroically dying. I wish to contribute my services to the glorious German army and herewith solicit the privilege of immediate induction into the *Wehrmacht*.[3]

Because the authorities in Berlin had more pressing business than checking the authenticity of such a request, the letter would likely be followed by immediate induction of the "writer."

ORGANIZATIONAL SUBVERSION

The General Plan

Undergrounds seek to infiltrate communication and transportation industries because agents therein can sabotage facilities needed for the mobilization of the military and police forces. Unions are also prime targets, as the control of these groups enables an underground to call strikes, weaken governmental control, or cause general social disorganization. Control of labor unions is also desirable because union funds can be diverted to underground activities.

Underground funds also may be concealed in union accounts by falsifying the records. Strikes, demonstrations, and riots also diminish the effectiveness of the government forces. Police, militia, and regular army troops may be required to control them, and this draws manpower from the units assigned to combatting the underground. Punitive measures taken by the police and injuries suffered by participants or onlookers are exploited by underground agitators to turn minor skirmishes into major incidents. By exploiting the resulting agitation, an underground may be able to rally the people to the revolutionary movement and disrupt government control.

Tactics of Subversion

Leadership Tactics

An underground may influence the actions of an organization if it can install its members in leadership positions. The underground member seeking a leadership post in an infiltrated organization represents himself as dedicated and loyal to the organization, takes the initiative in planning activities, and volunteers for any job no matter how time-consuming or unpleasant. He avoids any appearance of subversive activity. His candidacy for a position is supported by cell members in the rank-and-file, but close ties between the candidate and his cell collaborators are hidden from the general membership so that the candidate's support appears spontaneous and unsolicited. Thus, having demonstrated to the members that he is active, eager, and capable, and having the apparently unsolicited support of a number of other members, the underground agent rises to a position of leadership. It has often been noted that Communists who aspire to leadership in any organization are "the readiest volunteers, the devoted committee workers." [4]

Membership Tactics

By being the most vocal members at meetings and the last to leave, a small, articulate, and disciplined group can pass resolutions which the apathetic or outmaneuvered majority may not favor. One author has pointed out that most voluntary groups are composed of a small core of administrators and subleaders, a few faithful meeting-goers, and a large group of dues-payers who take little active part in its work. A small group working in concert can thus easily influence the direction of the organization and eventually gain control. [5] Subversive agents also attempt to gain control of recruiting. This enables them to draw in more of their own group and obtain information useful in screening future recruits. By installing one of its members as education officer, the cell can disseminate information on the underground movement. Editorship of the organization's newspaper permits the publication of subversive ideas, gives the underground access to printing materials, and permits it to establish its own distribution routes.

Rewards and Sanctions

If an underground can gain control of a labor union, it has at its disposal

Strategy, Tactics, and Countermeasures

a powerful system of rewards and sanctions by which to obtain strict obedience to its orders. If a man is dropped from a union he may not be able to get employment. If the union leader improves, by legitimate means, the lot of union members, they will be more willing to go along with political actions or to obey strike calls. Goon squads may be used to persuade reluctant members. Having all these instruments of persuasion and coercion, the leadership can call a strike which a majority of the members may not want. They will go along, however, either because they have faith in the leadership's ability to win higher wages for them, or because they know that if they oppose the leadership, they may be punished by loss of membership, or worse.

Front Groups

If it is unable to infiltrate existing organizations, the underground creates organizations which serve as an innocent façade to its actual work. These organizations usually espouse some worthy cause which will enlist the support of respectable members of the community, at least to the extent of permitting the use of their names, but the leadership is kept firmly in the hands of underground members.

Paper Groups

These are organizations with only one or two members, which obtain charters from some national or international body and then send delegations to its confederation or congress. Usually the paper group gets equal representation with larger, legitimate groups. If many paper groups send delegations that vote as a bloc, they can give the impression that there is wide popular backing for a particular individual or policy.[6]

United Front

When the subversive group is not large, it seeks to draw a number of legitimate groups into a united front, and thus gain the prestige of speaking for a larger group of people. Once in such a united front, the subversive group will seek to discredit the leaders and take control. Here, too, the technique of the paper organization is used in bargaining for leadership positions.[7]

PSYCHOLOGICAL OPERATIONS

Gen. Vo Nguyen Giap, on looking back at the Vietminh revolution, says that in preparing for armed insurrection, propaganda is the most essential task to be performed; and during the insurrection propaganda is even more important than fighting.[8]

The broad objectives of psychological operations are to affect by various means the attitudes, emotions, and actions of given groups within a society for specific political or military reasons. Under the label of psychological operations are communicative acts such as propaganda as well as physical acts of murder, assassination, or a simple show of force which are intended to influence the minds and behavior of men.

THE AUDIENCE

Psychological operations are directed toward six broad audiences, each of which requires specific appeals and may require the use of different communications media. These audiences are—

(1) *The enemy*. This may be a foreign occupier, a colonial government, or a native ruling group. The underground's psychological operations objectives are to harass and confuse the enemy, and reduce his morale and efficiency.

(2) *Persons sympathetic to the enemy*. The object here is to persuade the group to withhold assistance from the enemy and perhaps to win some persons over to the support of the underground cause.

(3) *The uncommitted*. The underground seeks to persuade this group to resist authority and support the underground, or at least not to cooperate actively with the government.

(4) *Persons sympathetic to the underground*. The object here is to provide moral support and tactical instruction on what is to be done and how to do it.

(5) *The underground*. Psychological operations directed at members of the underground are designed to maintain morale and unity.

(6) *Foreign supporters*. The objective here is to win financial aid, material assistance, and diplomatic recognition from foreign governments.

THE PROPAGANDA MESSAGE

When writing a message for the various subgroups within a society, the theme of the message, as well as the objectives of the propaganda campaign, must be carefully considered. One author had identified four types of propaganda messages: "conversionary," "divisive," "consolidation," and "counterpropaganda." In this schema, a conversionary message attempts to transfer the allegiance of persons from one group to another. Divisive appeals are designed to divide various groups under enemy domination and control. Consolidation propaganda hopes to bring about unified compliance by the population to directives of the occupying force. Finally, counterpropaganda messages are aimed at disrupting the images portrayed by enemy propagandists.[9]

Studies of past propaganda campaigns have brought to light certain guidelines for composing messages which, if followed, may increase the probability of success of propaganda operations. First, messages should be directed to *audiences* within the target country rather than directed to the public at large. Second, messages should exploit *existing attitudes* of the audiences rather than

(U.S. Army Photograph)

A Vietnamese soldier destroys a Viet Cong propaganda poster.

atempt to effect a complete change in attitudes. Third, claims made in messages should not exceed the limits of belief of the audiences. Fourth, messages should ask for responses which will promote not only an ideal but also the *individual's own well-being*, measured in terms of job opportunities, survival, etc.[10]

MEANS OF COMMUNICATION

In directing a message at a target audience, the underground worker must first determine the communications media available to that audience. For example, unless a message were intended for the educated elite, it would be foolish to print newspapers in a country where most people are illiterate; likewise, it would be manifestly unproductive to direct radio broadcasts to a country where there are few radio receivers, or none. Other devices, such as rumors spread by word of mouth, slogans on walls, or chants at mass demonstrations may be used.

Radio

Most radio broadcasts to an enemy-controlled country originate outside its borders. The BBC assumed the responsibility for such operations for European resistance movements during World War II. Within the country, the underground may be able to broadcast for short periods on popular channels. It also may interrupt the government's communications or prevent them from reaching the people. In Algeria, for example, the OAS frequently jammed government radio broadcasts. One important advantage of radio is, of course, that one does not have to be literate in order to hear and understand radio broadcasts. This medium also has some disadvantages, however, If the underground wants to be sure that the target group listens, it must find a way to inform the audience in advance of the time and channel on which the illicit broadcast is to be heard. Furthermore, broadcasts by low-power portable transmitters have limited range. Enemy radio-locating equipment can pinpoint the position of a transmitter, forcing the underground to change the frequency and the site of the transmitter frequently. One successful technique used during the war to counteract this was to set up radio equipment close to a government transmitter, and to broadcast close to the frequencies used by the government. This made it difficult to locate the transmitter and attracted the audience listening to the government broadcast, since they could hear the clandestine broadcast in the background.

Newspapers

Undergrounds have made extensive use of newspapers and leaflets. The advantage of printed material is that it can be used and reused by passing it on from person to person. In World War II, the clandestine underground press in Europe kept the people informed and rallied them to the resistance. They

A Norwegian underground worker prepares a news bulletin about Allied frontlines activities. She obtains the information over the radio.

carried on a fight against collaborators and traitors by publishing information about them and printed pictures of Gestapo and other agents. They also printed articles describing how individuals could engage in sabotage.[11]

Written materials do present certain problems. Large quantities of paper, ink, and other supplies are required for continuing publication, and usually the regime controls printing materials and presses. Also, the distribution of printed matter requires a complex and coordinated effort if the material and the distributors are not to be intercepted. Finally, the possession of subversive literature is hazardous to readers as well as distributors.

The most difficult problems in running an underground newspaper are staffing it with reporters, printers, and distributors, and finding a safe place to print it. In some cases, such as Algeria, newspapers were printed outside the country. News is often obtained from foreign broadcasts via shortwave.

Where presses and printing materials are licensed or under close surveillance, chain letters have been used effectively to communicate information to a large segment of the population. In Italy, *Alleanza Nazionale*, an anti-Fascist underground, employed a chain-letter technique, having each recipient make six copies and forward them to six other people, including two Fascists.[12]

Word-of-Mouth Communication

In countries where a large portion of the population is illiterate and few radio receivers are available, word-of-mouth messages are the principal means of communication. Agitators circulating in crowds, spreading rumors, and appealing for aid to the underground have often been very effective.[13] In South Vietnam the Viet Cong have set propaganda messages to music and traveling minstrels have gone from village to village singing revolutionary songs. Agitators who spread rumors usually seek out locations such as marketplaces or talk to travelers who are likely to pass the rumor along to the next village. Word-of-mouth communications have the advantage that the message is usually spread by people who know each other, and therefore it gains credibility. Another advantage is the fact that the messages, though subversive in content, may not sound subversive when presented by the agent, and there are no materials to incriminate the agent. A big disadvantage, however, is that the message may be distorted or may never reach the target group. Once the agitator transmits the initial message, he has lost control over where it will go and the form it will take. North Korean Communists in South Korea, by placing agitators in various territories, cities, and precincts, attempted by sheer numbers and distribution of agitators to get their message across.

Military units are often effective in spreading communications. In South Korea, an investigation by the ROK intelligence revealed that 50 percent of all rumors spread among the local population were disseminated consciously or unconsciously by military elements who came in contact with the civilians and whose opinions on military situations were considered "hot" news.[14] On the other hand, word-of-mouth communication is a two-way street: government messages and those of the underground may be equally credible and acceptable. For either side, word-of-mouth communication usually has the advantage of being viewed as "hot" news or information not normally available through other mass media.

Symbolic Devices

Another way to transmit information and harass the enemy is by symbolic devices such as slogans or symbols written on walls or in public places which are convenient to the target groups. Antigovernment slogans and messages can be displayed on walls in such a way that they cannot easily be eradicated. Jokes and cartoons carry great impact and are an effective way of conveying disrespect and resistance in a socially acceptable manner.

Agitators

North Korean agitators trained by the Russians for intelligence, sabotage, agitation, and propaganda were sent into South Korea. Communist directives ordered them to organize cells of four members and to set up chan-

(UPI Photo)

Secret Army Organization propaganda in Algiers (March 1962).

'nels through which propaganda materials and other equipment could be smuggled to partisans in South Korea. The agitators were to keep a close check on the attitudes of the populace within their precinct and to record all South Korean cooperation with U.N. forces. They were to prepare and distribute posters and leaflets and other propaganda, and compose slogans and appeals and post them at night in public places. They were warned not to speak openly in favor of North Korea, but to point out that the North Koreans did not destroy citizens' homes, bomb their children, or kill their cattle. The agitators were told to look for South Koreans who had lost all they owned or who had not received help from the government, and those rejected from U.N. jobs. They were instructed to help sick neighbors; South Koreans whose sons were serving with the ROK Armed Forces were to be offered a hiding place for their sons if and when the latter deserted. To utilize the war situation, the agitators were to invent and spread rumors to upset and frighten the people, and instill hatred against the Syngman Rhee government and the United Nations. They were also to exploit existing rumors, and to concentrate on themes which struck close to the hearts of the majority of the South Koreans regardless of their political creed. They were told to emphasize the sentimental over the rational and to utilize the elements of uncertainty, fear, and doubt which existed among the people.[15]

One analysis of speeches given by various agitators indicates that an agitator will concentrate on emotional appeals and attempt to exploit the frustrations of his public.[16] His function is described as bringing to flame the smoldering resentments of his listeners and then lending social sanctions to actions that might otherwise seem to be simply dangerous temptations. His themes are—

(1) *Distrust.* The agitator plays on his audience's suspicion of things they do not understand. He points out that the individual is being manipulated and duped by the government.

(2) *Dependence.* The agitator talks to the crowd as if they suffered from a sense of helplessness and offers them protection through membership in a strong organization led by a strong leader.

(3) *Exclusion.* The agitator suggests that there is an abundance of material goods for everyone, but that the crowd does not get the share to which it is entitled.

(4) *Anxiety.* He points to a general premonition of disasters to come, and plays upon the fears of the individuals and the general uncertainty of life in the community.

(5) *Disillusionment.* The agitator points to politics and alleges that the government and its leaders are guilty of fraud, deception, falsehoods, and hypocrisy.

The agitator does not invent issues, nor does he base his appeals on abstract, intellectual theories. He exploits the vagueness of his terms, playing not on facts but upon basic emotions of fear and insecurity. The agitator is not hindered by facts, and he needs none, since the themes he uses are emotional and common to all men.

(Wide World Photos)

A group of Venezuelan demonstrators await the motorcade of Vice President Richard Nixon and his party.

OVERT MASS RESPONSES

Mob Violence

Mobs and demonstrations have been used with great effect, particularly by Communists in recent years. For example, in 1958, when Vice President Richard Nixon visited Venezuela, top Communists—with the support of the local party organization—hired hoodlums, armed them with long wooden clubs and placards, and put on a demonstration against the Vice President. Gustavo Machado, the Venezuelan Communist Party leader, admitted that the party had organized the demonstration of 8,000 in order to ruin Nixon's trip. Sometime later in Japan, agitators organized a similar demonstration against Press Secretary James Hagerty. Before Hagerty's arrival, the agitators visited the unemployment offices to hire all applicants present. The hiring was so complete that the police were able to tell newsmen that the absence of lines at the employment office made it certain that demonstrations would occur later in the day. The Japanese security officials also estimated that the 5 weeks of violence against the Japanese-United States Treaty cost the Communists as much as 1.4 million dollars.[17] One observer has outlined the organization of a Communist demonstration and mob violence as follows:

(1) *External command.* The leaders who comprised the external command remained some distance from the scene of immediate action, observing and issuing orders from a place of relative safety.

(2) *Internal command.* Within the mob was the Communist cadre who implemented orders from the external command. Because the cadre was in the thick of the action, great care was taken to protect the leaders of this unit.

(3) *Bravadoes.* Bodyguards, or bravadoes, surrounded and shielded the internal command from the police and facilitated their escape if necessary. These men also flanked processions and guarded the banner carriers.

(Wide World Photos)

Demonstration at Tokyo's Haneda Airport, June 1960, where White House Press Secretary James Hagerty and companions had to be rescued by helicopter from their car besieged by some 5,000 demonstrators.

(4) *Banner carriers.* These men carried the banners and switched them upon instruction. At first, signs with slogans that expressed popular grievances were used, but as the mob became frenzied, the banners were exchanged for others bearing Communist propaganda. Key agitators were stationed near conspicuous banners so that they could easily be found by messengers bearing instructions from the leaders.

(5) *Cheering sections.* Demonstrators were carefully rehearsed on slogans to be chanted and the sequence in which they were to be used.

(6) *Messengers.* They ran messages between the external and internal commands. Generally, they rode their bicycles along sidewalks, keeping abreast of the moving demonstration.

(Wide World Photos)

Japanese demonstrate at Haneda Airport in Tokyo as Press Secretary James Hagerty arrives to arrange Eisenhower's visit to Japan (June 1960)

(7) *Shock guards.* These men carried clubs and accompanied the Communist cadre. They marched along the sidewalks, screened by the spectators, and rushed into the mob only as reinforcements if the Communists were engaged by the police. Their sudden and violent action was designed to provide enough diversion to enable the Communist demonstrators to escape from the area, leaving to the police the bystanders, unknowing excitement seekers, and sympathizers.[18]

Agitators frequently work on the assumption that bloodshed can be very effective in giving the proper impetus to the cause they are promoting, and that such bloodshed can turn an ordinary grievance into a holy commission. They may recruit and use women and children for acts of violence to inhibit the police in their use of countermeasures or at least to embarrass them. It is also common for women and children to be used by resistance groups to mock government occupation troops. The hope is that the enemy soldier will be so infuriated as to attack one of these people and hence to arouse the wrath of the populace.

Passive Resistance

Passive resistance is an important supplement to major underground activities, particularly since it allows sympathizers to aid the movement with relatively little risk. One advantage of passive resistance is that the offenses committed are so trivial that the governing authority will not take extreme countermeasures. Particularly if many engage in such activity, the government is unlikely to undertake any severe punitive action. A second advantage is that this type of activity camouflages an organized underground and hence enhances its effectiveness. The government arrests passive resisters and in turn finds no organized threat against the government; it exhausts itself searching the wrong quarter for organizations, plans, or activities. This acts as a distraction and provides a cover for the more dangerous underground activities.

Some of the activities of passive resistance are—

(1) *Boycotts.* By boycotting certain products, markets, or activities, the resisters show contempt for the governing authority and can thereby affect the morale of the government forces or supporters. The boycott of French cigarettes in Morocco in 1953–55, for example, deprived the French Government of revenue and also demonstrated the strength of the Istiqlal Party.[19]

(2) *Social ostracism.* Collaborators or people sympathetic to the governing authority or occupier are frequent targets for ostracism. To maximize its effect, ostracism is best employed against a particular group of people who are easily identifiable and small in number. Exploitation of existing prejudices toward minorities or special interest groups also helps to magnify the effect of ostracism.

(3) *Fear and suggestion.* Telephoned threats, bomb scares, and threats to contaminate drinking water are matters which the police must investigate and against which they must take precautionary action; the attention of the police is thereby diverted and their effectiveness is reduced.

(4) *Overloading the systems.* By following governmental instructions to report suspicious incidents and persons, large numbers of people can turn in false alarms or make unfounded denunciations of people who are "suspected of aiding the enemy" and in this way so overload the governmental authority that valid reports cannot be handled. This device has been especially effective against a block-warden surveillance technique used to counter underground activities.[20]

(5) *Symbolic acts.* In the Netherlands, when the Jews were told to wear armbands to identify them, many Dutch also wore armbands in defiance of the German authorities. This showed contempt for German regulations and had the effect of encouraging more resistance.

(6) *Absenteeism and slowdowns.* Production is hampered when workers fail to report for reasons of "illness," or when they go to work but create natural errors or work slowly.

COERCION, THREATS, AND TERRORISM

Coercion

While an underground will attempt to win support wherever possible through persuasion alone, it will often use intimidation and coercion to obtain

Strategy, Tactics, and Countermeasures

the support of undecided or uncommitted people. In Malaya, for example, the Communists would ask fellow Chinese labor union members to pay as little as 25 cents a month dues to the party. Rather than risk trouble over such a small amount, the workers would contribute. Then the Communists would ask for more money and threaten to expose the workers to the government as dues-paying Communists unless they cooperated further with the party. When such methods of simple coercion failed to produce the food, arms, and money needed to expand the movement, the Communists formed an armed terrorist group called the "Blood and Steel Corps" to carry out threats and punish antiparty people.[21]

Threats

In order to be effective, the demand part of a threat must be stated so clearly that it cannot be misunderstood by the individual or group to which it is directed, otherwise the threatened party might not comply simply because he did not know what was demanded of him. If a particular threat to an individual or small group is to have an effect on a great many people, it must be specifically related to a clearly discernible act so that the others will have no doubt as to why the individual or group was threatened. For the threatener is not always attempting only to get particular individuals to perform an act, he frequently wants large numbers of people to learn the lesson demonstrated by any enforcement of his threat. Indiscriminate and nonspecific punishment or mass terror does not lead people to act in the desired manner. Rather than threatening an individual or group because of some intangible such as being sympathetic to the enemy, punishment should be contingent on some observable act such as giving material aid to the enemy.

Secondly, to be effective the threatener must have means of administering punishment which are obvious to the threatened person. In counterinsurgency situations such means can be a group such as the Malayan Communist "Blood and Steel Corps," the Greek *Aftoamyna*, the Yugoslav Partisan O.Z.Na., and the enforcing squadrons of the Huks in the Philippines.

Thirdly, the threatener must discern whether or not the threatened persons did comply with his demand before he inflicts punishment, lest the situation of a specific threat be changed to a situation of general terror. For example, the Nazi threat to shoot all Yugoslavs who cooperated with the Partisans became ineffective because the Nazis shot randomly-chosen groups of Yugoslavs which included both persons who had and persons who had not complied with the demand. The Yugoslavs soon understood that they were liable to punishment whether or not they cooperated with the Partisans.

In making a threat, it is to the advantage of the threatener to choose for the demands in his threat actions which the people are predisposed to do. In Algeria the OAS, in order to demonstrate its displeasure against the French Government, demanded that all Algerians stay off the streets during the evening hours and turn out their lights, and threatened to punish anyone found on the streets during the evening hours. These demands on the populace were consonant with what an individual might do on his own during any kind of disorder, so it was extremely likely that the populace would comply with the demands.

There are ways to improve the effectiveness of a threat.[22] To keep the people from acting in an undesirable way, the threatener may try to prevent their taking the first step toward the ultimate action; if he cannot do this, he tries to prevent their taking the second step; etc. If drastic threats contingent on completion of the action are prefaced by a series of smaller, successively harsher threats relating to each step, and if the threatener shows that he is willing to carry out the intermediate threats, he may not have to carry out the ultimate one. Similar to decomposing the threat into a series is decomposing the punishment into a series of increasingly severe punitive acts. In this way the threatener, by inflicting lesser punishments, gives credence to the possibility that he might carry out the most drastic punishment. In France, for example, the OAS would send someone a note saying that an OAS tax representative would call to collect money for the organization. If the victim refused to pay, his car might be blown up with a plastic bomb. He was approached again and asked to contribute; if he failed again to pay the tax, his house might be damaged by bombs. If an additional refusal was received, he might be assassinated.

Terrorism

Terror is used to disrupt government control over the populace. It is probably most effective in a rising revolution where it can be used to express the discontent of many, and it is least effective where there is popular support for the government.[23] Terror is used to draw attention to the movement and to demonstrate in a dramatic way the strength and seriousness of the underground. The small strong-arm-unit, which most undergrounds maintain to protect its members, may also be used against informers and people who cooperate with the government.[24] In this way the underground demonstrates that cooperation with the security forces is a risky thing.

Because terror is a state of mind, the underground must carefully assess the probable reactions that may follow it. Such acts as murder, assassination, and bombing will usually produce fear, but that fear does not necessarily lead the people into the ranks of the terrorists; it may instead lead them to mass indignation and counteractions.

The Malayan Communists, unable to carry out and win a guerrilla war, attempted through mass terror to neutralize the elements of the population who were assisting the government. But soon the party received complaints from its underground, the *Min Yuen*, that the indiscriminate use of terror had made even routine *Min Yuen* work difficult and was making the populace more willing than before to cooperate with the government. The party then called for discriminate use of violence and greater political infiltration. However, as one source points out, the Malayan Communist Party had become a "prisoner of terrorism." [25] It could not use violence effectively; on the other hand, to

105

give up violence would have indicated that it was losing its effectiveness and its ability to enforce demands.[26] The discriminate use of terror is probably more effective than mass terror. The assassination of a key government official may lead some people to refrain from seeking political office.

A common strategy is to perform acts of terrorism which will drive the government to retaliate in kind and in this way drive people who were sympathetic to the government or neutral over to the underground. Yugoslav Partisans attacked villages knowing full well that the Germans would execute hostages. But they also knew that the remaining villagers would flee the villages and would come to the partisan movement voluntarily.[27]

In Algeria, the OAS committed acts of terrorism against Muslims in the hope that the Muslims would retaliate, thereby making it necessary for the French Army to fight Muslims and indirectly win the sympathy of the European population of the country. The FLN, having used similar tactics, was prepared for this maneuver and provided FLN agents to prevent any counter-demonstrations by the Muslims or counteractions against the French Algerians.

PUBLIC RELATIONS

Since an underground is attempting to win people to a cause, it must consider how best to present itself to the populace. Its leaders must decide what activities would be useful to the populace and make for better relationships between the people and the underground. In South Korea the Communists helped people who were homeless and who needed assistance. In Belgium the underground supported families of former underground workers who had been caught and nonunderground people who were evading the German conscript laws. This type of public relations is one of the most significant tasks the underground performs to enhance its cause.

Members of the underground are usually instructed to respect the customs and the moral standards of the major groups within the country. Failure to do so might provide the security forces with an effective propaganda tool. In Malaya, the British were successful in characterizing the Communist Party as "Communist terrorists" and as a "Chinese" organization, and thereby prejudiced the Malayan people against the movement. In Algeria, the OAS was successfully labeled as "Fascist." Such unpopular labels can destroy the underground's appeal and ability to recruit and expand.

SHADOW GOVERNMENTS

Shadow governments are reflections of the important objective of the underground to win control of the people, an objective more important than control of territory or military victories.

EXTERNAL SHADOW GOVERNMENTS

Many World War II resistance movements established governments-in-exile. Such a government acted as a rallying point for the resistance and conferred legitimacy on the acts of resistance against the occupier. It was felt that the existence of an exile government restrained people from collaboration. Also, funds were obtained in the name of the exile government within or outside the country.

In a revolutionary movement, the establishment of a provisional government and its recognition by foreign powers adds prestige to its cause. The provisional government can also act as a spokesman for the underground and make concessions to the other countries in return for aid.

INTERNAL SHADOW GOVERNMENTS

In spite of war and unrest certain public needs must be met in order to maintain social stability. Courts, schools, and police and public health services must be provided. In unoccupied territory, the underground resistance usually steps in and fills this vacuum by forming local and regional governments. In occupied territory, shadow governments are set up by the resistance. The underground in Poland was able to build a secret state complete with ministries, a parliament, and an army. It carried on underground schools and courts. The effectiveness of this state along with the threat of assassination for collaborators permitted the Poles to boast of having never produced a Quisling—a unique record in occupied Europe. In Yugoslavia, Tito organized National Liberation Committees for all levels of government and when the resistance ended, he was in political control of most of the country. Much of the postwar strength and success of the Communist revolutionary groups in Greece, Malaya, Indochina, and the Philippines was built upon the political control which they were able to build through local and regional governments during the wartime resistance.

The local governments play a military as well as a civil role. They collect taxes and supply food and intelligence as well as safe-houses for the guerrillas. Civil defense or militia units are formed to protect the villages and also to serve as sources for guerrilla recruitment.

Shadow governments are usually established under the protection of a guerrilla force, with "elections" of individuals sympathetic to the movement. The government usually maintains itself by providing needed public services. If the territory should be retaken by the government forces, underground agents, through terrorism and assassination, coerce people into refraining from collaboration and keep a record of whatever collaboration occurs.

INTELLIGENCE

SCENE-OF-BATTLE INTELLIGENCE

Underground workers assist associated military forces by providing valuable data about the enemy and the area of impending combat. This may include the number of enemy troops, their deployment, their unit designations, the nature of their arms and equipment, the location of their supply depots, the placement of their minefields, the pattern and routine of their patrols, the morale of the troops, and various topographical factors, such as swamps and ravines, that govern access to enemy emplacements.

Sometimes this information is obtained directly by underground personnel by visual observation of the targets. For example, members of the French resistance reconnoitered German coastal defenses in the preparation for the Allied invasion of France in June 1944. Such data may also be collected by the local populace, or "popular antennae," as these sources are described in one Vietminh manual.[28] The Vietminh used children playing near French fortifications as a source of information on troop arrivals and departures, the guard system, and other pertinent details which aided the guerrillas in planning attacks—all of which were easily observable by untrained children.

Intelligence activities are generally conducted under the guidance of outside governments or companion military forces in the field. These sponsors not only assign targets for reconnaissance but also give technical direction, because most underground personnel lack experience in this type of work. For example, in World War II specially-trained "Jedburgh" teams (composed of an American, a Briton, and a Frenchman) were sent into France to guide resistance workers in their intelligence surveys. These teams were also equipped to conduct the necessary radio communications with Britain. Likewise, Red Army personnel were assigned to the Soviet partisans to direct these activities.

When military personnel are not available to give instruction, underground members have been instructed by manuals. This was done during World War II in the Soviet Union where detailed booklets such as the *Guide Book for Partisans* were circulated for use in regions under German occupation. The following excerpt from a passage in this manual is a typical instruction:

> If you happen to encounter troops . . . do not show that you observe the enemy . . . ascertain the colour of their headgear, their collar braid, and the figures on their shoulder straps. If they have questioned the inhabitants about something, try to find out what the Fascists have asked. . . .[29]

This manual also gives tips about ascertaining enemy intentions: if an attack is planned, trucks will arrive loaded and depart empty; if the enemy intends to retreat, fuel and foodstuffs will be removed, roads and bridges will be demolished, telephone wires will be removed, and trains and trucks will arrive empty and depart full.[30]

SABOTAGE INTELLIGENCE

Reconnoitering transportation and communication facilities prior to sabotage attacks occupied much of the time of French resistance persons. Often working closely with Allied advisers, these people surveyed targets earmarked for sabotage on D-Day. In reconnoitering a bridge, for example, resistance members looked for such factors as (1) the guard system covering the bridge—if a number of permanent troops were evident, a step to eliminate them had to be included in the sabotage plan; when there was only an occasional patrol, the resistance would time an attack to avoid the patrol; (2) the bridge's construction—so that the size of the explosives could be calculated. By determining the schedule of enemy train movements, saboteurs were able to destroy a stretch of railroad track while it was in use, thereby compounding the wreckage and complicating repair work. Danish railroad saboteurs had an elaborate system to provide this information. Throughout Jutland, underground members were stationed near major terminals to note the departures of enemy troop trains. Whenever one was seen, the observer telephoned prearranged code phrases to the sabotage cell in the town next on the railroad line. Members of this cell then proceeded to predetermined spots on the tracks to lay their mines. With this advance notice, the mines could be placed at the last moment, preventing detection by patrolling guards. The train delayed by sabotage might eventually reach the next stop, but there observers would be waiting to repeat the process. Using these observers along a train's route, the resistance was sometimes able to slow a train's progress by days or even weeks.[31]

Production facilities are also surveyed by undergrounds in preparation for sabotage attacks. When possible, underground personnel are often aided in planning factory sabotage by outside intelligence experts, for these are best qualified to make the necessary technical judgments: it is a problem in itself to determine just which components in a plant should be incapacitated. Prior to the blowing up of a Norwegian heavy-water plant being operated by the Germans during World War II, the preliminary reconnaissance was done by an SOE agent parachuted into Norway. Details about the factory's equipment were obtained from a Norwegian scientist in London. Other data, perhaps about the guard system and access to the equipment, apparently were supplied by underground workers in the plant.[32]

SECRET SCIENTIFIC AND MILITARY INTELLIGENCE

Secret scientific and military data may be obtained by recruitment of employees in scientific and military installations, or by simple observation elsewhere. An example of the latter was the valuable data about the V–2 rockets obtained by the Danish resistance. During the summer of 1943, fishermen near the island of Bornholm began to report the crashes of unidentified objects into the sea; these were recorded by the resistance leader on the island. In August the island's police commissioner notified the underground leader of the crash of a flying craft in a nearby field. The two men rushed to the scene before the Germans and found the wreckage of what was clearly a new kind of aircraft. The only identification mark was a number: "V[1]-83." The men took photographs immediately before the arrival of the German investigators. From the skid marks the underground leader was able to determine that the device had come from the southwest. From the pictures, the underground chief drew a complete sketch of the weapon. This sketch, the photographs, and the notations as to the direction from which the missile had come were sent by courier to England, providing the British with perhaps their first technical data on the new German rocket.[33]

POLITICAL INTELLIGENCE

Underground agents also collect political intelligence. They note the statements and activities of persons to determine who favors the regime, so that these persons may be closely watched, or eliminated if their actions seriously threaten the underground. In Belgium during World War II, the *Mouvement National Belge* kept files on collaborators and campaigned by threatening phone calls and letters to dissuade these individuals from working with the enemy. If this failed, the collaborators were often executed. The list of collaborators was never made public in order to keep concealed the extent of cooperation with the enemy.[34]

In wartime the underground also notes the morale of the enemy soldiers. The Polish Home Army systematically collected data on German troops by reading their mail. There were too few Germans to handle all of the postal work; thus many Poles were employed. These workers would open letters and photograph the contents before sending them on. From these letters a fairly good estimate could be made of the enemy's morale.[35]

SABOTAGE

SELECTIVE SABOTAGE

The General Plan

A principal advantage of underground sabotage in wartime is that it can succeed in destroying a target not easily reached by conventional means. The former head of SOE activities in France said that this consideration often governed British action against targets in France. To launch an RAF attack on important targets in France required aircraft and trained flying-crews whose main task was to hit targets in Germany or Italy. The chance of success on a small target was very low, even with the refinements of precision bombing. One saboteur could—with more personal risk, but also with far less expenditure of total lives and money—obtain more certain results in a shorter time.[36]

The importance of this capability was demonstrated in France soon after D-Day. Fifteen days after the landings, the SOE's French section received a "Most Urgent" message from General Eisenhower. The problem outlined was the following: the Allied hold in Normandy was tenuous; all Allied forces were fully committed; any German reinforcement might swing the balance and prove fatal to the invasion; such a reinforcement, a Panzer Corps, was moving towards Normandy and was expected to cross the Eure River near Evreux at the only bridge the RAF had failed to destroy; it had to be destroyed before the Panzers crossed it. Could SOE saboteurs in France undertake this mission?

The operation was commenced—its importance increased when yet a second air attack that day failed to demolish the bridge—and the bridge was completely blown up that night approximately 3 hours before the tanks were to arrive. RAF reconnaissance the next day noted the success of the mission—the Panzer Corps drawn up at the river. It never crossed the Eure. The following day the SOE French section received a message of congratulation from the Supreme Commander.[37]

This type of sabotage is selective: it aims at incapacitating installations which cannot easily be replaced or repaired in time to meet the enemy's crucial needs. The length of delay required varies with the target. A tactical target such as a bridge, which might be crucial in the transport of troops and supplies to a battle area, need be removed from use for only a few hours or days. A strategic target such as a factory, however, must be incapacitated for a much longer period, perhaps months, if there is to be a telling effect. Since massive bombardment is often required to knock this type of target out of production for a long period, underground sabotage is most often directed at tactical targets—the enemy's lines of transportation and communication. By destroying transportation routes, the underground obstructs the flow of troops and supplies to battle areas, and by disrupting communications, it can interrupt or confuse the transmission of battle orders.

Timing, of course, is of the essence. To blow up a bridge without considering the immediate needs of the enemy might only put an unnecessary hardship on the populace, but to destroy it at a time when it is vital to the enemy's troop and supply movements would be of real tactical value.[38]

A series of sabotage acts that met this requirement of timing were those conducted by the Danish resistance against the Jutland rail system after the Allied landings in Normandy in June 1944. At this time the Germans began to transfer divisions stationed in Scandinavia to the Western Front. From Denmark alone about two and a half German divisions were sent immediately to oppose the landings. Intelligence reports indicated that these troops were to be followed in November by more than 12 divisions from Norway, and the

Troop train derailed by Norwegian saboteurs during World War II.

British asked the underground to disrupt this transfer of troops by systematic sabotage of the railroads.

Throughout the winter of 1944–45 about 300 resistance members engaged in these operations. They supposedly accounted for most of the following items: 92 wagons, 58 locomotives, 11 cranes, 14 water towers, 25 signal boxes, 8 bridges, 8 locomotive sheds, 9 turntables, and 31 level-crossings were destroyed; 119 trains were derailed; and 7,512 attacks were made along the tracks.[39] Other resistance movements during the war surpassed these figures, but failed to match this sabotage operation in terms of effectiveness. The Danes succeeded in carrying out a concentrated attack on the railroads which incapacitated items *immediately needed*. According to General Viscount Montgomery, these operations changed the entire tide of the Battle of the Ardennes. During the most crucial 2 weeks of the campaign, when it appeared that Allied troops would be pushed back if the Germans could get reinforcements, the Danish saboteurs worked so well that, for 2 weeks, every trainload of German soldiers was stopped.[40]

Less effective were the numerous acts of sabotage committed by the French resistance *prior* to D-Day against railways transporting supplies to Germany. According to one source, only about 500 of the 9,500 locomotives in the French railroad service were put out of commission by saboteurs, and those 500 locomotives constituted the margin by which French civilian transportation was provided for.[41] The time delays caused by derailments seldom exceeded 13 hours.[42] Highway sabotage was also easily repaired by the Germans with their powerful equipment.[43] These acts of sabotage were of little operational value because these installations were restored to usability in a matter of hours, and the brief delay did not impede the delivery of food and other goods to Germany. In time of battle similar delays in transportation would have been effective on the tactical level, and indeed were of some value at the time of the Normandy landings.

Sometimes undergrounds can successfully sabotage factories. One such case occurred in June 1944 in Denmark, where a resistance organization, the BOPA, destroyed the Globus radio factory. This factory, located just outside Copenhagen, was making components for the new V–2 rockets. A review of this operation reveals the extent of the planning and preparation that is needed in a sabotage operation of this scope.

(Courtesy of the Norwegian Information Service)

Remains of Shell building in Oslo after being struck by saboteurs in January 1945. The building had served as an oil depot for the Germans. About 200 tons of oil were destroyed in the fire which lasted for several days.

Three months before the raid, BOPA began planning; it received drawings of the Globus buildings and the deployment plan of the German guards from sympathizers in the police department. Several days before the attack, BOPA men, dressed as laborers and operating in full view of passers-by, planted a minefield in the road between the factory and Copenhagen. If necessary, the electronically-controlled mines could be triggered to blow the road and halt any pursuing German guards after the attack. On the day of the attack, over 100 underground members assembled on the outskirts of Copenhagen with knapsacks and bicycles. They could have passed as a cycling club, but in their packs were guns, ammunition, grenades, and mines. The men moved to the gardens of homes around the factory, from which they launched their attack. After using grenades and submachineguns to blast through the barbed wire and guards, the saboteurs placed their mines and withdraw to the factory yard where two buses were waiting to carry them quickly from the scene. The subsequent blasts so effectively incapacitated the vital machinery that the BOPA earned a radioed message of congratulations from General Eisenhower's SHAEF headquarters.[44]

Of course, selective sabotage is exercised with caution. Installations in regions which military units associated with the underground expect to occupy within a short time are usually spared, because they may be needed in an advance against the enemy or in production for the antiregime forces.[45]

Tactical Aspects

Explosives

A number of explosives can be used in sabotage, but only a few can be handled easily by underground members. These persons generally are not demolition experts and therefore require explosives that are relatively safe and easy to use. Nitroglycerine is potent but unstable; almost any jarring can cause it to explode. When mixed with sawdust or other absorbent material, making dynamite, it is safer but still very sensitive in warm temperatures. More suitable for use by undergrounds is trinitrotoluene, or TNT, which is so stable that even a piercing bullet might not cause an explosion. To set it off it is necessary to explode an embedded blasting cap of gunpowder. An improved explosive is made by mixing TNT with hexogene; the product is a malleable but equally powerful explosive. Although it is sometimes called cyclonite, or RDX, its popular label is "plastic."

This mixture is ideal for underground use because not only is it stable— it can be stamped upon, cut, frozen, or fried, and it will not explode—but it can also be molded for any use and readily stuck to surfaces, as putty. Like TNT, it must be detonated by a blasting cap, and also like TNT, a one-half pound charge of "plastic" can kill or severely wound a person standing a few feet away.

To have available such a handy explosive is not enough, however; a certain amount of expertise is still needed to use an explosive with maximum effectiveness. Would-be assassins failed in their attempt to kill President De Gaulle in his car on September 8, 1961, because the 66 pounds of "plastic" used were not skillfully tamped into the ground beside the road. As a result, most of the charge did not explode.[46] Detailed instructions in the use of explosives, therefore, are usually needed.

Training of Saboteurs

The problem of training underground saboteurs has been treated in various ways. They have been trained, when possible, by liaison personnel from military units. Manuals, directives, newspapers, leaflets, and even radio broadcasts have been used to disseminate technical instructions to units which do not have the benefit of personal instruction.

One Soviet partisan newspaper, *Red Star*, in World War II instructed readers on the following points: (1) the placement of charges for optimum effect in sabotaging steel or wooden bridges; (2) the best points for sabotaging railway tracks, e.g., at bends or on high embankments; and (3) the alternative ways of sabotaging tracks, e.g., by using explosives or removing rivets from joints.[47] Similar instructions were broadcast in a twice-daily, 10-minute program called "Course for Partisans."[48] and the *Soviet Handbook for Partisans* with detailed instructions was distributed extensively.

The following incident from the French resistance illustrates the practical effect of sabotage instruction. A directive on railway sabotage issued by the French Forces of the Interior departmental chief in Meurthe-et-Moselle instructed sabotage units to avoid breaking tracks before the passing of a train, as such ruptures "were always revealed before an accident was caused." The saboteurs were told "to cause accidents by provoking the break in the rail under the train in motion . . . [and] to place the charges in the curves or against the main switching points: these elements being more difficult to replace than straight rails."[49] According to the department chief, "the result [was] felt immediately. . . . The duration of traffic interruptions [rose] immediately to more than thirteen hours on the average instead of 6 hours and 50 minutes."[50]

Tactical Planning

Effective sabotage missions are generally preceded by a tactical reconnaissance of the target and surrounding area. The specific information sought by this reconnaissance will vary with the choice of targets. In general, however, it will include the exact location of the target and pertinent details such as the structure of a bridge, the number and positioning of guards, the routine of the guards, the paths of access to the targets, the routes of escape, and areas for regrouping in case of dispersal. Before a sabotage mission proceeds to the target, the commander briefs the unit members on the plan of operations. A large raiding party may be divided into three units—one in charge of eliminating the guards and securing the area around the target, one to execute the act of sabotage, and one to stand in reserve.

GENERAL SABOTAGE

The General Plan

An underground may undertake sabotage operations not only to hamper the enemy's war effort but also to encourage the populace to engage in general acts of destruction. Although the latter probably would not have much material effect, they could serve as "a form of propaganda among the population, a stimulant in the fight against the enemy."[51] By inducing people to perform minor acts of sabotage, the underground can weld them more firmly to its cause.

Tactical Operations

To foster this type of sabotage, undergrounds often instruct the population in the use of certain sabotage techniques. These instructions are usually limited to the use of simple devices which do not require technical skills or elaborate equipment. For arson, the public may be instructed in the use of homemade incendiary grenades, or "Molotov cocktails." According to one member of the Castro movement, "every man, woman, child, or aged person" can make an incendiary grenade by following a few simple steps: fill a glass bottle with gasoline; insert a rag into the bottle so that one end rests on the bottom and the other sticks out of the top to act as a fuse; plug the remaining space in the neck with a cork or some other stopper to prevent the liquid from spilling out. To use the grenade, one lights the fuse, after which he throws the bottle against a target. When the bottle breaks, the spilt gasoline will be ignited by the burning rag, producing the destructive effect.[52]

Fragmentary hand grenades can also be made with little trouble. The source referred to above suggests that an empty evaporated-milk can be used for the casing. After making sure that the can is dry, one begins by placing a dynamite cap on the bottom, and over that, a layer of nails or small iron pieces. The metal bits are tamped and the process is repeated until several layers of dynamite caps and metal bits are built up to the top of the can. A detonator and fuse are then laid on the charge. A wooden or metal lid with a hole for the fuse is then prepared. When this is put in place, with the fuse protruding, the device is ready for use. When the fuse is lighted, it sets off the detonator and the charge.[53]

There are many other easy-to-use devices. Fires may also be started by substituting incendiary solutions for nonvolatile fuels and by deliberately overloading machines. Mechanical interference may be produced by placing emery dust or sand in delicate bearings, or tossing bolts and pieces of scrap into moving mechanisms. Miscellaneous techniques for disrupting transportation include putting sugar into gasoline tanks, strewing nails in roads, blocking roads with stalled trucks and felled trees, and changing signs to misdirect traffic.[54]

In "passive" sabotage, the enemy is hampered when workers fail to lubricate machines, misplace spare parts, slow down production, practice absenteeism, etc. Supposedly, in the spring of 1949 a quarter of a million Italian workers used such methods in the metallurgical industries to cause a 16 percent reduction in output.[55]

ESCAPE AND EVASION

TYPES OF PERSONS GENERALLY AFFECTED

Escape and evasion operations assist the following types of persons to elude the authorities: underground workers whose identity has become known to the enemy and are being sought; members who are in imminent danger of being exposed, perhaps by captured coworkers; stranded military personnel (e.g., downed airmen, troops stranded behind enemy lines, and escaped POWs); refugees; and couriers.

ALTERNATIVE COURSES OF ACTION

Escape From the Country

One way to save persons from capture is to send them to a friendly or neutral country. This may be the only effective course if a large number of persons are in danger (as, for example, where the Jews in Nazi-occupied areas), if there are not enough guerrilla units or secret camps to absorb them, or if is impossible to house and feed them in the houses of underground workers. In Norway during World War II inclement weather and barren terrain made guerrilla activities infeasible and precluded the secret establishment of large camps. The only alternative was to take the refugees out of the country, mainly to Sweden.[56]

The direction of out-of-the-country escape and evasion routes is governed by political-military factors. Such routes usually cross borders into friendly (to the underground) or neutral countries where sanctuary is given, and avoid military fronts where enemy patrols and security measures are the heaviest. Since there are no fronts on the borders of neutral countries, egress routes often lead to these countries in time of war. The principal routes in Europe during World War II fit this pattern of operation.

From Denmark and Norway

Most escapees from Denmark were taken to Sweden, departing from the eastern islands, particularly Sjaelland, and traveling in private speedboats, in fishing boats, and as stowaways aboard commercial ferries. Before the fall of 1943, when the Nazi authorities stopped Danish travel to Sweden and initiated the pogrom against the Jews, the relatively few escapees, unassisted by an organization, made the crossing in small boats, even canoes and kayaks. After these Nazi measures were introduced, organizations were immediately formed to help the escapees, most of whom were Jews. Fishermen were co-opted, and the transport of Jews to Sweden began from the small fishing harbors on the east coast nearest to Sweden. Some 6,500 of the 7,000 Jews in Denmark were thus taken to Sweden. Speedboats made direct trips; the slower fishing and cargo vessels sometimes transferred their cargoes in the middle of the sound to boats which came from the Swedish side. Larger vessels in regular service were often diverted at gunpoint to a Swedish port to discharge the escapees, who had been hiding below deck. One cargo vessel had secret compartments which could hide 16 refugees.[57]

Escapees from Norway also found refuge in Sweden, crossing the border on foot at numerous points between the Norwegian cities of Halden and Elverum. During the first years of the Nazi occupation, they had to make their own way, but the increase in Nazi terror against the Jews fostered the growth of organized escape operations. Through the effort of an escape network known as the "Spider," 700 of Norway's 1,420 Jews were helped over the border into Sweden.[58]

From France and the Benelux Countries

From these countries, escapees were generally taken south by trains, cars, and on foot to the towns in the foothills of the French Pyrenees—Perpignan, Foix, Toulouse, Saint-Gaudens, Tarbes, and Pau. Thence, for a fee of 50 to 100 francs per person, they were conducted over the mountains into Spain by local French or Basque guides called *passeurs*. Once in Spain, they made their way with forged documents to Portugal or to the southernmost coast of Spain, whence speedboats, fishing boats, or submarines took them to Gibraltar or Algiers.[59] During the first months of the Nazi occupation, before the German coastal watch was tightened, many escaped across the Channel to England in fishing boats or other small craft.[60]

From Eastern Europe

Most Eastern European escapees came from Poland, Yugoslavia, and Greece. The major escape route from Poland first led south to the Carpathian Mountains on the Polish-Slovakian border. Then escapees followed mountain paths across German-occupied Slovakia and into Hungarian Slovakia. Until the fall of France in June 1940 and Italy's entry into the war, Poles who escaped to Hungary—which remained neutral until August 1940—were able to board trains and travel openly through Hungary, Yugoslavia, and Italy into France.[61] The most extensive use of this route was made by Polish soldiers following the collapse of Poland in September 1939, although this particular operation was not carried out by an underground. Approximately 140,000 Polish soldiers reached Hungary this way, and by June 1940 over 100,000 of them had traveled on to France to join the reconstituted Polish Army.[62] Even after this route was curtailed by the extension of the war, some Polish Jews still sought refuge in Hungary, where persecution was less severe than in most other countries under Axis control. It is estimated that nearly 70,000 foreign Jews—from Galicia, Slovakia, Rumania, etc.—sought refuge in Hungary until March 1944,[63] when the Nazis assumed full control of Hungarian affairs.

From Yugoslavia, most escapees were taken out by Mediterranean-based Allied airplanes, which used crude landing strips built by the Yugoslav guerrillas. Probably the most notable episode was the rescue of several Allied airmen who had been sheltered by underground and guerrilla units under the direction of General Mihailovic. Between August 9 and December 27, 1944, the U.S. Air Crew Rescue Unit evacuated from Mihailovic's territory 432 U.S. airmen who had been shot down during the war in various parts of Yugoslavia. They had been found and brought to several concentration points where U.S. aircraft could land. In addition to the 432 U.S. personnel, the Rescue Unit evacuated 4 British airmen, 2 Canadians, 2 Belgians, 30 Russians, and 76 Italians.[64]

The same technique accounted for many escapes from Greece. The August 11, 1943, flight of an RAF C–47 to and from a guerrilla "safe-area" near Neraidha is thought to have been the first Allied flight during the war to use a landing strip built by guerrillas in German-occupied territory. This particular flight picked up 12 passengers for the return trip to Cairo; and by late summer 1943 other landing strips were in operation in Greece.[65]

After the extension of the war in 1940–41, there were relatively few escapees from the Axis-occupied countries of Eastern Europe. Several factors not present in Western Europe to the same degree made out-of-the-country escape operations less feasible: the shortage of trains and cars made travel difficult; the terrain (e.g., the forests of Poland and the mountains of the other occupied countries) made in-country sanctuaries possible; and, except for Turkey, there were no contiguous neutral countries during most of the war.

Secret Lodgings

Couriers and persons facing capture may be hidden in safe-houses—quarters provided by underground sympathizers. Such a place is not necessarily a family's residence, but may be any building controlled by persons friendly to the underground. For example, Algerian doctors working for the FLN sometimes placed evaders in the Algiers Municipal Hospital as patients. Polish Home Army personnel sometimes asked prostitutes to provide secret lodging.

Regular underground workers sometimes allow their homes to be used as secret lodgings, but this compounds the risks already confronting these workers and poses an additional threat to the security of the organization. To avoid this, undergrounds have often preferred to house persons in hiding with sympathizers who know little about the underground's workings, but have consented to do their part for the movement by providing lodging on call.[66]

For a Short Period

Some situations call for only temporary housing. To facilitate the travel of persons along escape routes, undergrounds provide houses where escapees may stop for food, rest, and directions to the next place. The French resistance provided Allied forces with addresses of such places for use by Allied airmen shot down over French territory. Usually directions are given only to the next stop, to prevent the exposure of several safe-houses should an escapee be captured and interrogated.

Temporary lodging may be required also by underground workers who have reason to fear for their safety; there is always a possibility of exposure when an associate is captured and interrogated. In such circumstances, the endangered person goes into hiding until it can be determined whether he has been implicated. The experience of an anti-Nazi underground leader in Germany illustrates this need for safe-houses during periods of danger. On two occasions this leader felt it would be safer for him to stay away from his apartment for several nights: the first time, there was a risk of accidental detection because the Gestapo was watching his apartment building for another man; the second time, a good friend of the leader was arrested and there was risk of the leader himself being implicated. Both times the difficulty was where to go. In his words:

> You know that you can't walk in the door of a respectable hotel in Germany without having to fill out the regular police registration. A hotel with a doubtful reputation would be the surest way to be caught in a police raid . . . finding quarters for underground purposes is one of our most perplexing tasks.[67]

For an Extended Period

Persons sought by the authorities may require more or less permanent refuge. This situation arises when other alternatives are closed; that is, when such persons cannot leave the country or take refuge in hidden camps or with guerrilla units. This situation confronted most of the fugitives in the Netherlands during World War II. Escape by sea was effectively sealed by the Germans so that only 150 to 200 Dutchmen were able to leave by this route. There were some railroad lines to Spain via France, but they were not able to cope with the vast numbers. The number of fugitives ran into the hundreds of thousands, and most of them became *"onderduikers"* ("divers") and went into hiding. Because of the country's size, topography, and density of population, it was impracticable to organize large guerrilla bands or collective hiding places. The great majority of "divers," therefore, had to be hidden by relatives, friends, or others willing to run the risks of providing safe-houses.

The risks in providing safe-houses are often considerable. Neighbors may notice and report the evaders. There is also the problem of obtaining forged ration cards, forged identity papers, and even money with which to buy food and other necessities for the "guests." [68] Cover stories must also be devised to explain the presence of lodgers in case of inquiry by neighbors or the authorities.

The risks in providing refuge are greatly increased when a number of people rather than a single person has to be housed. A well-known instance of this was the effort to hide the Frank family altogether as a group in an Amsterdam attic.

In Remote Camps

Given wooded or mountainous terrain, camps may be established to accommodate evaders. Such camps were established by the Dutch in the spring of 1943 in the forests of Limburg and Gelderland; but the problems of feeding the fugitives and maintaining the secrecy of the camps soon led to their abandonment.[69]

With Guerrilla Units

Fugitives may evade capture by joining guerrilla units. Able-bodied persons become fighters or supply carriers. Those who are not able to fight—

aged persons, children, and the handicapped—may perform support tasks such as working in small manufacturing shops in the safe-areas.

In Indochina, a shortage of arms and ammunition did not cause the Vietminh to reject able-bodied evaders—they used them in the auxiliary service to transport supplies. The less physically fit could always be taken care of in safe-areas. On the other hand, the French Maquis was forced to turn away many persons because there was often a lack of arms and ammunition and there was no significant support activity in which they could be engaged.

SPECIAL CONSIDERATIONS

Supplies

An underground attempts to provide all escapees, evaders, and couriers with such necessary documents as birth certificates, identity papers, travel permits, ration cards, and work stamps, so that they may safely pass through police checks. Of course, this cannot always be done when there is a sudden demand for a large number of papers. Groups of Jewish refugees from Nazi-occupied areas often failed to receive covering papers because of their quick exit and the lack of adequate collections of forged documents. To prepare for such emergencies, an underground may accumulate large quantities of documents: in 1943 one underground in Germany was in possession of over 5,000 passports.[70]

Besides documents, an underground may give money to travelers for the purchase of tickets and food and clothes for disguises. Habib Bourguiba, then head of the illegal Neo-Destour Party, escaped from Tunisia in 1945 in the clothes of a Berber tribesman. He crossed into Libya and made his way on camelback across the desert to Egypt.[71] An underground also may provide those traveling on foot with compasses, maps, or food.

Counterinfiltration Measures

To guard against the infiltration of the escape and evasion network by enemy security personnel, strangers seeking the assistance of the underground may be subjected to tests. They may be questioned on their background to determine whether there are any discrepancies in their accounts of their places of residence, jobs, friends, and reasons for soliciting aid.

Persons identifying themselves as foreign military personnel can be checked by asking them questions whose answers only genuine nationals of the given country would be likely to know; for example, questions about geographical details, sporting figures, units of measurement. A former member of the French resistance reports that German *agents provocateurs* posing as Allied airmen were sometimes exposed by being asked their weight. An Englishman would reply quickly, but a German would hesitate, trying to convert his weight into stone. The British assisted by providing a detailed list of useful questions.[72]

Strategy, Tactics, and Countermeasures

Other Security Aspects

Despite precautions, a traveler may find himself in need of help. In such a case, he should appeal to someone who is alone rather than to a group whose members might hesitate to extend aid for fear there might be an enemy collaborator among them. Foreigners traversing an escape route have to cover their accents or inability to speak the country's language in case they are questioned by the police. British airmen downed in France were sometimes given papers certifying them as deaf and dumb or attributing to them some other disability that would account for their silence if questioned.[73] A Polish Home Army courier traveling through Europe as a Frenchman concealed his imperfect French by pretending to have a toothache and mumbling his words.

FOOTNOTES

1. T. Bor-Komorowski, *The Secret Army* (London: Victor Gollancz, Ltd., 1950), p. 79.
2. Ibid., pp. 78-79.
3. Jan Karski, *Secret State* (London: Hodder and Stoughton, Ltd., 1945), pp. 177-179.
4. Philip Selznik, *The Organizational Weapon: A Study of Bolshevik Strategy and Tactics* (Glencoe, Ill.: The Free Press, 1960), p. 250.
5. See ibid., p. 96.
6. See ibid., p. 164.
7. See ibid., p. 49.
8. Vo Nguyen Giap, *People's War, People's Army* (Hanoi: Foreign Languages Publishing House, 1961), pp. 78-79.
9. Paul M. A. Linebarger, *Psychological Warfare* (Washington: Combat Forces Press, 1948), p. 46.
10. See John W. Riley, Jr. and Leonard S. Cottrell, Jr., "Research for Psychological Warfare," in William E. Dougherty and Morris Janowitz, *A Psychological Warfare Casebook* (Bethesda, Md.: Operations Research Office, 1958), pp. 536-544; Daniel M. Lerner, "Effective Propaganda Conditions and Evaluation," *Propaganda in War and Crisis*, ed. Daniel Lerner (New York: G. W. Stewart, Inc., 1951), p. 347; M. F. Herz, "Some Psychological Lessons from Leaflet Propaganda," in *Propaganda in War and Crisis*, pp. 416-417; and E. A. Shils and Morris Janowitz, "Cohesion and Disintegration in the Wehrmacht," in *Propaganda in War and Crisis*, pp. 367-413.
11. See George K. Tanham, "The Belgian Underground Movement, 1940-1944" (unpublished Ph. D. thesis, Stanford University, 1951), pp. 316, 318.
12. Charles F. Delzell, *Mussolini's Enemies* (Princeton: Princeton University Press, 1961), p. 69.
13. See F. B. Barton, *North Korean Propaganda to South Koreans*, Technical Memorandum ORO-T-10, EUSAK (Bethesda, Md.: Operations Research Office, 1951), p. 3.
14. Ibid., p. 46.
15. Ibid., p. 110.
16. Leo Lowenthal and Norbert Guterman, "Portrait of the American Agitator," *Public Opinion Quarterly*, V (1948), 417-429.
17. Eugene Methvin, "Mob Violence and Communist Strategy," *Orbis*, V (1961), 166-181.
18. Ibid.
19. Personal interview, October 12, 1955, by Lorna Hahn (Special Operations Research Office) of Abderahmen Anegai, Istiqlal organizer in Tangiers and former head of imperial cabinet of Mohammed V.

20. See Bor-Komorowski, *Secret Army*, p. 79.
21. Lucien W. Pye, *Guerrilla Communism in Malaya* (Princeton: Princeton University Press, 1956), p. 88.
22. For an analysis and discussion of threats, see T. C. Schelling, *The Strategy of Conflict* (Cambridge: Harvard University Press, 1960).
23. See Jon B. Jansen and Stefan Weyl, *The Silent War* (New York: J. B. Lippincott, 1943), p. 236.
24. See G. Rivlin, "Some Aspects of Clandestine Arms Production and Arms Smuggling," *Inspection for Disarmament*, ed. S. Melman (New York: Columbia University Press, 1958), pp. 191–202.
25. Pye, *Guerrilla Communism*, p. 106.
26. Ibid., pp. 104–106.
27. See Julian Amery, *Sons of the Eagle* (London: Macmillan Co., 1948), p. 168.
28. Otto Heilbrunn, *Partisan Warfare* (New York: Frederick A. Praeger, 1962), p. 87.
29. Otto Heilbrunn, *The Soviet Secret Services* (New York: Frederick A. Praeger, 1956), p. 53.
30. Ibid., pp. 53–54.
31. David Lampe, *Danish Resistance* (New York: Ballantine Books, 1957), pp. 28–31.
32. See Ronald Seth, *The Undaunted: The Story of the Resistance in Western Europe* (New York: Philosophical Library, 1956), pp. 42–45.
33. Lampe, *Danish Resistance*, pp. 93–98.
34. Tanham, "Belgian Underground," pp. 157–158.
35. Bor-Komorowski, *Secret Army* p. 82.
36. Maurice J. Buckmaster, *Specially Employed* (London: The Batchworth Press, 1952), p. 84.
37. Maurice J. Buckmaster, *They Fought Alone* (New York: W. W. Norton, 1958), pp. 239–242.
38. Buckmaster, *Specially Employed*, p. 86.
39. Seth, *The Undaunted*, p. 129.
40. Lampe, *Danish Resistance*, pp. 28–31.
41. French Resistance Collection (Hoover Library, Stanford University), Folder 8, No. 1, p. 1.
42. P. de Preval, *Sabotages et Guérilla* (Paris: Berger-Levrault, 1946), pp. 35–36.
43. French Resistance Collection, Folder 8, No. 1, p. 70.
44. Lampe, *Danish Resistance*, pp. 33–34.
45. Ladislas A. Farago, *War of Wits* (New York: Funk and Wagnalls Co., 1954), p. 249.
46. See R. Alden, " 'Plastic' is Handy in French Terror," *New York Times*, January 28, 1962.
47. C. Aubrey Dixon and Otto Heilbrunn, *Communist Guerrilla Warfare* (New York: Frederick A. Praeger, 1955), pp. 73–74.
48. Ibid., p. 74.
49. de Preval, *Sabotages et Guérilla*, pp. 35–36.
50. Ibid.
51. Heilbrunn, *Partisan Warfare*, p. 89.
52. Alberto Bayo, *One Hundred Questions Asked of a Guerrilla Fighter*, tr. Department of Commerce, Joint Publications Research Service (Washington: Government Printing Office, 1962), Question 37.
53. Ibid., Question 30.
54. Farago, *War of Wits*, p. 240.
55. Ibid.
56. Seth, *The Undaunted*, p. 37.
57. Ibid., pp. 123–126.
58. Ibid., p. 37.
59. See Philippe de Vomecourt, *An Army of Amateurs* (New York: Doubleday, 1961), p. 48; Tanham, "Belgian Underground," p. 107; Seth, *The Undaunted*, p. 179; and Buckmaster, *Specially Employed*, pp. 50–51.

Strategy, Tactics, and Countermeasures

60. P. de Grande Combe, *The Three Years of Fighting France* (London: Wells, Gardner, Darton and Co., Ltd., 1943), p. 23.
61. For a personal account of the use of this route from Poland to France, see Karski, *Secret State*, pp. 91–97.
62. C. A. Macartney, *A History of Hungary, 1929–1945* (New York: Frederick A. Praeger, 1957), p. 368.
63. Ibid., p. 101.
64. David Martin, *Ally Betrayed* (Englewood, N.J.: Prentice-Hall, Inc., 1946), p. 245.
65. D. M. Condit, *Case Study in Guerrilla War: Greece During World War II* (Washington: Special Operations Research Office, 1961), p. 137.
66. de Vomecourt, *Army of Amateurs*, p. 50.
67. Jansen and Weyl, *Silent War*, p. 113.
68. Seth, *The Undaunted*, p. 179.
69. Ibid., p. 180.
70. David J. Dallin, *Soviet Espionage* (New Haven: Yale University Press, 1955), p. 93.
71. Lorna Hahn, *North Africa: Nationalism to Nationhood* (Washington: Public Affairs Press, 1960), pp. 31–32.
72. de Vomecourt, *Army of Amateurs*, p. 50.
73. Ibid., p. 48.

CHAPTER 4

COMMUNIST USE OF UNDERGROUNDS IN RESISTANCE AND REVOLUTION

INTRODUCTION

This chapter summarizes the basic elements of the Communist theory and practice of subversion, and describes the role of Communist undergrounds in (1) German-occupied Soviet territory, (2) the "wars of liberation" in the underdeveloped countries, and (3) the Soviet takeover of Eastern European countries. The chapter concludes with a brief description of the Soviet security and espionage systems and a summary statement of the activities of national Communist parties.

ELEMENTS OF COMMUNIST SUBVERSIVE ACTIVITIES

Although many 19th century advocates of revolution had propounded theories concerning the organization of subversive movements, Karl Marx was the first to analyze the historical processes and the political environment in which the Communists would operate, and to evolve his strategy and tactics accordingly. In his theory, modern capitalist society was divided into two major classes: the small bourgeois group, which owned the means of production and exploited the workers, or proletariat, and the latter, who were destined to revolt against the bourgeoisie and set up a society wherein the workers controlled the means of production and enjoyed the full products of their labor. The Communist Party, acting as the "vanguard of the proletariat," would clarify the issues for the masses and guide them toward revolution. Thus, while assuming a certain "class consciousness" and "activism" on the part of the proletariat, Marx carefully stressed the need for skilled cadres. Such cadres would be thoroughly versed in dialectics and trained in revolutionary techniques by which they would develop and channel properly the predominantly dormant revolutionary sentiment of the masses which he assumed to exist in all societies.

From this concept, Nicolai Lenin evolved the theory that the party alone would be capable of making a revolution; that the proletariat was incapable of any meaningful action. Whereas many of his comrades were interested in enrolling large numbers of people in the revolutionary movement, Lenin insisted on confining it to a small, hard core of dedicated militants who could be depended upon to maintain firm discipline and carry out orders precisely. Making a revolution, in other words, was to be the work not of idealistic amateurs but of a realistic elite professionally trained in the skills of subverting a government.[1]

Strategy, Tactics, and Countermeasures

The Leninist strategy for revolution encompassed three basic phases: organizing party nuclei capable of expanding; employing these nuclei, once they became strong enough, in such centers of power as unions, other political parties, and the *Duma* (the Russian parliament); and the final seizure of power. When his party was in the initial stage, Lenin expounded many precepts designed to increase its efficiency and protect it against governmental interference. The most important of these was the injunction to form both an open party organization and an underground apparatus, which could continue functioning even if the party was declared illegal. When illegal, he stated, the party must seek to obtain control of legal nonparty organizations. This duality of apparatuses was to become a cardinal rule of Communist organization throughout the world. The Communist International obliged all parties, whatever their status, to construct underground organizations.*

The second stage entailed not only infiltrating, "capturing," and manipulating smaller groups, but also finding temporary "allies" with whom to work in larger arenas of power, while trying to gain influence over the masses. Here Lenin cautioned his followers to conceal their party affiliation or, if this was known, to disguise the party's intentions.[2] He also advocated extensive use of propaganda designed to allay the fears of other organizations (including allies) regarding the party's true aims: to divert hostile attention away from the party and toward a common enemy, to create and spread disunity among the other groups and organizations, and to undermine the authority of the government.

Lenin believed that the Soviet Union must do everything possible to further world revolution, and to this end he supervised the founding of the Comintern in 1919. Through this organization, the Soviet Government was able to direct the activities of Communist Parties throughout the world and tailor their policies to suit its own interests. Lenin continued to concentrate on plans for making revolution in the more industrialized countries of Europe, but also became impressed with the political significance of the backward areas. In his *Imperialism: The Last Stage of Capitalism* (1915) he transferred the intranational class struggle between bourgeoisie and proletariat to the international arena, describing the colonized or underdeveloped nations as the victims of exploiting "imperialist" powers. After the war he stated:

> It is necessary to pursue a policy that will achieve the closest alliance of all the national and colonial Liberation movements with Soviet Russia, the form of this alliance to be determined by the degree of development of a Communist movement among the proletariat of each country, and of the bourgeois Democratic Liberation movement of the workers and peasants in backward countries or among backward nationalities.

These general ideas were subsequently developed by Josef Stalin, who added the concept of Communists being "scavengers of revolutions"—that is,

*In 1936 the Communist Party of Indochina was permitted legal existence by the French Government. Nevertheless, on orders of the Comintern, it still maintained clandestine networks.

capturing control of revolutionary movements started by other groups and by other leaders such as Mao Tse-tung and Vo Nguyen Giap. Under Stalin, the world Communist movement grew in size and strength. It tended to attract different types of members in different areas, depending on the given society. In the United States, for example, its members were drawn mainly from minority groups and from otherwise insecure people of various economic strata. In Europe communism tended to be identified with workers and trade union interests and leftwing intellectuals. In Asia it attracted nationalists who wished to fight colonialism.[3]

Organizational methods and tactics became highly developed. The concept of the professional revolutionary was maintained, and recruits were subjected to various trials and tests to determine their willingness to devote themselves completely to the movement and their ability to carry out useful tasks. Once accepted, members were given extensive and continuous indoctrination in the theories of Marx, and the theories and strategy developed by Lenin and Stalin. They were also given technical training for specific tasks, such as gathering intelligence, sabotage, and clandestine publication of subversive literature. The dual apparatuses were maintained. The legal parties concentrated on political agitation and propaganda and on the infiltration of such important groups as unions, professional associations, communications media, and even churches. They also organized front groups under names which would have appeal in the given environment; for example, the Thomas Jefferson Club in New York and the "agrarian friendship societies" in Indochina. Although organized to support popular reform, these groups used organizational pressure to exploit issues, government actions, and public opinion which coincided with the interests of international communism.

The clandestine parties followed a system of organization adapted from the overt party structure: they were organized in a hierarchical structure with numerous cells on the bottom and a central committee at the top. While concealing their party affiliation, members held, or tried to attain, positions of some importance in the country. It was mainly these people who collected intelligence on governmental activities, conducted sabotage operations, acted as couriers, and maintained contact with individuals who were cooperating unwittingly with the Communist Party. In countries with which the Soviet Union had diplomatic relations, an underground party member might serve as a member of the embassy staff protected by diplomatic immunity.

The Kremlin-controlled Comintern not only set the "party line" for the various Communist parties, but also ran training schools for Communists from all nations.[4] Such men as Tito, Mao Tse-tung, Ho Chi-minh, and Walter Ulbricht received instruction in Moscow not only in organizational methods and tactics but also in means of infiltrating governments and effecting coups d'etat. These trained leaders profited from World War II by attracting patriotic people to a resistance movement, training them for military operations, conducting a logistical and liaison operation, and winning foreign support. After the war these factors were critical in their winning control of the government.

Strategy, Tactics, and Countermeasures

In selecting the proper moment to launch a revolutionary offensive, Communist leaders were also taught to assess both "objective" and "subjective" factors in making their decision. The former consisted of the overall conditions prevailing in the country—for example, the amount of unemployment, degree of social discontent, degree of political instability and corruption. The subjective conditions included the factors pertinent to the strength of the party itself—the size and quality of its membership, its skill and experience in techniques of subversion and organization for guerrilla warfare, its ability to obtain personal and material support, etc. Given satisfactory conditions in both categories, the leadership could then launch a revolution.

ROLE OF UNDERGROUNDS IN GERMAN-OCCUPIED SOVIET TERRITORY

PREWAR FACTORS INFLUENCING RESISTANCE

In creating its anti-German resistance campaign, the Soviet Government was aided appreciably by the existence of several reliable organizations which could constitute a framework for an underground resistance movement. The most important of these was the Communist Party apparat; consisting of a hierarchy of echelons throughout the Soviet Union centrally controlled from Moscow. Supplementing this network were the Komsomol youth organization and the NKVD police force, both of which had chains of command generally paralleling those of the party.

The Soviet leaders were also able to profit from both prerevoluntionary and postrevolutionary party experiences. Clandestine antigovernmental operations during Czarist days and the overthrow of the moderate Kerensky regime had demonstrated that, under certain conditions, a well-directed underground movement comprising disciplined individuals could overthrow a regime by hammering away at its weak points and infiltrating its centers of power. The civil war and the subsequent Allied intervention provided lessons on how to mobilize the populace against a foreign enemy and how to weaken or destroy an occupying force. These experiences influenced the development of Soviet military doctrine, which always emphasized the importance of resistance activities.

Soviet military theorists, before and since World War II, have placed great stress on the "stability of the rear," a concept which they rank alongside such better-known tactical principles as surprise, mobility, and concentration of force. According to Marshal Klementi Voroshilov stability of the rear includes—

> all that constitutes the life and activity of the whole state—social system, politics, economy, the apparatus of production, the degree of organization of the working class, the ideology, science, art, the morale of the people and other things.[5]

The concept of "security of the rear" connotes both the strengthening and securing of the Soviet rear and the simultaneous weakening and undermining of the enemy rear. Stalin noted that as the Germans advanced into Russia they were getting away from their normal rear base, and that by operating in hostile surroundings, they were becoming an increasingly vulnerable target.[6] He welcomed the Western European resistance movements because they further weakened the German rear and helped undermine the entire German military effort.[7]

In operating against the German rear, the Soviets relied mainly upon partisan warfare. This combined large-scale fighting by guerrilla bands in rural areas with political activities was designed to obtain or maintain the allegiance of people in the German-occupied areas. The underground movement, then, was essentially an adjunct to the guerrilla forces in this effort to disrupt the "stability of the German rear."

Although partisan warfare was not a new concept in the Soviet Union, the population and even the party cadres in the German-occupied areas were not sufficiently prepared initially for underground resistance against an occupying power. Years of operating as a ruling elite had caused many older party members to lose the feeling for conspiratorial work, while younger members had never participated directly in clandestine operations. Also, probably because prewar Soviet military doctrine envisaged a future war as an offensive campaign, there was a lack of detailed contingency planning at the regional and local levels.

Early Soviet plans for resistance existed only at the All-Union and Republic levels, and envisioned organizations paralleling the territorial structure of the party and the NKVD. In each *oblast* (region) an "underground secretary," secretly appointed by the party, was to remain and direct the entire partisan effort. A major component of this effort was to be a network of "diversionist" groups, whose most important activities would take place in the urban areas under the direction of remaining NKVD cadres.

On its lowest levels, the resistance network was to consist of groups of three to seven persons trained for sabotage and miscellaneous missions. Members were to continue working at their normal occupations in order to cloak their clandestine activities, and, to ensure security, each member would know only his group leader and the others in his immediate band. In addition to the diversionist groups, a considerable *ad hoc* party network under the direct control of the underground secretary was to carry out sabotage, armed attacks, and propaganda activities.[8]

The diversionist groups were to be supplemented by territorially based *otryad*, or "destruction" groups, which were full-time, overt partisan bands.[9] Theoretically there was to be no organizational connection between the diversionist network and the partisan bands, in order to reduce the risk of discovery. In practice, however, many underground diversionist groups later found it necessary to rely on the partisan bands for assistance and refuge. The partisans were to engage in hit-and-run tactics against small isolated German posts and acts of terrorism against collaborators. They were to avoid any lengthy or major engagements with the enemy, and to make no attempts to control any areas through physical presence. The emphasis was to be on stealth and surprise, with the underground members resuming their normal activities after the completion of an operation.

The initial plans for underground and partisan activities, however, were generally swept aside by the swift advance of the German Army during the summer and early fall of 1941. During the winter of 1941–42 the surviving party underground organization led a precarious existence, and concerned itself primarily with self-preservation rather than with actions against the Germans. Nevertheless, its ability to survive at all must not be minimized.

DEVELOPMENT OF UNDERGROUNDS DURING THE WAR

During early 1942, significant changes were made in the Soviet partisan structure which reflected a fundamental reorganization of the entire resistance effort. Instead of remaining a relatively small elite movement based on loyal party members and NKVD agents, the resistance became a mass movement utilizing all available sources of manpower, including peasants, escaped prisoners of war, and soldiers who had been cut off from their units by the German advance. Tasks were assigned to individuals or to groups, not on the basis of their party affiliations or proved political loyalties, but on the basis of their ability to operate effectively behind German lines.[10]

The partisan bands were constantly reorganized into larger units, and were assigned to occupy large rural areas. Since a manpower shortage restricted the physical presence of the German occupation forces to the major towns and road junctions even in regions where the guerrillas were active, the countryside fell under the control of the partisan bands. While the partisans were able to gain control of large regions and to operate with some effectiveness in many "twilight zones," these areas contained only a minority of the population of the occupied territories. Nevertheless, the guerrillas were able to give support and refuge to the underground networks concentrated in the cities, and thus to contribute to their effectiveness.

Both the partisans and the underground networks pursued the dual aims of disrupting the German-oriented local administrations and spreading the idea of Soviet invincibility. The former, by constantly harassing the enemy, were able to demonstrate to the populace that Soviet power was not extinct, and to discourage the development of anti-Soviet movements. The latter, through extensive propaganda and psychological warfare, spread the news of partisan successes and discouraged collaboration. Both partisans and underground workers, in other words, fought political and psychological battles in order to maintain or recapture the political allegiance of the people in the occupied zones.[11]

The urban underground was assigned two main tasks: psychological warfare and intelligence operations. As was previously mentioned, the Communist Party planned originally to leave in the occupied areas an underground apparatus which could not only harass the enemy, but also conduct such propaganda activities as publishing clandestine newspapers. This party organization was to be aided by the Komsomol, which had an extensive network in the occupied territories. As it turned out, more Komsomol cadres than party members actually remained behind. Therefore, the government made them the nucleus of the urban underground. Literate, physically fit, and highly motivated, the Komsomol members were ideally suited for psychological actions directed at both the Germans and the indigenous population.

Both the partisans and the urban underground employed sabotage, terror, and intimidation. Since the latter group operated in closer contact with occupation forces, collaborators, and large numbers of people, assassination and terrorism became its special domain. (Partisan bands specifically formed for these tasks operated occasionally, and were particularly active during the summer and fall of 1943.) [12] The urban underground also concentrated on infiltrating and subverting German and collaborator organizations. Agents would infiltrate spontaneously when the opportunity arose, or when ordered to do so by a superior, or when recruited by someone already working as an infiltrator. Not only clerks and interpreters but also such officials as police and municipal authorities and chairmen of *kolkhoz* units were recruited.

The crucial task of penetrating and subverting indigenous puppet administrations was given to special agents usually recruited from the party or Komsomol. They were to plan activities designed to cause the breakdown of the local administration, sabotage German reconstruction and rehabilitation work, and discredit collaborators. It was not unusual for agents to infiltrate and capture from within a police or civil organization and thus reduce considerably the efficiency of the German administration.[13]

Agents assigned to intelligence activities sought information on three basic subjects: German operations which might affect the security of local partisan bands; German operations which might affect the Red Army; and political, economic, social and cultural matters. Although some of these agents worked only for the urban underground, the majority were affiliated with local partisan bands. These informants worked as railroad men and hospital staff; others volunteered for jobs as German interpreters, helpers or servants. As agents who came in daily contact with the Germans, they provided the partisan groups with a constant source of information.[14]

In recruiting intelligence agents, a premium was placed on innocent appearance and freedom of movement. Thus women, children, and older people were used extensively.[15] According to one observer, 70 percent of all Soviet intelligence agents working behind German lines were women. Employed as clerks, laundresses, attendants, and translators, they could not only obtain useful information, but also help recruit other agents and guerrilla fighters and perform occasional sabotage missions.[16]

The organization of underground intelligence networks varied considerably in different areas. Each network was under the direction of the local partisan intelligence officer who controlled four types of agents: (1) informants among the population, such as former activists, party members and candidates,

```
                    ┌─────────────────────────┐
                    │   Central Committee     │
                    │      of the all         │
                    │  Russian Communist Party│
                    └─────────────────────────┘
                              │
                    ┌─────────────────────────┐
                    │ National Defense Committee│
                    └─────────────────────────┘
```

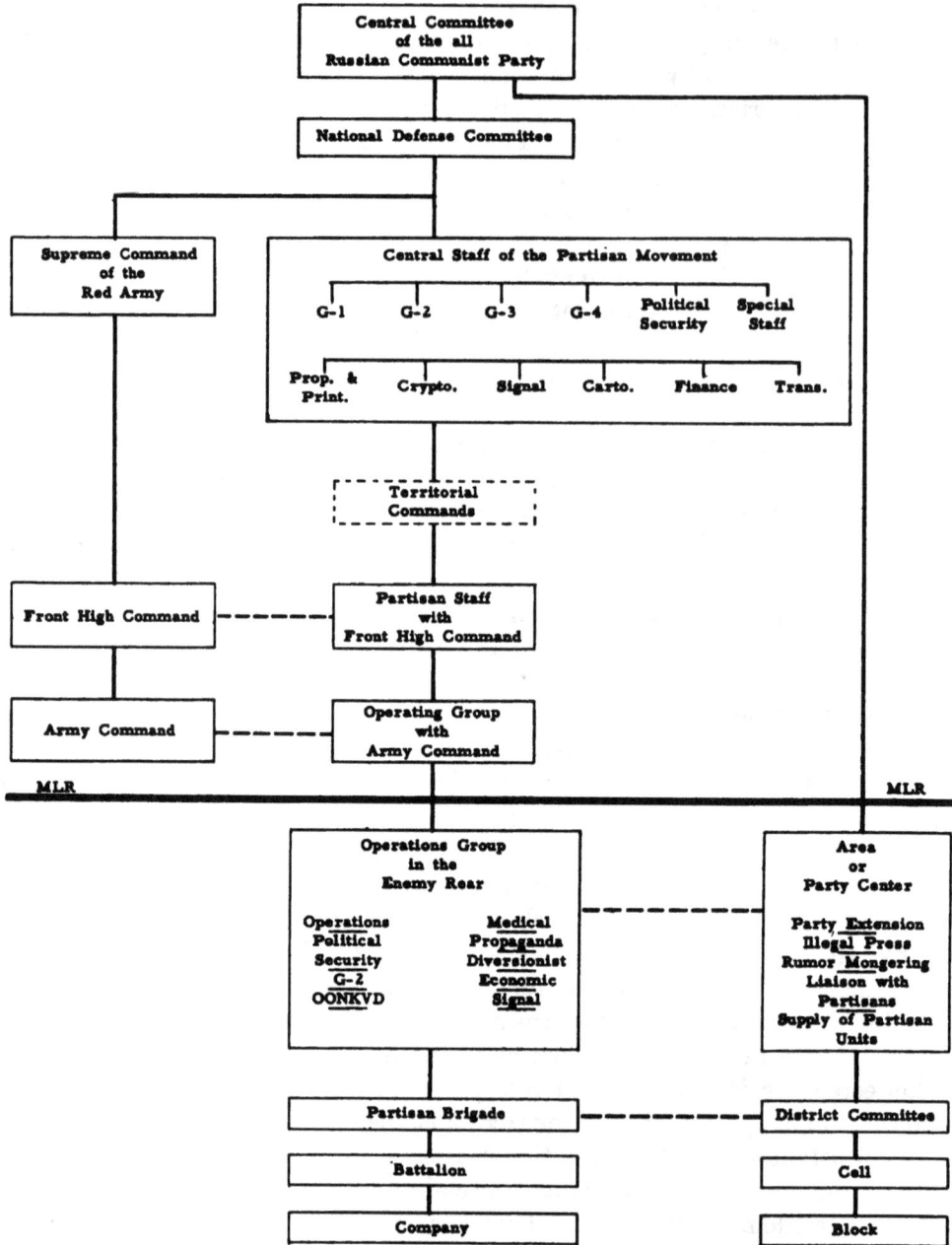

Figure 3. Organization of Partisan and Soviet Army Units.

and wives of soldiers, (2) resident agents, (3) agents recruited from groups working for the Germans, (4) information agents recruited from those persons who had special passes and were thus able to move freely between localities.

In view of the risks assumed by full-time informants, only the most active and patriotic citizens volunteered for these activities. There were others who occasionally participated to provide insurance for the future, expecting that the Soviets would eventually reestablish control of the occupied areas. From the majority of the population, however, the partisan intelligence network could expect only passive neutrality or, at most, timely warning of German operations.[17]

IMPORTANCE AND EFFECTIVENESS OF SOVIET UNDERGROUNDS

The undergrounds affiliated with the partisan movement became so important during the war that by 1943 the Kremlin relied almost exclusively on their intelligence networks for information on German activities and the attitudes of the population. The effectiveness of these networks, according to one source, was probably due less to the quality of the agents than to the quantity of intelligence activities and the large amount of *ad hoc*, informal reporting on German activities.[18] As the partisan forces grew in number and the successes of the Red Army presaged a Soviet victory, more and more civilians cooperated in underground activities, including the gathering of intelligence.

Besides its intelligence activities, the underground also performed effectively in psychological warfare. In evaluating their importance in this field, one must remember that the underground networks were employed by the Soviet Government not only to create difficulties for the German occupiers, but also to represent and maintain its own authority over the people in occupied areas. Their help in propaganda operations, purges, psychopolitical operations, and the establishment of governing bodies behind German lines was a valuable asset [19] in maintaining or regaining for the Kremlin the allegiance of the populace. It must be noted that as the Soviet counteroffensive moved westward in 1944, and anti-German activities decreased, the underground networks posed a problem for the Soviet authorities. For although they had on the whole maintained close connections with the party elite and had been staffed by people of known political loyalty—in contrast with the partisans, who had armed large numbers of politically untested men—the undergrounds presented a potential danger to the security of the regime because of the revival of conspiracy and the "legitimization of illegality." The Soviet Government succeeded in destroying this threat by imposing tight party controls over the underground and by disseminating propaganda which stressed that the anti-Fascist resistance activities were a bona fide aspect of Soviet Communist policies.

THE ROLE OF UNDERGROUNDS IN "WARS OF LIBERATION"

In World War I the Russian Communists seized power by organizing strong urban networks and effecting a rapid coup d'etat. The conditions which had permitted the Bolsheviks to enter and control key centers of power, however, did not exist in most other countries prior to World War II. Leaders of Communist movements in industrialized countries saw no alternative other than to bide their time and wait for the proper moment to make a bid for power. Leaders in underdeveloped countries, however, concluded that there might be a direct path to power through a "war of liberation" or "people's war," in which guerrilla forces led by the Communist Party and backed by the masses would eventually oust the incumbent government through military victory.

The most important and the most characteristic feature of "liberation wars" is the use of the peasantry rather than the proletariat as the main source of Communist strength.[20] In China, after 1927, the Kuomintang was attemping to crush the Communists and the Communist-controlled labor unions in the major cities. Rather than continue trying to organize the urban groups and to subvert Kuomintang-controlled institutions, Mao went into the countryside, where Kuomintang administration was either weak or nonexistent, and organized the peasants into rural soviets. He succeded in organizing millions of people into such soviets and his success set a pattern for other countries where the bulk of the population was rural. Thus, in Indochina in 1930, when all nationalist activities were illegal and the Communist Party had not yet gained prominence among the nationalist underground groups, rural soviets were organized in northern Vietnam by Chinese agents from Hong Kong.[21] These initial soviets spread their influence among the peasantry and later became the bastion of Communist strength in Indochina.

Another characteristic of liberation wars is the creation or control by the Communist Party of an army which serves as the vanguard of the revolution. Such an army, usually termed a "People's Liberation Army," combines military with political activities.[22] Under the direction of the party, each soldier is indoctrinated and trained for political as well as military duties, so that he can help convince the masses that this is their struggle and persuade them to support the revolution.

A third feature is the creation by the Communists not only of "bases," which serve as supply depots, but also of large areas of "liberated territory" which serve the liberation army as sources of intelligence and supplies, sanctuaries for escape, evasion, or recuperation, and centers of political strength.[23] There is no set rule concerning how or when such liberated areas are organized. In China, the Red Army had been active for years before Mao began forming the "anti-Japanese war bases" in 1937. In forming them, he reorganized the administration and functions of rural soviets he had already set up; also, his troops occupied by force large areas hitherto controlled by the Kuomintang or the Japanese, and then organized the people.[24] In some cases, particularly during the Japanese occupation of Manchuria, Mao sent "armed working teams" consisting of small party cadres into an area with the ostensible purpose of helping the people, and then organized them. In Indochina, on the other hand, Ho Chi-minh usually won over villagers first, engaged in small-scale actions designed to make small isolated units of the government forces leave the area, and then created "liberated" zones in which "shadow governments" operated.[25]

Another noteworthy point is that urban undergrounds play a secondary or supplementary role in most liberation wars. In China it was not until the Japanese were defeated and the Communists entered the "Third Revolutionary War," or final drive against the Kuomintang, that concerted efforts were made to extend Communist influence in urban areas. Efforts were centered in the cities and communications centers of Manchuria, where Communist agents were instructed to turn the people against Chiang, form underground networks, and recruit workers and intellectuals for construction work and military duties in the rural base areas.[26] In Indochina, where clandestine urban organizations had operated since 1927, they played a minor role, their main task being the extortion of money from merchants and other well-to-do people in order to help the guerrillas.[27]

The general overall strategy in liberation wars calls for a long "protracted war." [28] This permits the guerrilla army, which may be very small initially, to gain new recruits, and also gives the Communists the opportunity to establish and consolidate rural soviets and bases. From the military viewpoint, a long guerrilla war wears down the enemy both materially and psychologically, and thus facilitates its defeat. The conflict is conceived as having three main stages: an inital phase, when the guerrilla forces are weak, of terrorizing or harassing the enemy; a second phase, when the guerrillas have developed strength, of extensive guerrilla operations; and a final stage of regular warfare against the enemy's military forces,[29] which may culminate in a decisive battle. The basic plan is to gain control of the countryside first, then encircle or otherwise gain control of the major cities.

While their plan for revolution was new, the advocates of liberation wars generally followed Marxist-Leninist tenets in their overall approach and in their specific tactics. Thus the Communist parties in China, Indochina, and other Asian countries remained the tight elite groups Lenin had advocated. Each of these parties maintained firm political control over its army, with its cadres playing roles in the guerrilla forces similar to that played by party cadres in the Soviet Army. When village soviets and bases were created, party cadres played dominant parts in the governing structure.

Standard Communist techniques of forming front groups and infiltrating organizations have been used in the new milieu. Thus Mao formed a united front with the Kuomintang to fight the Japanese in 1937; in 1941 the Indochinese Communist Party organized the Vietminh as a front organization in which other nationalist groups participated but which the Communists dominated.[30] Deception regarding affiliations and goals is also practiced extensively. A war is not termed a war for "the triumph of communism" but for "agrarian reform." In Indochina, when the Communist Party was illegal, it formed "rural friendship societies" and a front group known as the "Indochina Democratic Front." [31] In other words, every effort is consciously made to present the movement and its goals in terms which attract the masses and at the same time deceive outsiders.

Specific tactics to win popular support vary according to circumstances, but some general patterns have been followed in every liberation war. A typical procedure is for the army, before entering a village, to dispatch a team of soldiers which includes a few party members. These men act as friends of the people, and do something useful such as helping to harvest crops. After establishing a social indebtedness for the services performed, they disseminate propaganda—usually by word of mouth, sometimes accompanied by pamphlets—describing the evils of the existing regime and the benefits which the liberation force wishes to bring to the people. They hold elections and "elect" officials favorable to the liberation force. They then form villagers into groups of interlocking organizations according to occupation or interest— for example, farmers and merchants associations, sporting clubs.[32] Particular attention is paid to organizing people, such as women, who previously have had little opportunity to play an important role in public life.

Next, a village soviet is created in which party members, or persons known to adhere to the party line, hold dominant positions. If a base is to be organized, the government structure extends above the village level to include the entire area, with at least one party member in control. The political leaders then proceed to operate a "shadow government," collecting taxes, conducting regular propaganda sessions, etc. During these preliminary operations, "liberation army" soldiers are required to conduct themselves in an exemplary manner and to respect the personal rights and property of the people as much as possible.[33]

The Communists seek more than passive acceptance of their control. Their doctrine teaches that a guerrilla force must "swim among the people like a fish in water";[34] accordingly, the people are organized to assist in what is allegedly the "people's war." Villagers are required to provide money, food, clothing, and other supplies—but preferably not in amounts that would arouse resentment. Those unwilling to cooperate are threatened with severe reprisals.

Local residents are expected also to collect intelligence concerning the enemy. All villagers are usually instructed to report items of information gathered at their places of work, social meetings, etc. Since political and military actions are considered inseparable, information is sought not only on the positions and strengths of troops, but also on political and economic developments. Usually one or more specialized intelligence groups—perhaps one for civilian and one for military information—are formed; these devote most of their time to gathering information.

The interlocking "associations" and other groups provide an excellent means of maintaining security. Members are instructed to report at group meetings any activities of other members which might be harmful to the Communists. They are also asked to confess their own misdeeds and shortcomings—a procedure proved extremely effective.

Villagers are also expected—and often forced—to contribute soldiers to the guerrilla army and to provide for the defense of the village and the base. The Communists thus usually have three levels of armed forces at their command: the regular army, regional troops composed of men from the locality, and the village-level "self-defense" units, or home guards. When the armed forces are expanded or replacements are needed, the regular forces draw from the regional forces, the regionals from the self-defense groups, and the last-named from the villagers.[35]

In addition to the regular army, which maintained a war of movement and fought only on selected battlefields, the Vietminh maintained strong regional troops. It was their duty to protect an area and its population, as well as harass the enemy and ambush reinforcements. Many times the French could not locate the regular troops but came in contact with the regional troops. The popular troops under them consisted of two groups, the *Dan Quan* and the *Du Kich*. Although the members of both groups served only in their spare time without uniforms, the *Dan Quan* included everyone and was essential as a labor force, while the *Du Kich* was a guerrilla group. Although they occasionally committed acts of sabotage, their main duties were to collect intelligence, act as guards (during the daytime they station themselves from one half to one mile away from the village, posing as peasants in the fields), build roads, and fortify villages. They also infiltrated villages prior to a Vietminh assault. Their main purpose was not military, but political: to demonstrate that the entire population was participating in the national struggle.[36]

Since logistical problems nearly always hamper guerrillas, making it difficult for them to maintain sustained operations and forcing them to rely on surprise attacks at suitable places and times, the Vietminh took great pains to prepare the chosen battlefield, etc. The commander would be assigned a particular target and be given regular units, some regional units, and popular troops that were in that zone. The popular troops would spend a long period of time "preparing the battlefield" with stockpiles of food and munitions, gathering intelligence, sometimes preparing fortifications. The regular troops if necessary would break up into small units, infiltrate through the French lines, and assemble at the prepared battlefield prior to the attack.[37]

In urban centers the Communists concentrate less on intelligence collection and harassment than on psychological warfare aimed at creating or exacerbating antagonisms between the people and the government. If an underground does not already exist, agents are sent to penetrate unions and student groups. These agents distribute propaganda and organize meetings and demonstrations; they also foment strikes and riots to force the authorities to take countermeasures which will further alienate the populace. An existing underground or the special agents may also collect funds and information and commit acts of sabotage, or encourage citizens to do so. These urban activities, which Mao termed the "second front" in his struggle,[38] have been helpful in undermining regimes and in convincing groups of people in other nations that the rebels, not the incumbents, represent the will of the people.

COMMUNIST TAKEOVERS: EASTERN EUROPE

Before World War II, the Soviet Communist Party was able to dominate the world Communist movement for two basic reasons: first, the fact that the

Strategy, Tactics, and Countermeasures

Soviet Union was the sole example of Marxism gave its party unchallengeable prestige and authority, while the obvious factor of the size and strength of the state the party had created lent weight to its decisions. Second, the formation of and control over the Comintern [39] by the Soviet Government permitted it to tailor the activities of other Communist parties to fit the current dictates of its own foreign policy.* These material and political factors were to help the Soviet Communists to bring their proteges to power in Eastern Europe after the war.

Stalin and his Kremlin colleagues had anticipated a second global conflict, and had instructed their followers in neighboring countries to prepare to bring Communist order out of the subsequent chaos.[40] Prior to the German attack on the Soviet Union in June 1941, the position of the Communists in Eastern Europe was difficult. Although they opposed nazism ideologically, they followed the official Soviet policy of "neutrality" and did not engage in any anti-Nazi activities. Remaining underground, they avoided clash with the Nazis and, like the Soviet state, collaborated with them whenever such a collaboration might result in immediate advantages either to themselves or to the Soviet Union. Hence in Poland and Yugoslavia, the Communists preserved "neutrality" when Germany attacked their countries; in the former, they actually collaborated with the Gestapo in uncovering or denouncing the anti-Nazi underground resistance.[41] At that time the Soviet and the Nazi Governments were bound to cooperate on the basis of the secret protocol of the Nazi-Soviet pact of 1939.[42] Nevertheless, numerous Communists, particularly from Poland and Czechoslovakia, left the Nazi-controlled territories and escaped to the Soviet Union to train for an ultimate seizure of power.

The situation changed radically the day Germany attacked Russia. Without delay the Communists all over Eastern Europe turned against the Germans and began to organize undergrounds and partisan detachments. The Communists were determined to form their own underground organizations, independent of democratic, anti-Nazi resistance movements. In German-conquered areas they also refused to recognize the authority of the governments-in-exile, which as a rule exercised leadership and control over the "official" underground movements. This was particularly true in Poland, Yugoslavia, and Greece.

While preserving their own organizational independence, the Communists tried hard to conceal their identity. The term "Communist" was avoided; they generally described their organizations as "democratic," "liberal," or "people's." They also tried to form "democratic fronts" or "blocs" with non-Communist groups. Soon individual Communists established working contacts with the official non-Communist undergrounds and thus became privy to their organizational structures, activities, strength, and, what later proved extremely important, the identity of their rank-and-file members.

Throughout 1941 the Communists were generally left to shift for themselves. Beginning in 1942, however, the Soviet authorities organized material

---

*The Communist International, or Third International, was formed in 1919 as a brotherhood of all Communist parties in the old Marxist tradition. Although representatives from other countries usually held the presidency, the Comintern had its headquarters and most of its bureaucracy in Moscow and was actually directed by the Soviet Foreign Ministry.

support for them, and by the following year that support consisted not only of equipment and arms, but also of instructors and staff sent from Russia. Radio transmitters supplied by the Soviet Union as well as special Soviet radio programs broadcast to the occupied countries were particularly important in strengthening the position of the Communists among their countrymen.

Notwithstanding their propaganda allegations, the Communists did not concentrate their energies on fighting the Nazis. On the contrary, in Poland, Yugoslovia, and Czechoslovakia they concentrated on preserving their own forces until the end of the war. As a rule they avoided actions which might result in human losses, especially among their political elite, while trying to weaken the non-Communist undergrounds as much as possible. They carried on propaganda campaigns accusing the latter of "collaboration" and "fascism" in order to undermine their prestige, and often provoked Nazi reprisals against the local population, later claiming the reprisals to be the result of "official" underground cowardice or irresponsibility. Sometimes they denounced non-Communist underground workers to the Gestapo.[43] In Poland, for example, a special Communist cell was organized for that purpose.[44] Their overall strategy was to preserve their own forces, eliminate competitors for power, then seize the power for themselves.

COMMUNIST ACTIVITIES PRIOR TO 1944

Soviet diplomacy, East European Communist activities, and worldwide Communist propaganda were closely coordinated. Soviet wartime diplomacy aimed at and achieved two important goals: (1) Allied secret recognition of the 1939–40 Soviet territorial acquisitions in Eastern Europe and (2) Allied recognition of Soviet special "interests" in Eastern Europe.

After 2 years of secret negotiations, British Foreign Secretary Anthony Eden and President Franklin D. Roosevelt agreed in Washington in March 1943 that recognition of the territorial acquisitions would have to be accorded. At the Tehran Conference Stalin was finally satisfied.[45] And on the second point, the British Government in August and October 1940 offered Stalin "leadership" in the Balkans in exchange for a "benevolent" Soviet attitude.[46] In October 1943, at a British-United States-Soviet conference of foreign ministers meeting in Moscow; it was agreed that each of the three Allies would act in the territories liberated by its own forces according to its own judgment.[47] In June 1944 Churchill agreed in principle to partitioning the Balkans into Soviet and British spheres of influence, and the following October an unofficial but binding Soviet-British agreement was concluded in Moscow.[48] At Tehran, as well as at Yalta, it was secretly agreed that the postwar governments of the East European countries would have to be "friendly" to the Soviet Union, and thus would include some Communists.[49] To supplement these secret negotiations, a worldwide public opinion campaign was mounted in support of the "democratic," or "liberal," Communist-led undergrounds and against the "undemocratic," "Fascist" undergrounds, led by the governments-in-exile.

Strategy, Tactics, and Countermeasures

Beginning in 1942 the Soviet Government also started to supply the Communist underground groups with arms, equipment, and training personnel. Simultaneously, East European refugees and deportees in the Soviet Union formed several "national committees" challenging the legality of the governments-in-exile and preparing their own cadres for the postwar seizure of power. Whenever possible they also organized "liberation armies." In time, fairly sizable groups were organized for each of the East European countries. Through pressure, rewards, promises, or persuasion they obtained supporters from such groups as Polish and Rumanian deportees; German, Hungarian, Polish, or Rumanian prisoners of war; and Czechoslovakian or Rumanian escapees. They organized radio programs for Eastern Europe in the languages of the various ethnic or national groups.

COMMUNIST ACTIVITIES IN THE "LIBERATED" AREAS, 1944–48 [50]

The Red Army approached the prewar western frontiers of the Soviet Union in January 1944, starting the process of "liberation" which lasted until Germany's surrender in May 1945, when all of Eastern Europe, except Greece, was occupied by the Red Army. No British or United States forces were present and the Western Allies exercised no control in the area; thus the Red Army commanders as well as the Soviet civilian authorities were free to impose any measures they wished under the cover of military exigency. Confusion, fear, unprecedented destruction, and above all, hunger, were rampant. In all of the countries except Poland, relatively large segments of the population were liable to accusations of collaboration with the Germans. Except for Czechoslovakia, no country had had a background of successful democratic government.

The Initial Steps Taken By the Soviet Authorities (Red Army)

In most of Eastern Europe the Red Army entered an administrative vacuum. Most persons who held government positions, public offices, or even significant business posts preferred to resign and either withdraw with the *Wehrmacht* or go into hiding. There was fear of the Red Army and fear of being accused of collaboration with the enemy or with the puppet governments set up by the Germans. The Red Army commanders and Soviet civilian authorities had to fill the vacuum, particularly on the lower levels, in order to maintain security and public order. They assigned administrative positions to trusted individuals—to local Communists, if possible. In addition, national councils were soon created to fill the urgent need for local lawmaking authorities and for safety and administrative measures. From the outset, these were controlled or dominated by the Communists.[51]

From the very beginning, the Soviet authorities were careful to entrust the ministries which controlled the police forces, justice, and propaganda to the most experienced Communist elite. Thus before the war ended and long before the public realized it, the actual power in Eastern Europe was in the hands of Soviet officials or local Communists.

Liquidation of the Non-Communist Resistance Groups

As soon as the Red Army occupied an area, the local non-Communist underground fighters found that both the Soviet authorities and the local Communists were hostile. No help could be expected from any quarter. The governments-in-exile received less and less support from the Western Allies and were practically powerless. Most of their members decided to remain refugees, rather than return to their homeland and subject themselves to the new regime. In the first months, the Soviets pursued the policy of a ruthless liquidation, especially in Poland and Yugoslavia. Guerrilla fighters were shot and underground workers arrested. For some time the non-Communist resistance groups tried to remain underground, hoping that the situation would change, that the Western Allies would intervene. It soon became evident, however, that the Western Allies either would not or could not do much, and that it was impossible to maintain the underground structure any longer. By late 1945 or early 1946 the underground members revealed themselves. Those who agreed to cooperate with the Communists saved their lives and careers; the uncooperative were destroyed.

Communist Means of Mass Pressure

The Communists were in an advantageous position to exert mass pressure and restrict or paralyze the activities of opponents. Through the Red Army and Soviet civilian authorities they controlled all railroads, rolling stock, housing, paper, printing plants, radio, etc. They were free to use security measures as a political weapon. Travel permits, passes, accommodations for public meetings, transportation, and printing and radio facilities were restricted to persons or groups "friendly" to them. They exerted pressure over numerous individuals who had collaborated with the Nazis or Fascists, supported or served in the undemocratic prewar governments, participated in anti-Communist activities, etc. They were also able to dispose of the property formerly owned by Germans and Jews and other emigrees.

While using their power to stamp out actual or potential opposition, the Communists sought to enlist the support of the population through persuasion. They used propaganda slogans adroitly. A few of many successful slogans were the following:

Let us all concentrate on reconstruction. Politics now is not essential.

We need the Soviet friendship, regardless of whether we like it or not.

We have been abandoned to the Soviets and their leadership is unavoidable.

Before the war our industry and a great part of our commerce were controlled by foreign capital. This nationalization cannot harm the people and can do much good.

Our peasants are land hungry and our countryside is overpopulated. The land reform will be useful.

143

Strategy, Tactics, and Countermeasures

Political Organizations

According to the Yalta agreement, the only political parties which were supposed to operate freely in the Soviet Union's sphere of influence were those which were "democratic," "anti-Fascist," "friendly to the Soviet Union," and willing to "collaborate" with the already existing Communist-controlled governments. The terms were vague; there was no agreement on interpretation and the Communists interpreted them to the advantage of themselves and the Soviet Union: any organization which opposed Communist leadership or Soviet policies was branded as "undemocratic," "Fascist," "unfriendly to the Soviet Union," or unwilling to "collaborate" with the government. The prewar rightwing parties, such as the National Democrats in Poland or the Agrarians in Czechoslovakia, were banned altogether. The Peasant and Socialist Parties were officially allowed to operate freely. Soon, however, numerous splinter groups formed which challenged their anti-Communist leadership. Receiving considerable help from the Communists, these groups subsequently weakened the democratic forces within the parties. Confusion was increased by the emergence of new "liberal," "democratic," "progressive" parties, most of which were actually controlled by crypto-Communists.

When political maneuvers failed, anti-Communist democratic leaders were arrested and public trials were staged. Invariably convictions followed, on charges of "collaboration with the enemy," "supplying foreign powers with intelligence" (Great Britain and the United States), "anti-Soviet activities," currency crimes, etc.

Coalition Governments

The Yalta agreement provided for the formation of coalition governments in which the Communists would participate but not have a controlling voice. The practice was much different. Usually only an insignificant minority of democratic leaders were admitted into the Communist-controlled government, where they were outvoted on each important issue and were unable to prevent the Communist-led persecution of their own followers. Then they themselves were arrested and put on trial. A few of them, like Ferenc Nagy of Hungary, Stanislaw Mikolajczyk of Poland, and Georgi Dimitrov (not related to the Communist leader of the same name) of Bulgaria succeeded in escaping. The rest perished or chose to collaborate with the Communist regimes.

Both the British and United States Governments intervened. Their diplomatic notes, however, were rejected as an "interference in the internal affairs" of "sovereign" states. They did not go beyond diplomatic intervention.

General Elections (1946–47)

According to the Yalta agreement, general elections were supposed to be "free," "democratic," "unfettered," and "secret." Freedom of speech and freedom of the press were to be respected. Again the Yalta agreement was broken by both the Soviet Government and East European Communists. Although the "national fronts" which were organized included splinter groups and pseudo-non-Communist parties, they were controlled by the Communists. The members of the front agreed between themselves on the division of parliamentary seats and presented the voters with one synthetic ballot. The elections were not secret, mass terror prevailed. Opposition parties had no chance; they were ruthlessly persecuted. When in Poland, Hungary, and Bulgaria the Socialist or Peasant Parties received large votes, the results were falsified by the Communists. Special secret organizations had been set up just for that purpose.

Merger of the Socialist and Peasant Parties With Communists

Following the elections and the liquidation of the democratic leaders, the Communists initiated a merger of the Socialist and Peasant Parties with the Communists or Communist-dominated groups. By that time (1948–49), opposition political parties had virtually ceased to exist. The independent Socialists joined the "sister" Communist parties, and Peasants merged with the Communist-controlled splinter groups. With this formation of the uniform national fronts unchallenged by any organized political forces, the Communist seizure of power was concluded.

In 1917 the Communists had taken advantage of the chaos bred by World War I to seize power in Russia. A generation later, they and their disciples profited by World War II to gain control of other countries. There were several factors in Eastern Europe in 1945 which facilitated the takeovers. The fact that the Germans had worked with capitalists, aristocrats, feudal landowners, and other conservative or middle-of-the-road elements made these groups easy prey to charges of collaboration, and thus decreased or destroyed their potential political strength. The presence of large peasant groups who had ample reason to desire changes in the *status quo ante bellum* provided a good potential base of support. The fact that, save in Czechoslovakia, there was no tradition of real representative government or of people's rights meant that curbs on freedom of speech, assembly, etc., would be likely to draw little immediate or effective protest.

The international situation also helped the Communists. Had the Russian forces simply attempted by brute force to install puppet regimes, this would have been understood and actively opposed by the Allies, who had promised free elections to the peoples of Eastern Europe. With the subtler techniques of takeover, however, most Western statesmen were unfamiliar. Rather than incur the wrath of an ally who had suffered grievously from the war and whose cooperation was sought to ensure the peace, the Western Allies permitted the Russians and their followers to enter a political vacuum they themselves had made no plans to fill.

While the milieu was new, the basic tactics employed by the Communists were not. The formation of united fronts, the infiltration of groups, the concealing of affiliations and intentions, the dissemination of propaganda, the use of terrorism, threats, and blackmail to discredit, divide, and otherwise immobilize the opposition, were familiar tools invented by Lenin and refined by Stalin and his contemporaries. Some of the refinements used to actually seize the centers of power, however, are noteworthy because they set the pattern for later Communist efforts in other countries.

The Communists agreed to participate in coalition regimes rather than trying to create their own governments, but insisted on controlling the ministries which exercised physical, economic, and psychological power—that is, Defense, Interior, Agriculture, and Information. In other words, they were not concerned with making an immediate show of great political strength but with controlling organs which could mold the minds and actions of people, by coercion if necessary. Also by placing their agents in positions where they could discredit non-Communist officials, they made it extremely difficult for opposition groups to campaign effectively or unite against them.

THE SOVIET SECURITY AND ESPIONAGE SYSTEMS

SOVIET SECURITY SYSTEM

The Soviet security system and some of its techniques are discussed below, in order to understand better how the Soviet Union reduces the potential for insurrection against it. The Soviet Union is a highly centralized, authoritarian state in which the Communist Party controls the instruments of power. In such a system, security—against both internal traitors and foreign spies—is of extreme importance. The two major instruments of these operations are the elaborate state police systems and the international espionage network.

The present police system has several major branches.[52] The Committee for State Security (KGB), directly responsible to the Council of Ministers, performs the functions which are handled for the United States by the FBI, CIA, and the military counterintelligence agencies. It also controls censorship, border troops, the protection of state secrets (archives), and special KGB troops provide honor guards, protect sensitive communications and installations, etc.

The KGB is also responsible for counterintelligence operations in the Armed Forces. In addition to uniformed personnel, it has networks of plain-clothes police and informers, and has important functions in the field of foreign intelligence gathering. (The *Glavnoye Razzedyvetelnoye Upravlenize*—GRU, as the Chief Military Intelligence Directorate of the Ministry of Armed Forces, is concerned only with positive intelligence or police functions with the Armed Forces.) The Ministry of Internal Security (MVD) no longer exists on the all-union level but only on the republic level. It appears that the KGB now directs many of the activities of the republic MVD.*

*The MVD was abolished on the national level in 1960. Apparently the Government feared putting too much power into the hands of one individual as had been the case with Beria and former police leaders.

Each of the Soviet republics has a Ministry of Internal Security which controls three types of military units: the general militia, the transport militia, and departmental militia; (1) the general militia perform all normal police duties, and units are subordinate both to the local officials at republic, *oblast*, *kray*, *rayon*, or city levels, as well as to commanders of the next higher rank in the militia; (2) transport militia protect river fleets, ports, and railroads. Since railroads and airlines frequently employ their own guards to protect property, transport militia often share responsibilities with them; (3) departmental militia protect state enterprises or buildings, again sharing responsibility with guards hired to protect property. The militiamen perform another important function: supervision of the all-union passport system by which the government controls the movements and actions of Soviet citizens. The militia also enforce the licensing system for the control of firearms, explosives, poison, etc. Uniformed militiamen enjoy certain unique privileges; for example, a militiaman above the rank of sergeant is immune to arrest.

In addition to the regular militia there are such auxiliary forces as the people's guards, rural marshals, and volunteer brigades, whose main task is to curb disorderly conduct, such as drunkenness. Members serve part-time on an allegedly voluntary basis and are usually commanded by a party secretary. The Ministry of Internal Security also maintains security troops to suppress large-scale disturbances, maintain rear areas during wartime, etc.

In maintaining control over the population, the police have in recent years tended to substitute for the terrorism of the past a more subtle—and in many ways more effective—form of economic and social control. Threats to fire a person from his job and prevent him from finding another, social ostracism, and similar sanctions are now the rule instead of torture, forced labor, and other forms of terror, which were used more extensively in the past.

The government also controls closely the movements of all citizens. People are forbidden to travel without special permission in border areas or in certain restricted areas such as those containing relocation camps, military or security establishments, and strategic industries. All citizens must also have internal passports, and when traveling from one place to another, must check with authorities upon both departure and arrival. People must have work permits, and must obtain governmental permission before changing jobs. Housing permits are another effective check upon both the whereabouts and the actions of individuals, since unsatisfactory behavior can influence the type of living quarters assigned to them.

The legal system, which places the welfare of the state above the rights of the individual, permits the government to censor all mail, internal as well as external. It also gives the police great latitude in seizure and search. The citizen's knowledge of the premises of Soviet law and the special powers of the police, coupled with his fear of reprisals, constitute an effective deterrent to subversive statements or deeds.

In gathering information, the KGB and MVD units not only rely on their professional agents and *provocateurs*, but also depend heavily on informers. Chief among these are the "activists" who regularly supply intelligence to the police in order to achieve better jobs or positions of influence. Each secret police agent generally has his own group of informers, and maintains in a fireproof safe a personal file on each one. Each dossier contains such material as—

(1) a signed statement by the informer obligating him to work for the secret police;
(2) a photograph of the informer;
(3) the cover name of the informer;
(4) data on the personality and activities of the informer, usually including some compromising facts which could be used to prosecute him under one of the articles of the Criminal Code;
(5) a list of all the persons with whom the informer is acquainted;
(6) copies of all the informer's reports;
(7) data on rewards or punishments given to him.

Once a year the secret police operator reports to a superior concerning the work of each informer.

The number of secret informers assigned to each secret police operator varies according to the operator's need and his ability to make effective use of them. The quality of informers is usually very low; hence the police must rely more on quantity than quality. Whenever the informers report an important piece of information or a case requiring thorough investigation, professional agents from police headquarters are assigned to the job.

The secret informers are most numerous in administrative and industrial centers and in the border zones. The number of informers in each factory plant or organization depends upon the political, economic, or military importance of the establishment. A large number of informers are recruited among people who work in such public places as hotels, restaurants, barbershops, clubs, and railroad stations. In the villages, informers and "activists" also cooperate with the press by informing the newspapers of achievements or defects in work being done by the populace, etc.

The police try to recruit as informers people who, because of their past, will not be suspected by their fellow citizens. As a rule, a newly recruited secret informer is usually asked to watch persons whom he himself has named as being suspect; only after he has been tested in some way is he given real assignments for spying. Persons who have voluntarily denounced others or informed the police of subversive activities almost always receive a formal offer to become informers.

This elaborate security system, coupled with a long tradition of authoritarian rule and secret police activities, has served to check any serious indigenous threats to the security of the Soviet state. Several effective techniques have been developed for countering any subversive activity, whether internally generated or instituted by foreign agents.

One device is to undertake provocative activities designed to test the loyalty of Soviet citizens. A person may be contacted by somebody claiming to represent a subversive movement, but he would be wise to assume that the "resistance movement" is one directed and controlled by the KGB. Thus, even if he is strongly against the regime, he will hesitate to join a subversive movement because he will doubt its authenticity. Furthermore, since all Soviet citizens are required to show "active loyalty" to the regime, he must report the attempted contact to the authorities or face severe punishment. Any clandestinely organized resistance movement is therefore very likely to be reported to the police within a very short time.

Another practice is to form a clandestine anti-Communist movement, permit it to enroll members, and conduct actual operations against the regime so as to attract the support of real enemies of the state. Once the ring has grown to a size where it may be assumed to include a large number of real anti-Communists, the police arrest the entire membership.

The communes and collective farms are another means of checking potential revolutionary operations. By centralizing food supplies, rather than leaving them in the hands of individual peasants, the government makes it extremely difficult for guerrillas to obtain food from the populace and maintain themselves. Similarly, the practice of periodically issuing new currency makes it difficult to collect the funds necessary for conducting subversive activities.[53]

In addition to these negative checks on subversive activities, there are also positive factors which prevent Soviet citizens from participating in, or seeking participation in, antigovernment operations. The educational system and the steady flow of government-controlled propaganda teach the merits of the Soviet system and inculcate a feeling of loyalty to the regime. Pride in the material and cultural accomplishments of the Soviet union, and increased benefits in the form of better housing, clothing, and recreational facilities, contribute greatly to a feeling of contentment, particularly among older people who remember the hardships of the past. Furthermore, there are many immediate benefits to be obtained by becoming a party member or active sympathizer, or otherwise supporting rather than opposing the regime.

SOVIET ESPIONAGE

Within any country Soviet espionage and intelligence work are carried out through four major groups: the Communist Party, operating openly or through front groups; the Soviet Embassies, overseas delegations of the Ministry of Foreign Trade; Tass news agency personnel, all of whom enjoy diplomatic immunity; and the illegal underground espionage network.

The two most important foreign intelligence units are located within the GRU and the KGB. The GRU group, the army intelligence branch, is in charge of military espionage in other countries and directs information gathering, subversion, and terror.[54] The KGB unit is responsible for nonmilitary espionage in foreign countries, although its activities often parallel or rival those of the GRU group.[55]

As is the case in general party organization, the Russians differentiate between legal and illegal apparatuses. The former consists of a network of agents operating under a member of the Soviet Embassy staff, who enjoys the benefits of diplomatic status and immunity. The illegal apparatuses, on the other hand, are networks operating under a person other than a member of the Embassy staff.[56] Soviet espionage agents generally are organized for gathering four types of intelligence: (1) political intelligence, including information on the activities of any fifth column operating in the country; (2) scientific discoveries and atomic development, armament and military intelligence; (3) economic intelligence and supply of materials; (4) special tasks and control of Soviet officials abroad.[57]

The illegal apparatus is headed by a "resident director," who, despite his title, does not usually reside in the country against which his net is operating but lives in and directs his organization from a neighboring country. His activities in the neutral country are of a normal nature and he is usually forbidden to carry out work against that country. As a rule, he is not Russian nor a native of the country in which he is residing. The major responsibilities of the resident are the control of communications and finances, and the sorting out and evaluating of information supplied by his agents. He evaluates his information, ciphers it, and transmits it to the Center in Moscow. He is usually unknown to his agents, couriers, and radio men, being in contact with them only through middlemen,* who transmit messages and information to him by means of maildrops or letterboxes. This ensures the security of both the resident director and the agents.

In every Communist Party headquarters there is one highly placed agent whose main job is to supply the resident director with information acquired from party members about potential recruits or sources of information. Generally the party man looks for recruits who have little or no association with the party, and discourages them from joining the party so as not to jeopardize their potential usefulness at some later date.[58]

In developing their network and their sources, the agents seek information which facilitates: (1) the identification of persons who have direct or indirect access to security information; (2) the preparation of personality records on such persons in order to evaluate their vulnerability to persuasion or pressure, or the likelihood of their imparting information unwittingly; (3) the discovery of persons who are willing and able to identify potentially useful persons and supply the particulars needed; (4) the discovery of persons willing and able to contact and use persons indicated by the preliminary investigations as having worthwhile access to channels of information.[59]

Personal data are extremely important in Soviet espionage. Agents collect all sorts of directories, reference books, and official lists in order to get information about individuals in a target country.[60] The Soviet espionage system probably has one of the largest biographical directories in the world; it includes information on any person who has ever been contacted by the Communist organization and who may be even remotely regarded as helpful

---

*The middleman acts as a link between agents or between agents and higher-ups. He acts as a safeguard within the ring to prevent agents or residents from coming in contact with each other and thereby conceals the identity of each agent for security reasons.

150

SOVIET ESPIONAGE APPARAT

Moscow Center

Resident Director

Communist Party

ILLEGAL APPARAT

Soviet Embassy
LEGAL APPARAT

Network Chief

C.P. Members

Agents

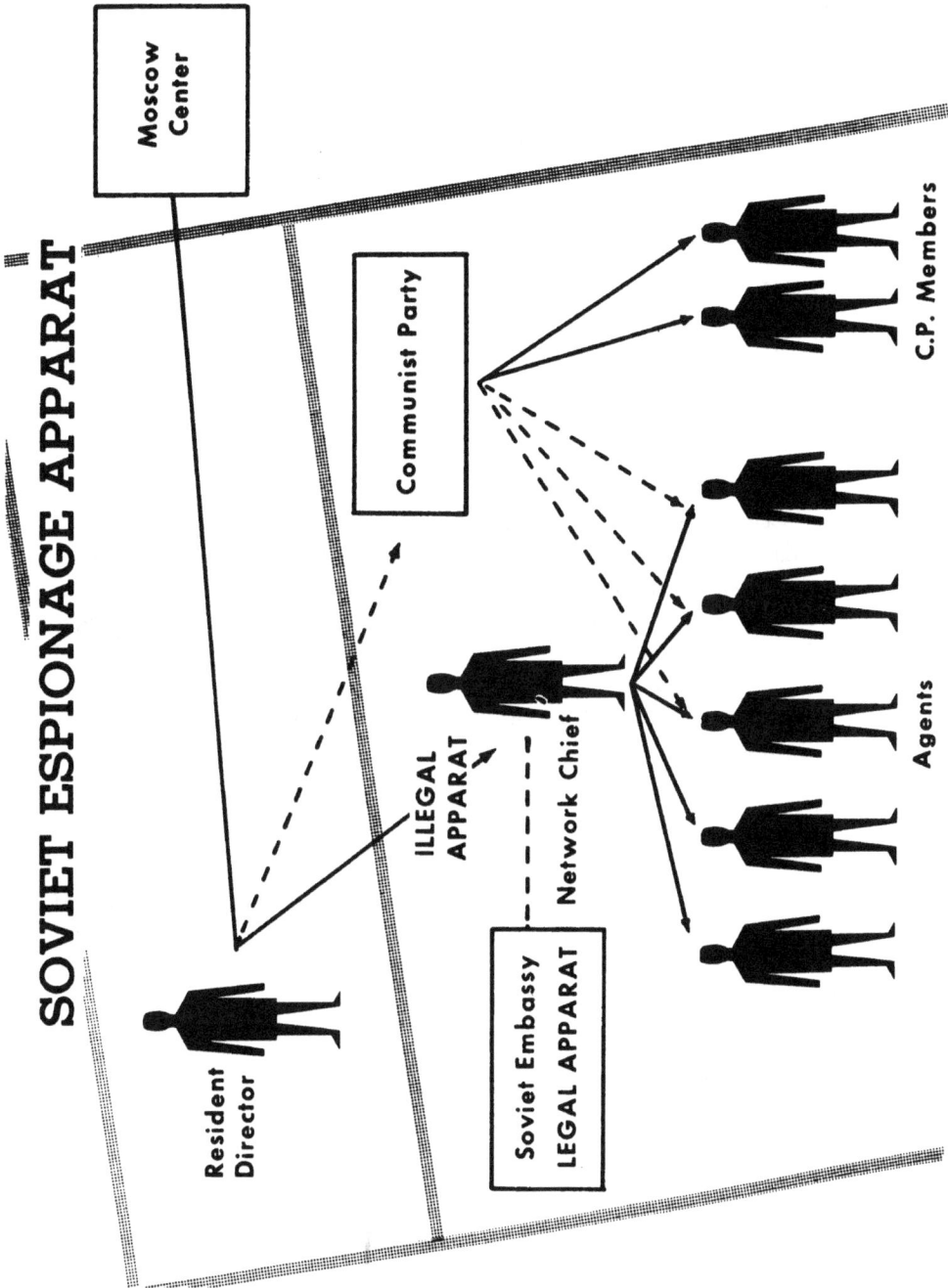

Figure 4. Soviet Espionage Apparat.

or as a sympathizer.[61] The directory contains such items as the present position, biographical and educational data, attitudes, financial position, and positive and negative characteristics of the individual.[62]

According to two experts on Soviet affairs, the foreign intelligence section of the KGB is organized as follows:

First section, responsible for espionage and counterintelligence in the entire Western hemisphere.

Second section, responsible for the United Kingdom, its colonies, and British Commonwealth members.

Third section, responsible for Austria and Germany.

Fourth section, responsible for all other Western European countries and Yugoslavia.

Fifth section, responsible for Russian emigrees. Its functions are to infiltrate, sabotage, and destroy all active organizations outside of Russia, and to persuade these emigrees to return to the Soviet Union. It is responsible for tracing emigrees regarded as traitors to the Soviet Union, penetrating their associations and societies, and recruiting them for possible use as Soviet agents.

Sixth section, responsible for Asia.

Seventh section, responsible for the surveillance of satellite countries from North Korea to Czechoslovakia. East Germany is handled separately. Officers from this section direct trained security forces in each of the satellite countries. These security forces also collect intelligence on local Communist leaders. In the past this information was used in Communist purges.

Eighth section, the "illegal apparatus," whose agents do not have diplomatic immunity.

Ninth section, concerned with Soviet nationals who make foreign visits. It is the duty of this section to make sure that no one escapes from these delegations.[63]

Special Bureau Number I (*Spetsburo*), an independent part of KGB, is charged with terror, murder, and assassination. Using local gangsters and other foreign nationals to do its work, this group is responsible for eliminating important people as well as Russian defectors.

SUMMARY OF THE STRATEGY OF COMMUNIST PARTIES

As the foregoing pages have shown, the international Communist movement has participated in a wide variety of resistance and revolutionary operations. The Soviet Union during World War II demonstrated how a Communist government organized resistance activities against an enemy invader. Under the general guidance of the Red Army, citizens were organized into partisan guerrilla bands and underground networks charged with intelligence, sabotage, and psychological warfare operations. Although most citizens were asked to participate for patriotic rather than ideological reasons, and although few of the guerrillas or underground workers were Communist Party members, party officials within the army or the government remained in control and directed their activities.

In other countries, wartime resistance movements became the vehicles by which Communist parties launched or completed revolutions. In Yugoslavia, for example, the partisan movement gradually widened its control over the country during the war and eventually assumed the reins of power. In other parts of Eastern Europe, leaders of Communist-led resistance movements seized power through coups d'etat, in most cases with the backing of the Red Army. In China, the movement already started by Mao Tse-tung gained so greatly in size and strength through its anti-Japanese activities that in 1949 it was able to consummate successfully a two-decade drive for power. In Indochina, Malaya, and the Philippines, Communist-led groups capitalized on their anti-Japanese actions in order to start revolutionary wars.

Regardless of their locale or method of operations, Communist parties have followed a general pattern in their attempts to gain power through revolution. Leaders or would-be leaders were recruited or selected by Communist agents for special training in Moscow. Tito of Yugoslavia, Walter Ulbricht of Germany, Maurice Thorez of France, Evangelista in the Philippines, Lai Teck in Malaya, Ho Chi-minh in Indochina, and many others were trained in the Soviet Union. After returning home, they formed (or re-formed) small elite groups which in turn were sometimes the controlling core of large mass parties. Membership within the elite was sharply restricted: only individuals who were willing to place their dedication to the party before all other interests were admitted. Parties were organized on hierarchical structures headed by a central committee with a politburo, and discipline was strict. Preparations were also made for operating legally and illegally.

Whatever their size or strength or the degree of governmental opposition they encountered, Communist parties everywhere gave their members experience in clandestine operations, and tried to acquaint them with conspiratorial behavior. Cell meetings, for example, were often held secretly so that members attending them would learn how to travel and how to make contact with other members without arousing suspicion. Members were often assigned some sort of minor intelligence-gathering or sabotage missions which in themselves were of little or no practical use, but which both tested and trained members in the performance of clandestine activities. For the same reason, a clandestine press was often operated, even if local censorship was not severe. In addition, the members' sense of dedication was reinforced by the standard practice of self-criticism and criticizing fellow members at cell meetings. In this way, they were urged to higher goals of self-improvement and loyalty.

Even where the parties themselves remained small, Communists everywhere tried to increase their influence by organizing or infiltrating front groups. Popular causes were selected around which to build an organization. Humanitarian issues and, in the colonial countries, "imperialism" were the main reasons for organizing new groups or trying to expand and reorient existing groups. Particular efforts were made to form or infiltrate labor unions, which were a major instrument of power or potential power. For not only would control of union dues provide Communist leaders with a steady source of funds, but more important, it would give them a steady source of manpower to carry out other programs, and provide them with grievances and issues to exploit. In addition, Communist parties, when legal, sometimes cooperated with other leftwing parties in "united fronts" attempting to direct the activities of the fronts to further their own plans. By participating in these various social, economic, and political organizations, Communists were able to use popular discontent to foment strikes and demonstrations, and thus contribute to the general unrest necessary for revolutionary conditions.

In striving to obtain popular support for the party or for the organizations it controlled, Communists seldom stressed ideological matters. Instead, they played upon individual personal grievances or needs. Thus they clamored publicly for better working conditions generally, while helping unemployed individuals find jobs or houses. They often posed as the understanding friend of anyone who was discontented, indicating that their movement could use the services of the individual who had been unappreciated or maltreated under the existing system, and could at the same time serve the individual's needs. If possible, they sought to obligate individuals to the party for personal services rendered so that whenever the party made a bid for power, the individual could be asked or compelled to support it.

Wherever the Communists seized power through peaceful means, as in Eastern Europe, they followed a pattern of forming united fronts with other parties, obtaining posts in governments, steadily eliminating or absorbing opposition elements, and then seizing power. Where they have turned to revolutionary warfare, it has been because theoretically both "subjective" and "objective" conditions were right for it—that is, the party itself was strong and the country was filled with unrest and thus ripe for revolt. In Greece after the war, the party itself was fairly strong and found itself in a situation where economic conditions were poor and government efforts inadequate. After testing their strength, they attempted to seize control of the government by a popular revolution. However, there seem to have been circumstances which led Communists to attempt open warfare against a legal government: in the Philippines and Malaya, for example, the party was losing ground among the populace and it evidently concluded that if it was ever to have an important, let alone the most important, role in directing the country it would have to make a dramatic bid for power.

When they have decided upon revolutionary war, Communist leaders then follow a fairly standard pattern of action. The main Communist leaders themselves leave the major cities where they would probably have been arrested, and find sanctuary in various regions from which they can safely direct the war effort. At the same time, they usually destroy party records and other evidence which could directly incriminate the party. By having their front group conduct strikes and antigovernmental demonstrations, they try to augment existing unrest and thus weaken popular respect for the government.

The Communist Party then becomes part of a broader revolutionary movement. Under Communist influence or control, two organizations—a guerrilla army and a civilian underground—are created to carry on revolutionary activity. Both of these organizations assume the name of goals of popular causes such as resistance, liberation, or independence. Communist goals and objectives are seldom mentioned. In Malaya, the Communist Party organized the MRLA (military) and the *Min Yuen* (civilian); in Yugoslavia, the National Liberation Army and the National Liberation Committee; in France, *Francs-Tireurs et Partisans Françaises* and the *Front National*.

Many recruits do not understand communism nor do they know that the movement is Communist-led. Often such recruits join because they want independence or freedom and do not care who helps them gain it. However, it is the party members who hold decision-making posts within these organizations, and simultaneously indoctrinate their members along Communist lines. No matter how large the guerrilla armies of the undergrounds become, the number of party members remains small.

Membership in the guerrilla army or civilian underground is generally open to anyone, provided the recruit has no commitments to other political groups. He is not required to swear allegiance to communism. The party has always sought to absorb or eliminate all opposition or potential opposition elements in order to unify resistance into one movement. In Poland and Yugoslavia they informed to the Germans on rival movements. Sometimes they have tried to form united "liberation" fronts, as in Greece, but again with the Communists themselves firmly in control. As their strength increases, they have established "liberated" areas in which they have set up governments and exercised *de facto* control, and thus further strengthen their claim to represent "the people."

Where the Communist-led movements have succeeded in ousting the former government, leading party members have usually installed themselves in the key posts. As they have been extremely wary of any possible counter-revolutionary activity, the movements have been particularly careful to gain control over the police and security forces. In virtually every case where Communists have obtained power—whether through coup d'etat or war—external cadres from the Soviet Union have entered the country to reorganize and, in many cases, to staff the police and security forces.

While the Communists have often seized power, it must be noted that in Greece, Malaya, and the Philippines, for example, they failed. Either they misjudged the "subjective" or "objective" conditions, or lacked sufficient external support, or—perhaps most important—found themselves in later stages confronted by a government which was both willing and able to fight back on all fronts.

FOOTNOTES

1. Although Lenin first stated this in 1902 ("What Is To Be Done?") when the party was outlawed, he repeated it after 1905, when the party was again permitted legal existence. V. I. Lenin, *Collected Works* (New York: International Publishers, 1943).

Strategy, Tactics, and Countermeasures

2. In 1915, for example, when striking workers were preparing to present a petition to the Winter Palace, Lenin instructed Bolshevik agitators to address the crowd without announcing themselves as Bolsheviks. In 1920 he warned Communists, in *Left-Wing Communism: An Infantile Disorder*, against being too aggressive or too frank in announcing their revolutionary goals, and urged them to work wherever possible in united fronts. Ibid.

3. See Gabriel A. Almond, *The Appeals of Communism* (Princeton: Princeton University Press, 1954), pp. 230–257, for a thorough discussion of susceptibility to communism.

4. For an extensive account of the Comintern, see Günther Nollau, *International Communism and World Revolution: History and Methods* (New York: Frederick A. Praeger, 1961), particularly pp. 134–177.

5. Raymond L. Garthoff, *Soviet Military Doctrine* (Glencoe, Ill.: The Free Press, 1955), p. 287. This concept of "stability of the rear" has figured heavily in the planning of other Communist military leaders. Giap, for example, stresses "the building and consolidation of the rear" and the need "to combine economic and cultural needs with those of national defense." See General Vo Nguyen Giap, *People's War, People's Army* (Hanoi: Foreign Languages Publishing House, 1961), p. 146.

6. Joseph Stalin, *The Great Patriotic War of the Soviet Union* (New York: International Publishers, 1945), pp. 23–24.

7. Ibid., p. 51.

8. C. Aubrey Dixon and Otto Heilbrunn, *Communist Guerrilla Warfare* (New York: Frederick A. Praeger, 1957), pp. 67–69.

9. Nikolai Khokhlov, *In the Name of Conscience* (New York: David McKay Co., Inc., 1959), gives an account of his training and activities as a member of an NKVD destruction squad behind German lines in World War II and later activities in postwar Europe.

10. Otto Heilbrun, *The Soviet Secret Service* (New York: Frederick A. Praeger, 1956), p. 12.

11. See Dixon and Heilbrunn, *Communist Warfare*, p. 67.

12. Heilbrunn, *Secret Service*, pp. 52ff, cites the *Soviet Handbook for Partisans*, printed in 1942, which gives elaborate instructions on how partisans were to be organized and to operate.

13. Ibid., p. 54.

14. Ibid.

15. Dixon and Heilbrunn, *Communist Warfare*, p. 70.

16. *Readings in Guerrilla Warfare* (Fort Bragg, N.C.: Psychological Warfare School, 1954), p. 46.

17. Ibid., p. 52.

18. Ibid., p. 53.

19. See F. O. Miksche, *Secret Forces* (London: Faber and Faber, 1950). Author notes that underground and resistance organizations can have important effects after a war; for by destroying the soul of a nation and systematically encouraging violence and disrespect for law and order, they can disrupt the nation's social and political structure.

20. This separation of the Chinese Communist Party from the proletariat was at first a source of deep concern to both the Comintern and the Chinese themselves. Both the Maoists and their Comintern supporters tried to excuse this apparent sharp deviation by explanations to the effect that the Red Army and the territory it controlled were, by virtue of being Communist, speaking for the proletariat. For a good discussion of this point, see Benjamin I. Schwartz, *Chinese Communism and the Rise of Mao* (Cambridge: Harvard University Press, 1952), pp. 189–199.

21. Donald Lancaster, *The Emancipation of French Indochina* (New York: Oxford University Press, 1961), pp. 78–79, 82–83.

22. The writings of Mao Tse-tung make constant reference to the political-military nature of his revolution and to the dual role to be played by his soldiers, particularly the

bona fide party members. See *The Selected Works of Mao Tse-tung* (London: Lawrence and Wishart, Ltd., 1954), particularly I, pp. 106, 107.

23. Otto Heilbrunn, *Partisan Warfare* (New York: Frederick A. Praeger, 1962), pp. 44–45, summarizes Mao's conception of "bases" as follows:

(1) The base fulfills the same function for guerrillas as the rear does for a regular army. In addition to its logistical role it permits military power to be consolidated and expanded, and guerrillas and semiregulars to be recruited and trained.

(2) The base is a political testing ground for the mass appeal of the Communist program and it is here that they mold the masses into organizations, inquire into their opinions, "co-ordinate and systematize" these opinions, and then explain and spread them until the masses accept them as their own and translate them into action—a process known as the principle of "from the masses to the masses." If the mass response is insufficient, Communist policies can temporarily "be changed."

(3) By expanding they deny the enemy more and more territory until he is finally expelled.

(4) The base area, in conjunction with other base areas and the semiregular forces, acts as the encirclement line of enemy-held territory. While each isolated base area is encircled by the enemy, neighboring guerrilla bases and the regular forces' front lines in turn surround the enemy. The encircling enemy thus faces encirclement itself.

Lucien W. Pye, *Guerrilla Communism in Malaya* (Princeton: Princeton University Press, 1956), pp. 30–31, discusses how "People's Liberation" parties in Asia consider the rural areas as the weakest link in the enemy's defenses, and exploit "the limited power of unstable governments and the lack of effective communications common to much of Asia."

24. *Selected Works of Mao Tse-tung* (Peking: Foreign Languages Press, 1961), IV, pp. 83ff. This book contains the writings of Mao during the 3d Revolutionary Civil War Period, and is not to be confused with the writings of Mao published in 1954 by Lawrence and Wishart (see footnote 22).

25. See Heilbrunn, *Partisan Warfare*, p. 160. George K. Tanham, *Communist Guerrilla Warfare* (New York: Frederick A. Praeger, 1961), pp. 16, 24, 25, notes that following the principles of Mao, the Vietminh after a defeat were willing to sacrifice territory, people, and economic assets in order to preserve a hard core. They never relaxed, however, in their efforts to win over the population, and their clandestine cells and guerrilla and propaganda agents proved increasingly damaging to the French. In order to create insecurity in French-held areas, they organized guerrilla groups who not only harassed the French military forces, but also indoctrinated the people and undermined their loyalty to the government.

The Vietminh organized their bases according to their definition of "a closely integrated complex of villages prepared for defense; a politically indoctrinated population in which even children have their specific intelligence tasks; a network of food and weapons dumps; and an administrative machine parallel to that of the legal authority, to which may be added at will any regular (army) unit assigned to operations in the area."

26. *Selected Works of Mao Tse-tung* (Peking), p. 84.

27. For a discussion of the Vietminh underground, see Lancaster, *The Emancipation*, pp. 84–85, and Tanham, *Communist Warfare*, p. 21.

28. This term was originated in 1938 by Mao, "On Prolonged War." *Selected Works of Mao Tse-tung* (Lawrence and Wishart).

29. See ibid., passim, for Mao's discussion of strategy.

30. Lancaster, *The Emancipation*, pp. 14–16.

31. Ibid.

32. This has been termed "parallel hierarchies" by French authorities. See Colonel Lacheroy, "A Lesson In Revolutionary Warfare," Ximenès et al., *Revue Militaire d'Information* (France), p. 4.

33. *Selected Works of Mao Tse-tung* (Lawrence and Wishart), particularly II, pp. 246–248.

34. Ibid., particularly IV, pp. 84–85.

35. For a detailed discussion of how to mobilize the people, see the excerpts from the *Vietminh Manual on Partisan Warfare* cited in Heilbrunn, *Partisan Warfare*, p. 90. See also Ximenès et al., *Revue Militaire*, especially the case study on Vietnam; and Tanham, *Communist Warfare*, pp. 21–31.

36. Tanham, *Communist Warfare*, pp. 48–50.

37. Ibid., p. 25.

38. See *Selected Works of Mao Tse-tung* (Peking), pp. 83–138.

39. See Nollau, *International Communism*.

40. Ibid.

41. Władysław Pobog-Malinowski, *Najnowsza Historja Polityczna Polski* (*The Most Recent History of Poland*) (3 vols.; London: Gryf Printers, 1960), III, pp. 343ff.

42. Department of State, *Nazi-Soviet Relations 1939–1941* (Washington: Government Printing Office, 1948), p. 78.

43. Malinowski, *Recent History*, III, pp. 397–412; see also T. Bor-Komorowski, *The Secret Army* (New York: Macmillan Co., 1951), pp. 71, 118–122, 171–172.

44. See Joseph Swiatlo, *Za Kulisami; Bezpieki i Partji* (*Behind the Corridors of the Security Police and Party*) (New York: National Committee for a Free Europe, 1955), p. 15. Colonel Swiatlo was head of the key department in the Polish Security Police. He defected in 1954. The invaluable material in the above pamphlet is only a part of his knowledge of that subject.

45. For detailed information covering the Anglo-American-Soviet dealings on Eastern Europe, see: Winston S. Churchill, *The Grand Alliance* (Boston: Houghton Mifflin, 1951), p. 531, 628–631; *The Hinge of Fate* (Boston: Houghton Mifflin, 1950), pp. 327–339; *Closing the Ring* (Boston: Houghton Mifflin, 1951), pp. 324–407; Cordell Hull, *The Memoirs of Cordell Hull* (2 vols.; New York: Macmillan Co., 1948), II, pp. 1166–1173; Robert E. Sherwood, *Roosevelt and Hopkins: An Intimate History* (New York: Harper and Brothers, 1948), pp. 709–710. On U.S. foreign relations see: Department of State, *Diplomatic Papers, the Conference at Tehran, 1943* (Washington: Government Printing Office, 1961); Department of State, *Diplomatic Papers, the Conference of Malta and Yalta, 1945* (Washington: Government Printing Office, 1955); Herbert Feis, *Churchill, Roosevelt, Stalin: The War They Waged and the Peace They Sought* (Princeton: Princeton University Press, 1957), p. 285.

46. Department of State, *Nazi-Soviet Relations*, p. 167.

47. Feis, *Churchill*, p. 208.

48. Winston S. Churchill, *Triumph and Tragedy* (Boston: Houghton Mifflin, 1953), pp. 73–79, 227; Feis, *Churchill*, pp. 338–343; Hull, *Memoirs*, II, pp. 1951–1958.

49. Churchill, *Triumph*, pp. 346–402; also, James F. Byrnes, *Speaking Frankly* (New York: Harper and Brothers, 1947), pp. 21–45; Edward R. Stettinius, *Roosevelt and the Russians: The Yalta Conference* (Garden City, N.Y.: Doubleday, 1949), pp. 79–292.

50. Many excellent books have been written on this subject. The authoritative information pertinent to this study was supplied by Stanislaw Mikolajczyk, former Prime Minister of Poland; Ferenc Nagy, Former Prime Minister of Hungary; Stefan Korbonski, the last head of the Polish underground; Reuben H. Markham, distinguished scholar in the field of the Balkans. See Stanislaw Mikolajcyk, *The Rape of Poland: Pattern of Soviet Aggression* (New York and Toronto: McGraw-Hill Book Co., 1948); Ferenc Nagy, *The Struggle Behind the Iron Curtain* (New York: Macmillan Co., 1948); Stefan Korbonski, *The Fighting Warsaw* (London: Boleslaw Swiderski, 1958); Reuben H. Markham, *Rumania Under the Soviet Yoke* (Boston: Meador Publishing Co., 1949); Reuben H. Markham, *Tito's Imperial Communism* (Chapel Hill: University of North Carolina Press, 1947); David Martin, *Ally Betrayed* (New York: Prentice-Hall, 1946).

51. See "The Politics of Takeover," in *The Soviet Takeover of Eastern Europe* (Cambridge, Mass.: Center for International Studies, Massachusetts Institute of Technology, 1954), passim. For further details on the Soviet plans for takeover, see Hugh Seton-

Watson, *From Lenin to Malenkov* (New York: Frederick A. Praeger, 1956), pp. 248–260; Leland Stowe, *Conquest by Terror* (New York: Random House, 1951), p. 19, states: "By 1945–46,, M.V.D. officers and native "Muscovite" Communists had monopolized key police and government posts in Poland, Rumania, Bulgaria, and Hungary. Everywhere Communists were ministers of defense and interior, controlling both the armies and police; also of national economy, information, and communications. These Moscow-trained Red leaders were also Soviet citizens: Dimitrov and others in Bulgaria; Emil Bodnaras, Ana Pauker and company in Rumania; Rakosi, Zoltan Vas, Erno Gero in Hungary; more of the same stripe in Poland."

52. U.S. House of Representatives, Committee on Un-American Activities, "Facts on Communism," Volume II, *From Lenin to Khrushchev*, 86th Cong., 2d Sess. (Washington: Government Printing Office, 1961), p. 310.

Since the dissolution of the OGPU in July 1934, the Soviet secret police has been repeatedly reorganized and renamed. This constant reorganization complicates any discussion of secret police agencies after 1934.

Whereas the old Cheka and OGPU were separate Soviet Government agencies assigned solely to "secret police" or "state security" work, starting in 1934 such work was periodically assigned to a subdivision of a larger governmental apparatus dealing with Soviet "internal affairs" in general.

Thus, after the OGPU was dissolved in July 1934, its tasks were assigned to GUGBEZ, a section of the NKVD (People's Commissariat for Internal Affairs). The NKVD was a ministry which included, in addition to a secret police section, many other departments dealing with routine police work (e.g., crime investigations), as well as fire protection and the recording of birth and death certificates. Although from 1934 to February 3, 1941, secret police tasks were assigned only to the GUGBEZ section of the NKVD, Westerners commonly used the term "NKVD" to apply to the Soviet secret police apparatus.

From 1941 to March 1954, the secret police functions in the Soviet Union alternated between a separate agency devoted solely to security work and a subdivision of the Ministry of Internal Affairs.

On February 3, 1941, the GUGBEZ section of the NKVD became a separate agency under the new name: *Narodnyi Komissariat Gosudarstvennoi Bezopasnosti*—NKGB (People's Commissariat for State Security). On July 20, 1941, it reverted to a department of the NKVD, but in April 1943 it emerged once more as a separate organization, the NKGB.

The independent NKGB was renamed *Ministerstvo Gosudarstvennoi Bezopasnosti*—MGB (Ministry of State Security) in March 1946. At the same time, the NKVD was renamed *Ministerstvo Vnutrennikh Del*—MVD (Ministry of Internal Affairs). On March 15, 1953, the MGB reverted to a subordinate position as a department of the MVD. Westerners now, however, popularly applied the term "MVD" to the activities of one of its branches assigned to security work.

On March 13, 1954, the MGB once more became independent of the MVD, leaving the latter ministry with only routine internal affairs duties, and was at the same time renamed *Komitet Gosudarstvennoi Bezopasnosti*—KGB (Committee for State Security).

53. See Franklin A. Lindsay, "Unconventional Warfare," *Foreign Affairs*, 40 (January 1962), 264–274.

54. Peter Deriabin and Frank Gibney, *The Secret World* (Garden City, N.Y.: Doubleday, 1959), p. 243.

55. Ibid., p. 60.

56. *Report of the Royal Commission on Espionage* (Commonwealth of Australia, August 22, 1955) .(Available in the general collection of the Library of Congress, Washington, D.C.). The terms "legal" and "illegal" have no reference to lawful or unlawful activities but refer instead to the status of the directors. Thus Valentin Gubechev, who enjoyed diplomatic immunity as a member of the Soviet U.N. delegation, was merely expelled from the United States when apprehended in espionage activities. The Soviet

espionage agent, Rudolph Abel, on the other hand, having no diplomatic immunity, was imprisoned.

57. E. H. Cookridge [Edward Spiro], *Soviet Spy Net* (London: Frederick Mullen, Ltd., [1955]), p. 54.
58. Alexander Foote, *Handbook for Spies* (London: Museum Press, Ltd., 1948), p. 52.
59. *Report of the Royal Commission on Espionage* (Commonwealth of Australia, August 22, 1955).
60. Cookridge, *The Soviet Spy Net*, p. 70.
61. Ibid., p. 61.
62. Ibid.
63. Deriabin and Gibney, *Secret World*, pp. 184–187.

CHAPTER 5

GOVERNMENT COUNTERMEASURES

INTRODUCTION

OBJECTIVES OF COUNTERMEASURES

At the beginning of an underground movement government countermeasures are limited by lack of information about the nature of an enemy which is coming into being. Although the ultimate aim of all government countermeasures is to destroy the leadership and organization of an underground, initially the government must find out who the enemy is. Therefore the government's first objective is to identify the underground leaders, usually by infiltrating the movement. Next the government tries to prevent growth of the underground by restricting its access to the populace and to supplies. To do this the government may seek the cooperation of the people for intelligence purposes, offering them both protection from threats by the underground and evidence that the government measures are in their best interests.

In the second stage of development of the underground the objective of government countermeasures depends upon whether the underground is a resistance or a revolutionary movement and on circumstances external to the underground itself. The aim may be either pacification or control. Pacification entails obtaining a large amount of popular support and willing cooperation. Control does not require such a high degree of popular support; if the government's security forces* control resources and production facilities, and the lines of communication and transportation in strategic areas, that may be sufficient. In both resistance and revolutionary situations pacification is preferable, because a progovernment populace requires a minimum of physical restraint and permits the government to use security forces for other duties. If an occupying government aims for pacification in a resistance situation, a great many troops will be needed originally for occupation duty. In practice it has proved expedient during a military campaign for an occupier merely to establish control without attempting to achieve pacification.

Both the Germans and the Japanese attempted pacification, but when the demand for troops increased, they limited their efforts to control of strategic points. In China the Japanese were overextended, and learned to live with the problems of resistance, being satisfied to control raw materials, industrial and commercial centers, the capitals of the various provinces, and the major seaports and transportation centers. This situation required from one-half to one-third fewer troops than would have been required for the occupation of the entire country.[1]

---

*Security forces are those armed groups actively engaged in maintaining civil order and suppressing subversion and insurrection. In a given country, these may include any or all of the following groups: (1) the full-time national defense establishment, including the army, navy, air force, marines, etc., (2) the police, both uniformed and nonuniformed, at the national, regional, and local levels, and (3) the militia.

Strategy, Tactics, and Countermeasures

In revolutionary war, however, control alone cannot be a sufficient aim for the government. The ultimate objective must be pacification even though the government may be required to restrict personal freedom to such an extent that martial law is invoked. Such restrictions may cause resentment and aid the revolutionary movement by adding credence to its claims of government persecution. On the other hand, failure to undertake prompt and effective countermeasures may permit the illegal organization to grow rapidly. In dealing with revolution, a government typically works under several handicaps: (1) the revolution is usually well underway before control measures are applied, and therefore the security forces are on the defensive at the offset; (2) security forces are often subject to legal restraints; and (3) the government faces conflicting goals—to suppress the revolution and gain the active support of the people.

THE APPLICATION OF MILITARY COUNTERMEASURES DURING PHASES OF UNDERGROUND DEVELOPMENT

During the initial stages of underground development, military combat units are usually not involved, since nascent subversive activity is generally a matter for police action. The military usually does not take action until some event precipitates a crisis. Its role in suppressing underground activities may be outlined as follows:

(1) During the first phase of underground organization the role of the military is primarily one of gathering political intelligence.

(2) When the underground launches its psychological offensive—which may include demonstrations and acts of terrorism—the military role is to support the police in riot control, protect life and property, and reestablish law and order.

(3) As the underground expands and the intensity of strikes and riots increases, the military and police forces continue to restore order, and to identify and keep under surveillance suspected underground leaders.

(4) As the underground moves into its militarization phase and establishes "liberated" areas, the military usually begins a campaign of denial operations against the insurgents and a civil program designed to win the support of the populace. Usually the first step is to cut off potential outside sanctuary and supply. Next, action is taken to isolate the guerrillas from the underground to cut off internal aid and supply. Checkpoints are established to restrict the movements of underground agents and suspicious urban areas are cordoned off and searched for arms and subversive material. People in rural areas who are threatened by the guerrillas are sometimes moved into protected areas. Providing protection for the general populace helps the security forces to gain the active support of the people and establish intelligence sources. Once underground support to the guerrillas is cut off, guerrilla units are compelled to attack the security forces in order to obtain food and arms. Since the

security forces are usually better trained and equipped, they have a tactical advantage in these encounters.

This chapter presents a discussion of countermeasures against revolution and resistance in four historical cases, a brief discussion of the strategy and deployment of security forces, and an analysis of the military and political elements of countermeasures.

COUNTERMEASURES IN RESISTANCE AND REVOLUTION

This section presents the countermeasures employed against two resistance movements—the anti-Nazi resistance in Europe and the Chinese resistance to Japanese occupation, and two revolutionary movements—those in Algeria and Malaya.

GERMAN COUNTERMEASURES IN EUROPE

German Occupation Policy

During World War II, the German occupation policy varied in different parts of Europe and changed from time to time. Initially, the Germans attempted to make Denmark and Norway and the unoccupied part of France into model areas by maintaining the indigenous governments and police forces and using German troops as little as possible. In territories which the Germans appended to the *Reich*, a civil administration was established and a German *Reich* commissioner appointed. In areas which were of strategic or economic importance, a military administration controlled by the *Wehrmacht* through the various military commands was established.[2] In the Balkans, for example, the German interest, as defined by Hitler, was to maintain the security of supply routes and communications to German bases, to safeguard the mineral producing areas, and to protect open shipping. The Germans did not attempt to dominate the entire area, but limited their efforts to controlling major towns, transportation nets, and strategic industries. In this way they were able to extract economic resources without committing excessive numbers of troops to pacification and occupation duty.[3] Generally, the German occupation policy could be characterized as one of control rather than pacification in the occupied countries and battle zones.

In many cases, the Germans encountered initially little organized resistance from the civilian population. As the occupation progressed, however, and as the Germans suffered defeats on the military fronts, the attitude of the populace toward the German troops changed. To obtain manpower for their military production, the Germans began trying to recruit workers in the occupied countries by offering higher wages and better living conditions in

Germany. Although they had some success, they were still unable to get the manpower they required, so they resorted to conscription. This aroused bitterness and drove many persons into the underground.[4] The Germans also exploited the economies of the occupied countries, reducing the civilian populations to a starvation level. In many cases, relief agencies from neutral countries had to be called in.[5] This suffering, of course, also helped swell the ranks of resistance movements.

Command

German countermeasures were handicapped in some areas by a divided responsibility of command. In the Balkans, three separate agencies were responsible for the control and administration of the occupation. They had overlapping responsibilities and were often in conflict. The SS and police agencies in occupied countries reported to the German chief of police, Heinrich Himmler, and acted independently of the army. The military commanders were responsible for security, and various civilian agencies, including the German Foreign Office, were also represented.[6]

In Eastern Europe, responsibility for antipartisan measures was divided at first between the Supreme Command in the operational areas and the *SS Reichsfuehrer* in the *Reichskommissariat*. It was later centralized, however, under the *Reichsfuehrer*. Subordinate to him were the commander in chief of the antipartisan forces and the senior SS and police commanders.[7] The centralized command proved more effective in coordinating and carrying out a unified program of countermeasures.

Intelligence and Security

There was considerable confusion and wasted effort in antiresistance operations, particularly in the clandestine intelligence field. The SS and military organizations were attempting to accomplish the same missions independently, while the German Foreign Office, also working independently through its High Commissioner, was attempting to achieve political aims not always consistent with the directives given by the commanders. There was often considerable friction among these units.[8]

German security measures in the Balkans were extremely lax. Large groups of local civilians were hired to work in German areas of troop concentration and in military installations with only a minimum of security investigation. These workers were able to provide the guerrillas with information about the German troop movements and installations.[9]

The Germans used a system of informers organized into districts, cities, localities, down to units as small as apartment houses. They recruited local residents to watch the other inhabitants and report regularly to the German security forces. A block-warden system was used to check the movements and activities of the civilian populace. One individual in each dwelling was responsible for reporting to a warden on the movements of those who lived there. The Germans also used *agents provocateurs*, who would represent themselves as Allied pilots, guerrillas, or simply as patriots who wanted to perform some subversive act. These agents would attempt to get others to join them in order to test their support of the German regime and to ferret out illegal activities. In Belgium, a "false" escape and evasion network was set up to help Allied pilots out of the country. The Germans let a few escapees get through in order to keep up appearances and deceive the underground and then captured the rest.[10]

Administration of Countermeasures

Puppet governments were installed to lighten the Germans' administrative burden, and native police and security forces were also organized to protect officials and reduce the number of occupation troops required. In many countries German agents and sympathizers provided the occupiers with political intelligence necessary to select pro-German officials to staff various governmental agencies. The Germans also organized motor guards and police to control traffic across borders, and formed state guards to support the city and rural police. In many of these units a German cadre directed operations.[11]

Persons suspected of underground activity were removed to concentration camps and detention centers. The Germans maintained a policy of collective responsibility and often exacted reprisals against the civilian population for sabotage or guerrilla attacks. To discourage resistance, they took hostages indiscriminately, warning that they would be executed if the activities continued. While this may have been effective at first, the indiscriminate execution of hostages and the destruction of villages meant that pro-German families suffered as well as anti-German families. This aroused much resentment among the inhabitants and demonstrated to the people that their actions had no effect on their fate at the hands of the Germans. This led many persons to leave their villages and join the guerrillas.[12] Other measures such as the control of printing materials and censorship of the mails were also used. Radio receivers were confiscated and loudspeakers were set up in public places for listening to radio broadcasts. Individuals were forced to register with the security forces and people's movements were controlled; ration cards and work permits were issued. Many forced-labor battalions were formed.

Measured in term of control, German antiresistance measures were generally effective: the Germans were able to maintain control over the needed resources and manpower; they maintained reliable communications and transportation without serious interruptions; and through a good intelligence and police system, they were able to keep resistance activities in most areas to a tolerable level. On the other hand, such policies as the indiscriminate taking of hostages and reprisals as well as the depleting of local economies led many persons to join the underground.

JAPANESE PACIFICATION MEASURES IN CHINA

During their occupation of China in World War II, the Japanese were constantly harassed by guerrillas operating in rural areas. Having insufficient security forces to deal with them, the Japanese inaugurated a program of "rural purification" which they hoped would reduce their troops requirements.[13] The first phase of this pacification program called for sending troops into a region where guerrillas had been active, wiping out or driving out the guerrillas, and conducting mop-up operations. In the second phase the region was to be made a model—autonomous or ostensibly independent of Japanese control, with courts, militias, and other instruments of local power controlled by local authorities. The Japanese organized local police forces, established training schools, provided large rewards for obedience, and imposed severe punishment for criminal violations. Japanese officials supervised the administration from behind the scenes, and strong Japanese forces were stationed outside the region but close enough to counter any serious guerrilla activity which might develop. Passive defenses were established to protect the villages from attacks, and regular patrols were made outside the villages. The third phase consisted of intensive propaganda, rebuilding of schools, and a general amnesty to guerrillas who surrendered.

The purpose of the pacification movement was to increase control over the twilight areas between guerrilla-controlled territory and Japanese-held territory. The Japanese wanted to put the guerrillas on the defensive by establishing good relations with the population and strengthening the authority of the local governments. They hoped to expand the model areas. They assumed that by instituting a campaign of intensive propaganda and reeducation and by removing the severe rationing and other restrictive control measures, they could persuade the people in the guerrilla-controlled or twilight areas to welcome Japanese rule and withdraw their support from the guerrillas. Thus, the basic aim of the pacification program was to set up an economic blockade which would prevent the guerrillas from getting food and clothing from the villages and also from obtaining recruits. Simultaneously, the Japanese carried out aggressive counterattacks using small, mobile antiguerrilla units.

Capitalizing upon the strong family ties within the oriental culture, the Japanese developed a variation of the block-warden system. In order to extend their control over the smallest unit in the Chinese community, the Japanese used a system called *Pao-Chia:* 10 families who lived close together were organized into a *pao*, and five of these *pao's* into a *ta* or a grand *pao*. The Japanese then selected the most influential or respected person, appointed him Great Pao, and placed him in charge of all of the families. The Great Pao in turn would appoint a head for each *pao*. The Great Pao was completely responsible for the actions of all the people under his jurisdiction. The Japanese expected him to carry out such government activities as collecting taxes and disseminating information, and to maintain watch over all activities of the individuals under his control. Within these family units, self-defense corps for militia or guard duty were formed to maintain law and order. If any member left the family unit or assisted the guerrillas, the Great Pao was expected to turn him over to the authorities. If the Japanese did not get the cooperation they expected, they would execute the Great Pao as an example and appoint another one. The people within the units, respectful of the Great Pao, would not engage in subversive activities or aid the guerrillas for fear of provoking reprisals against him. In turn, the Great Pao would cooperate with the occupying authorities to prevent reprisals against his people.

Despite the careful planning and the effectiveness of the *Pao-Chia* system, the Japanese pacification was not very successful. One basic fault was in the administration of the program. The Chinese who were selected as administrators in the various villages usually lacked training for administrative duty, and many of them participated in the puppet governments purely for personal gain. For example, when food restrictions were lifted to win over the people, they often used the additional food allotments for their personal use. Consequently, the people developed contempt rather than respect for them and the Japanese, and many remained willing to aid the guerrillas.

FRENCH PACIFICATION IN ALGERIA

To quell the Algerian independence movement (1954–62), the French undertook a military campaign designed to defeat the guerrillas, and, after some delay, instituted a pacification program to bring the Algerian people closer to the administration.[14]

In 1954 General Challe introduced the "spot-of-oil" strategy in an effort to pacify the Algerians. Troops were concentrated in one section of the country at a time in sufficient strength to overcome the insurgents. Once subversive activity had ceased, they moved on to the next adjoining area. Although these areas were pacified militarily, many were not won politically, and as soon as the military departed for another area, the underground began its activities anew and guerrilla forces reinfiltrated the "pacified" zones.[15] The French, by concentrating their forces in certain areas, left many areas without military or civil control, and the insurgents simply moved in and established shadow governments without opposition.

Not until nearly 2 years after the initial uprising was the political pacification effort begun in earnest and even then it was done on a small scale. French soldiers were trained for political as well as military duties, and were to acts as "agents of pacification." Special *Sections Administratives Spécialistes* (SAS) were formed to help build schools, homes, and hospitals. The French also began to carry out large-scale reforms in education, health, and sanitation throughout the country. To show their respect for the "Algerian personality," they introduced for the first time instruction in Arabic in the French-operated schools.[16]

In order to better direct the operations, the Governor General was placed in charge of all political and military activities, thus ensuring a unified command for this area. In 1957 the Maurice line along the Tunisian border and another line along the Moroccan border were established to cut off the supply of food and arms to the *Front de Libération Nationale* (FLN). To deprive the guerrillas of internal support, over a million people were relocated into camps.

169

Strategy, Tactics, and Countermeasures

In the meantime the FLN had organized an army and an underground, and had established through persuasion and terror a strong base of popular support. Although the French had contained the FLN militarily, they could not destroy their organization. By the time the French began their pacification program, Muslim support for independence had crystalized to such an extent that it was very difficult at best to obtain any popular active support against the FLN. The French probably could have had control over any area they chose had they concentrated their forces. However, they could not stop the small guerrilla raids nor could they locate any large FLN units to engage in battle. The military situation became a stalemate. The external command of the FLN was out of reach in Tunisia, and the meager French pacification effort did not win sufficient active support among the populace to provide the intelligence sources required to destroy the clandestine organization within their midst.

Finally, the Algerians received psychological and diplomatic support from Tunisia and Morocco as well as from groups in other countries, including France, which for one reason or another felt Algeria should be independent. In addition to international political pressure, the drain on France's economy had its effect upon the decision to grant independence to Algeria.

BRITISH COUNTERMEASURES AND PACIFICATION IN MALAYA

In Malaya (1948–60), the British, and later the independent Malayan Government, succeeded in carrying out an effective countermeasures program against the Malayan Communist Party.

Emergency Powers

After the outbreak of subversive activities, the British instituted emergency measures. These gave the police summary powers to detain suspects for as much as 2 years without trial and to search without warrant. They also had the right to control food supplies and travel and to impose curfews and to close roads. A very effective measure was the detention and banishment law, under which the British could arrest anyone suspected of subversive activity and deport him. From 1949 to 1953 the regulations provided that the inhabitants of a village could be detained for questioning concerning terrorist activities, and could be subjected to collective punishment if they refused to provide such information. The British also instituted national registration and required everyone over 12 years of age to carry an ID card with his photo and thumbprint on it.

The Briggs Plan

When terrorist activities increased and police efforts failed to check them, the British called to Malaya Lt. Gen. Sir Harold Briggs, who drew up the

Briggs Plan in 1950. Its objectives were: (1) to control the populated areas and build up a feeling of complete security among the populace which would in turn result in an increasing flow of information from all sources, (2) to break up the Communist organization within the populated areas, (3) to isolate the insurgents from their food-supply organizations in the populated areas, and (4) to destroy the insurgents by forcing them to attack the security forces on the security forces terms.[17]

Command and Control

As the British had trouble coordinating the army and police efforts, they organized a war council, which was headed by a director of operations and comprised the chief officials of the civilian administration, the police, army, and air force. The police were to have priority in fulfilling their normal functions in towns and villages. The army was to maintain a permanent detachment of troops deployed in close conjunction with the police, cover the towns and villages which the police could not control, and act as a striking force ready to move against guerrillas anywhere within 5 hours marching distance from the towns and villages where its units were stationed. The day-to-day planning and coordinating sessions among the military, police, and civil authorities did much to bring about concerted political and military action. The British succeeded in cutting off outside sanctuary and supplies: the army and police closed off the Thai border and navy patrol boats prevented supplies from arriving by sea.[18] Although the jungle areas provided an internal sanctuary, it was difficult to grow food there and the guerrillas had to rely almost entirely upon the Communist civilian underground, the *Min Yuen*, for supplies.

Intelligence

A dossier was kept on every known terrorist in the area. If one was killed, the British would identify him and close the record. By offering rewards for information on subversive activity, they were able to secure a ready supply of informers. Through informers, defectors, and prisoners of war, they obtained most of the information needed to combat the insurgents. In order to develop a countrywide intelligence system, Briggs instituted the "Briggs map"—a massive chart of the Federation on which were shown the areas in which the terrorists were operating, the distribution of Chinese and Malays, and the areas in which squatters were living. Containing information from every conceivable source, the map also located targets which the terrorists were attacking, such as lumber camps, rubber estates, and mines, and showed where aboriginal tribes lived and traveled. Drop zones for aircraft were also indicated, and sites were located for resettlement of squatters on the basis of the information gathered.[19]

Resettlement

The major goal of the operation, in both its civil and military aspects, was to resettle people out of reach of the terrorists and to control the supply of food. Since the people living close to the jungles were susceptible to guerrilla terror and intimidation, and some willingly supported the guerrillas, they were resettled in defensible areas in order to protect them, make it more difficult for the guerrillas to gain recruits, and to deny food to the guerrillas. To prevent food from getting outside the residential areas, the British attempted to limit the village food supplies to a week's ration for the normal population.

The new villages were built in good farming areas. They were surrounded by barbed wire, and booby traps and mines were placed around the outer perimeter of the villages. They had police posts and search lights at strategic points, and maintained a dusk-to-dawn curfew. To supplement the activities of the regular police, the people had to provide their own home guard and perform police duty.

Those who were moved to the villages were compensated for the loss of the possessions they could not bring with them, and were given an allowance and building materials to build their own homes. They were given a plot of land up to one-sixth of an acre and were instructed how to build a house and plant a garden. If they were dependent solely upon agriculture for their livelihood, they received two to three additional acres.[20] Going to these villages afforded advantages other than protection: the villages had schools, shops, electricity, medical services, and other conveniences which were not obtainable at the jungle edges. In many cases, resettlement of the landless Chinese squatters in the new villages improved their lot economically and offered them hope for permanent improvement, and ultimately led to an actual desire to cooperate with the British.

To run the settlement camps, the British usually appointed Chinese, or Chinese-speaking civil servants selected for their ability to make friends, to be persuasive, and to organize and administer.[21] Every effort was made to obtain intelligent, sympathetic officials. To develop a sense of community fellowship, the administrators arranged competitive games between teams drawn from the government, police, army, and relocated settlers.

Information Campaign

The British carried on an intensive information program to convince the inhabitants that the struggle was not between Malayan Communists and security forces but between the Malayan people and subversive Chinese Communists. They were able to win the support of the Malays by capitalizing upon their long-established antagonism toward and suspicion of the overseas Chinese. They sent teams of captured and converted Communist Party members throughout the countryside lecturing to the local populace. They used demonstrations, films, and radio broadcasts to appeal to the terrorists and offer them amnesty.

Labor Unions and Infiltration

Since the Communists had taken the lead in organizing the Malayan labor movement and had infiltrated and taken control of many union groups, it was difficult to destroy the Communists without destroying the labor movement. Instead of abolishing the unions, the British controlled them by registration

Under the provisions for registration a union could perform all labor activities but no political ones. If it ventured into political activities or participated in demonstrations that had political overtones, its registration could be revoked. In addition, a union member was required to have at least 3 years of experience in the industry or trade before becoming an officer. (At this time, Communist cadres were creating unions and assuming leadership positions in many industries and trades without having ever worked in that trade.) Through these measures the British avoided antagonizing loyal citizens by eliminating the unions, but did control and minimize Communist infiltration by controlling union activities.

Retraining

The British also attempted to retrain Communists in institutions such as the Taipin School. The purpose was not to convert them to democracy, but to give them educational and vocational training that would enable them, after resettlement, to earn a living and to function as respected members of their society.[22]

In summary, the success of the British in Malaya probably resulted from a variety of effective countermeasures. They were able to close off the Thai border and deny sanctuary to terrorists. Deprived of supplies from outside and unable to convert jungle areas into base areas as prescribed in Mao's strategy, the terrorists were forced to depend exclusively upon the *Min Yuen* for supplies. The British also succeeded in isolating the rural populace from the terrorists, and thus ultimately destroyed the latter's food supply and source of recruits. Through the use of good administrators, they were able to win the support of the relocated inhabitants. They developed an efficient intelligence system, and their information program was effective in making the populace understand that the terrorists were Chinese Communist and not Malayan nationalists. The discriminating use of restrictive controls on Communist organizational activities limited Communist underground expansion in the cities. Finally, the promise and granting of independence did much to counter Communist propaganda against the British and bring an end to Communist activity.

MILITARY AND POLITICAL ELEMENTS OF COUNTERMEASURES

It is extremely difficult to combat underground activities without imposing restraints upon the populace. To win or hold the support of the populace, the government may have to permit greater freedom of action to the uncommitted, which freedom the underground can utilize to its advantage. Both the underground and the government compete for the support of the uncommitted; at the very least each wants to deny that support to the other.

Strategy, Tactics, and Countermeasures

The first task of the security forces is to prevent the expansion of the underground movement, since it will present a threat only if it can muster strong popular support. One way to ascertain underground activities and membership is to infiltrate the organization. Even if an underground attempts to maintain security by minimizing the information any one member can obtain, it is impossible to restrict all information concerning activities and membership. The final step is to capture the leadership of the underground movement and thus destroy its organization.

DENIAL OPERATIONS

To contain the underground and prevent expansion of its activities, the security forces must deny it both the human and the physical resources which it requires to expand. Some of the more common techniques for checking the underground are described below.

Access To Mass Organizations

To prevent an underground from manipulating legitimate organizations and mass meetings a government may impose control on the activities of large-scale organizations and on public meetings. In Malaya the British forbade political meetings and required unions and union members to register; thereafter, if the unions engaged in political activities, the British revoked their charters. When an underground attempts to make use of orderly assemblies, the government may employ the police to curtail the demonstrations, using tear gas to disperse crowds and taking photographs to aid in identifying underground leaders.

To hamper the propaganda activities of an underground the government may require licensing of all printed materials and communication facilities and may initiate censorship of newspapers and radio or television broadcasting. Also, radio detection equipment will locate clandestine transmitters enabling the government to capture them or to force them to move so frequently that they lose much of their effectiveness.

Movement and Sanctuary

Undergrounds frequently do much of their planning, maintain their records, and accumulate and store their arms and supplies outside the borders of their own country. To cut the insurgents off from such areas of sanctuary, the government endeavors to impose strict controls at the borders of the country. In countries with long borders, such as Algeria, this is difficult. However, in Malaya, where the border is short and access by sea was made difficult by British gunboats, it was possible to deny the terrorists access to outside help. To control the frontiers, regular guards are maintained on all roads and waterways and at all the normal approaches to the country while roving patrols in jeeps, helicopters, and airplanes cover larger areas.

A British corporal of an armoured car road patrol inspects the identity card of a lorry driver in the Selangor area of Malaya.

Curfews and travel restrictions are imposed to limit the movements of the underground members, with severe penalties for violations. Passes or permits are required for any travel, and identification papers are checked frequently along the route.

In some countries, security forces have instituted the block-warden system. Cities have been divided into areas, sectors, and blocks, and a reliable person appointed in each block to report on the movements of all persons within that area. Residents are instructed to check in and out with block or apartment leader, who in turn reports to the sector leader. Such data as arrival and departure times at jobs, schools, or stores provides the police with up-to-date information on the movements of people within the zones and areas. Other informants, working covertly, check on the block and sector leaders to determine whether they are conscientiously carrying out their duties.

Strategy, Tactics, and Countermeasures

Economic Measures

Government control of food supplies works a great hardship on underground and guerrilla activity. In Algeria, stores were permitted to stock only one week's supply; thus groups defecting to the guerrillas or the underground could not take a large quantity of supplies with them. Rationing helps control black-market trade and makes for a more equitable distribution among the populace. The use of scrip* makes it difficult for the underground to maintain large cash reserves for buying arms and food. To deny food to the enemy, it is also necessary to guard food storage areas from guerrilla raids.

When food, clothing, medicine, and other necessities are under tight control, there may be occasions when the supplies are not sufficient to meet legitimate needs. So that loyal citizens may not be subjected to unnecessary hardship, special supply units are usually formed to provide for these emergencies, as well as to aid towns and villages which have been raided by guerrillas or struck by natural disasters.

Passive Measures

Passive measures are required to protect bridges, underpasses, tunnels, power stations, radio stations, airports, rail stations, water towers or reservoirs, and critical points along important roads—all of them likely targets for sabotage. Repair crews also need protection. Areas around factories or other installations are made restricted areas, to which only authorized personnel are admitted at checkpoints. The area is usually encircled with barbed wire, and the adjacent terrain cleared of foliage. The clearings and open areas are illuminated, and alarm equipment is provided at critical points. Mines and booby traps are often placed along forbidden aproaches. In other areas, roadblocks and checkpoints are set up, and all traffic is stopped for identification. Static security posts are established at strategic points, with both regular and roving patrols traveling between them.

Active Measures

Security forces usually operate with a minimum number of personnel. If so few men are available that spreading them throughout the area would generally weaken the forces, it is preferable for maximum effectiveness against the underground to divide the area into subareas, and to control and pacify one area before moving on to the next. In this way the underground's shadow governments can be eliminated as well as its channels of supply.

The presence of government troops among the populace does much to reduce the state of terror and the effectiveness of underground threats and coercion. Population controls maintained by government troops in any area can be broken only at the risk of detection by the security forces themselves or

---

*Scrip is an artificial currency used for legal tender. It may be issued or recalled at any time; and since all currency must be redeemed for new scrip, the government can identify and control any large amounts of currency held by any one source.

Vietnamese soldiers exercise their dogs at the dog center near Saigon. The dogs receive 4 to 6 weeks of training prior to being assigned to a field unit patrolling against Viet Cong rebels.

by informers. But if the populace is not converted to cooperation with the government, then after the military forces leave the area the underground will reestablish itself. If pacification cannot be achieved, then resettlement of the people into controllable areas—as was done in Malaya, Greece, and Algeria— enables a small military force to protect and control large populations.

Patrols made regularly through populated areas also add to the effectiveness of controls over the populace and make overt underground action more difficult. Other patrols, timed at random intervals, can perform surprise searches. The size of the patrols depends upon the size of the security force and the usual strength of the raiders. Dogs are frequently used for patrol duty. The police and army often maintain a group of reaction troops which can respond instantaneously to surprise attacks.

To deny the underground weapons, search and seizure laws are usually enacted. These permit army and police personnel to search and detain suspects without a warrant. Successful search and seizure operations can result in the arrest of suspected members of the underground and confiscation of arms, communications equipment, medicine, and supplies. This can seriously cripple the underground. In populated areas, a cordon is thrown around a city block cutting off ingress and egress, and house-to-house searches for suspects and subversive materials are conducted. The cordon technique has the added advantage of bringing the police in direct contact with the population and emphasizing the hazards of harboring or assisting the underground in any way. In Malaya and Palestine, registration of weapons was required to keep them from falling into the hands of the underground.

STRATEGY AND DEPLOYMENT OF SECURITY FORCES

In using security forces, the military commander has several alternative strategies. He can, for example, concentrate his forces in sufficient strength to defeat the insurgents in one area and then move to the adjoining area. This was the aim of the Japanese rural "purification" and the French "spot-of-oil" strategies. If the area is pacified and the populace is willing to cooperate with the security forces, the underground shadow governments can be broken and the insurgents beaten. However, if the area cannot be won politically, the underground will resume its activities as soon as the security forces leave. In this eventuality the next step is the relocation to controlled villages of rural inhabitants and others vulnerable to the underground, as was done in Malaya. This provides an alternative means for destroying underground supply and access to the populace.

Another strategy is to establish a static defense of strategic centers. In Burma during World War II the British developed the "stronghold" concept, in which static defense posts were placed at key road junctions, railways, and important villages. They were always established in threes in triangular arrangement around the key position, and were spaced close enough together so that if any one of the forts was attacked and the alarm signal given, the other two forts could send reinforcing units to attack the flank or rear of the enemy. The Germans and Japanese found that by assuming a strategy of static defense they could not destroy the insurgents, and the price for controlling strategic centers was constant guerrilla attacks. Another alternative is to divide the forces available between static defense positions and a mobile reserve force which can be stationed and moved as necessary. The mobile reserve force is a reaction unit assigned to an area of responsibility which can reinforce or counterattack in case of a raid on static points.[23]

As noted earlier, the government usually faces the dilemma of being forced to impose restrictive measures upon the populace in order to destroy the underground at the same time that it must seek to maintain or win the active support of the people. To reduce the effects of this conflict of aims, a central authority is usually appointed to direct both military and civil activities. In the countries under German occupation, constant dissension between split commands and overlapping authority of the political, military, and internal security forces made concerted government efforts difficult. The same type of organization existed in the early stages of the Malayan Emergency; later, however, cooperation between military and civic authorities through joint planning meetings produced coordination of activity. The centralization of effort was carried throughout all of the levels by setting up committees of military, police, and administrative officials, which met daily to plan the campaign against the terrorists. The police performed their normal functions within the cities and villages, concentrating on identifying potential subversive elements within the community. Army units, which supported the police in actions against larger groups and maintained control of rural areas and strategic points outside the villages, were ready for aggressive antiguerrilla campaigns. The navy and air force were used for patrol duty along borders or over large uncontrolled areas.

The size of units and patrols usually depends to a large extent upon the security forces available and the size and composition of the underground and guerrilla units. Both the army and the police usually maintain reserve and reaction units which can be called upon if needed. Communications between patrols and reserve units are of critical importance.

Although centralized direction of countermeasures is necessary, each area commander must have a certain amount of tactical autonomy. Often an area commander can act swiftly and aggressively to counter underground measures only if he is permitted to act on his own initiative. Since underground activity and the amount of popular support may vary greatly from area to area and from time to time, administrative control is usually kept flexible. Area commanders are authorized and encouraged to effect civic improvements, pay informers, or take on-the-spot corrective action to adjust any deficiencies in either the military or civic programs.

INTELLIGENCE

Kinds of Intelligence

One of the first tasks of intelligence is to set up a system which makes it possible to identify every inhabitant of an area and, if desired, keep track of his movements. National registration, which involves the issuance of identification cards bearing the photograph and fingerprints of the bearer, is the usual procedure. Further, by requiring everyone who enters certain areas to check in and out, the security forces can ascertain who was in the vicinity at the time a subversive act was committed. Undergrounds try to evade the identification system by counterfeiting identification cards, ration cards, work permits, travel permits, etc. If the government frequently changes documentary procedure, underground agents may be caught through their attempts to use out-of-date credentials.

Census information has been used successfully to catch guerrillas hiding in a village. The security forces counted or photographed the inhabitants, and when the count exceeded the census, investigation usually led to the apprehension of guerrillas who were hiding in the homes of underground members. The disappearance of a particular age group from a community or sudden shortages of materials and food are good indications of guerrilla recruitment and underground supply network, and a prediction of future guerrilla action in that area.

The particular nature of captured underground members often provides valuable clues to uncaptured members, because local underground leaders tend to seek recruits among persons in their same profession or craft or from the same ethnic group. In Malaya, for example, the British were greatly helped by the fact that the Communist movement was composed principally of Chinese; thus the British could concentrate their efforts on the Chinese element, which

179

comprised about 50 percent of the total population. Further, an analysis of the activities of known individuals and of guerrilla operations may establish patterns of behavior which aid in the detection of other members of the underground. Captured personnel or defectors often supply information on how the movement developed, who its leaders are, what groups within the populace it depends upon, how it is organized, how members communicate with each other, the size and composition of the cells, the background of its members, where supplies are kept, where safe-homes are located, and who arranges for escape from the country. The government also tries to find out whether there is any rivalry among the members of the movement, what hardships members face, how effective government appeals for amnesty and punishment have been, what motives led members to join the organization and whether political ideology played a role. All this information is kept in central files where it can be studied by the authorities to discern a pattern in the underground's organization, activities, and membership. This information, combined with dossiers on leaders of the movement, can lead to the identification and capture of key underground members.

An underground is always desirous of enlisting the support of leading political figures who may have opposed members of the government prior to the insurrection. About such leaders the government requires as much information as it can gather—their political views, the organizations to which they belong, their finances, families, friends, and any circumstances which might make them vulnerable to blackmail or coercion. Political intelligence also has a positive role. It can be used to identify capable and loyal citizens who would be willing to undertake assignments in the government or to serve on commissions and other civic bodies.

Usually the underground attempts to infiltrate the security forces by blackmail, bribery of the people working with security units, or by planting secretaries, janitors, or other employees who through their work come in contact with officials who have direct or indirect access to security information. Careful security measures to seek out these covert elements are critical to any intelligence effort. Civilians employed by the military are screened for loyalty and periodic checks are made on their activities. Security checks and background investigation of all personnel in sensitive assignments are essential.

Obtaining Intelligence

Although something can be learned through interrogation of prisoners, defectors, and friends and family of known underground members, a government must develop reliable sources of information by using aggressive measures.

Paid Informants

The paid informer is probably the best source of information. Informers are usually recruited from various groups throughout the country so as to provide wide geographic and ethnic coverage. The underground uses terror, threats, and reprisals to ensure that no one reveals its clandestine activity or the anonymity of its members. Therefore, before adequate sources of information can be developed, security forces must reduce the level of risk to the individual and offer some inducements for this dangerous role. Contacts with the informer are usually kept to a minimum and his true identity is usually known only to his security agent contact. In some cases, payment is made to the informer through a bank account which is kept for him by the government so that he will not be betrayed by a sudden increase in wealth. If the informer becomes known to the underground, provisions are made to relocate him in some other part of the country.

One of the most risky parts of the informer's task is transferring information to the security agent. Many of the techniques used are similar to those used by the underground, such as the use of intermediaries or prearranged signals. In the Philippines, the transfer of information about the movements of the Huk guerrillas was accomplished through the use of a set of prearranged signals which the farmer-agent communicated to a government air reconnaissance plane flying a set pattern over the region. The number, location, and direction of the Huk units were indicated through such devices as tying a cow to a particular post, placing the barnyard gate in a particular position, or opening a certain window. Since the signals used were common everyday actions, they could not give away his role of informer, and thus the informer could feign cooperation with the Huks or even act as a sympathizer, thereby

(Courtesy of the Natural Rubber Bureau)

Gen. Sir Gerald Templer (left) empties information boxes brought in from several communities during the "Emergency." The communities' leaders watch the proceedings.

181

protecting himself. This system provided the security forces with a rapid, large-scale intelligence system which could pin-point the movements of insurgents.[24]

The Populace

During an insurgency people are generally under a great deal of stress and many seek to avoid becoming involved with either side. To open up these sources of information, measures must be taken to provide security for the informants and inducements to encourage them to perform acts which involve risk. In Malaya, soldiers and police went to each house and gave the occupants a sheet of paper on which to write, anonymously, any information concerning subversive activity. The next day the security forces returned with a sealed ballot box and collected all papers, blank or not. Since everyone was required to hand in a paper, the underground could not determine who in the village was informing to the authorities.[25] During a routine cordon and search operation in Palestine, everyone in the area was interrogated individually in an enclosed booth, and if information was obtained the identity of the informer was thereby protected.[26]

Even when citizens report information to local officials, the information is often not passed on to government headquarters because the local officials are afraid to oppose the underground which frequently directs its terrorism against such local authority. Also citizens are reluctant to give local officials information lest the officials be unknown members of the underground. In order to overcome such a situation, Ramón Magsaysay, then Secretary of Defense in the Philippines, directed the people to send information directly to him through a special mailbox, thereby preventing disloyal officials from leaking the information or the identity of the informer to the underground.[27] The effectiveness of this system was due mainly to the fact that Magsaysay commanded much respect from the people and had a reputation for being a man of the highest integrity. In the Philippines, when security forces had to use informants to identify underground members in a village, a tent was set up with a peephole, or "Magic Eye," cut in it. All the villagers were then lined up and made to pass before the "Magic Eye," behind which hid the informer. This type of operation not only produced intelligence, but had a strong psychological effect upon the community.[28]

In Kenya a similar technique was the use of the "hooded men." The informer met with his government contact at a prearranged location, and was covered with a sheet from head to foot to conceal his identity before coming in contact with other informers. All the informers were assembled and seated in a row of chairs, a security agent standing behind each one. The suspects passed before the hooded men and if one recognized a terrorist he whispered the information to the agent behind his chair; if more than one informer recognized a man, it was certain that he was a genuine terrorist.[29]

If the populace are so frightened of the insurgents that they continue to withhold information in the hope of not becoming involved, the threat of group punishment may be used. The Germans and Japanese in World War II and the British and French in postwar Malaya and Algeria used the principle of group responsibility.[30] The entire community was made responsible for the actions of any of its members and all suffered sanctions unless information concerning subversive activities was reported to the authorities. Some system of surveillance, such as the block warden, and *Pao-Chia* or parallel hierarchy systems, was imposed to increase certainty of detecting violations. It then became a crime for the individual or the group to withhold information, and active checks were made to enforce the system. The Germans used *agents provocateurs* who pretended to be allied pilots in distress or underground workers seeking aid from the community.[31] This made it doubly dangerous for the populace and greatly increased the risk involved in remaining silent.

In some cases fear of underground reprisal is so great that even the principle of group responsibility fails to produce the necessary information. One method of obtaining information without directly threatening the populace was used in the Philippines. The security forces leaked information that they were about to raid a village; on receiving this news the active insurgents immediately left the area. Instead of sending armed troops, government officials sent in a small unit with a camera and took a group photo of the village inhabitants. The next time the security forces received intelligence information that the insurgents were active in the area, they went to the village and arrested anyone not in the original group picture. Ninety percent of the suspects taken by this method proved to be members of the Huk organization.[32]

Agents

One way to penetrate an underground apparatus is to persuade or coerce a captured underground worker into working for the security forces, while ostensibly remaining a loyal member of the movement. Another way is to use the underground's own recruiting process: placing a government undercover agent in a critical job with access to classified government information, which makes him a likely target for underground recruiting.

In the Philippines, security forces used relatives of known insurgents to infiltrate the movement. A cousin of a Huk commander agreed to cooperate with the government by infiltrating the ranks of the underground. To ensure his well-being as a counteragent, the government prepared a series of incidents which helped convince the insurgent high command of the agent's "loyalty." The agent's brother was placed in jail and other members of his family were relocated. This contrivance persuaded the Huk leaders that this man joined their ranks because of "obvious grievances" against the government. In time, the Huk command had such confidence in this man that he was made an official in the National Finance Committee, and later a bodyguard to Taruc himself, a high Communist leader in the movement.[33]

An example of infiltration of underground organization in another part of the world occurred in Norway during World War II. A Norwegian named Henry Oliver Rinnan, sympathetic to the Germans, formed his own organization to disrupt and infiltrate Norwegian underground activities. With sea routes closed by German military activity, the Norwegian underground sent

Strategy, Tactics, and Countermeasures

an agent to organize overland supply routes to Sweden. Rinnan, posing as a member of another underground organization, made contact with this agent, but the agent was suspicious and would not work with Rinnan. Shortly thereafter, the Gestapo captured the underground agent and then permitted Rinnan to "arrange" the escape of the man from the Gestapo. This established the underground's confidence in Rinnan and permitted him to gain vital information. When Rinnan's true identity became known later, his knowledge of the underground organization was so great that the central leadership decided to abandon all activities in the area.[34]

Government infiltration of undergrounds by double agents is not only a source of information, it is a disruptive influence compelling the underground to tighten security and creating fear which makes the underground membership less aggressive in its actions and less likely to recruit people.

Pseudo-Gangs and Large Unit Infiltration

The Germans used "dummies" on pseudo-gangs (individuals disguised as terrorists or guerrillas) in Russia and the British used them in Kenya and Palestine.[35] The British used ex-terrorists and European military men who disguised themselves as Mau Mau to penetrate this terrorist organization and gain intelligence. They dressed and acted as terrorists. The Europeans used burnt cork or bootblack to color their skin, learned Mau Mau handshakes, songs, and used terrorist clothing and equipment. The pseudo-gang used homemade guns, wore ragged overcoats and Mau Mau beads. They contacted other Mau Mau terrorist gangs and gained valuable information about the organization and its leaders.[36] In the Philippines the security forces used an organization called "Force X" to carry out "Large Unit Infiltration." The prime objective of this force, as was that of the pseudo-gangs in Kenya, was to obtain intelligence. By penetrating the insurgent organization, they were able to gain information about the security, communications, and supply systems as well as the nature and extent of civilian support and liaison methods. They would go into villages disguised as Huks and in this way determine which officials and civilians were working with the insurgents.[37] Still another way to infiltrate is to create a group which is independent of the underground but which appears to be operating against the government. For its own security (because the group could draw the attention of security forces to other subversive activity) the underground will contact this group, with the intention of restricting its movements or of absorbing it into the movement.

Uses of Intelligence

When the government identifies a member of the underground, it is often more useful to keep him under surveillance than to arrest him immediately. By shadowing him, the identity of other underground workers may be learned and the *modus operandi* of the apparatus determined. Wiretaps on telephones, hidden microphones, and other eavesdropping equipment are useful aids in the surveillance of suspects.

184

The nature of the conflict in insurgency situations imposes a very short lifetime on much, if not most, of tactical intelligence, particularly when compared to conventional war situations. In conventional war a commander has direct contact with the enemy or knows his whereabouts; in unconventional war the commander of security forces is fighting an enemy who avoids contacts and whose whereabouts are often unknown because his combat units are dispersed except during engagements. Therefore, in unconventional war tactical information needs to be converted into counteraction chiefly in the area and at the time that it is gathered.

The centralization of long-term intelligence concerning the underground organization and its leaders is essential for effective countermeasures. The underground is based and relies on a national organization. A centralized collection of information can provide patterns of underground activities which could not be detected on a local level. However, information for tactical, immediate use is frequently not coordinated for fear of leaks in the security system or because underground agents may deduce government intentions from their preparatory actions.

CONTROL AND PACIFICATION

To develop intelligence sources among the populace the security forces must provide the people with a reasonable degree of protection against retaliation by the insurgents. The government also attempts to mobilize suitable people into the armed forces or into labor projects. It further attempts to mobilize the attitudes of the people because, normally, a large portion of the populace is indifferent to political strife and avoids involvement with either insurgent or government forces. To reach these uncommitted people the government usually undertakes information campaigns and civil action programs to draw attention to the existing civic advantages and to the obligation of each citizen to support the government. In this endeavor the role of the individual government soldier is an important one; he must know not only how to fight the underground but how to conduct himself in contact with the people. The French in Algeria recognized that the soldier was the agent of pacification and that his actions are a major determinant of the success or failure of any pacification effort.

Relocation and Retraining

Because it is difficult for people to cooperate with the government forces in areas where guerrillas are active, the inhabitants are often relocated to an area where they can be protected from guerrilla raids and their activities controlled. The government may build a new, fortified village located near a highway so that security forces can use their mobile reaction units to come to the aid of local militia in case of attack. An alarm system is set up to ensure instantaneous help from the nearest military installation. It is important to

(U.S. Army Photograph)

A deep moat imbedded with bamboo stakes and surrounded by barbed wire encircles the Vietnamese village of Cu Chi. This is but one of a system of fortified villages designed to protect the villagers from the Viet Cong.

(U.S. Army Photograph)

Bamboo is used instead of barbed wire to parallel this moat in Vietnam.

recognize that relocation may cause hardship and resentment even if properly administered. However, this is a calculated risk in the attempt to cut off internal supplies and underground support for the guerrillas.

In most of the areas where both guerrilla and underground Communist cadres have had some success in recruiting individuals into front organizations, they have drawn from the dispossessed or the "have nots" of the society. Farmers who have been driven from their farms to the city and cannot find employment, the unskilled, the unemployable—all are likely to be potential recruits for subversive movements. In Malaya the British successfully provided a retraining program, intended not to convert the ex-terrorists from communism to democracy, but to provide them with skills which ensured subsistence and a future in a modern society. The EDCOR program in the Philippines is another example of how effective rehabilitation of insurgents played a significant psychological role in the defeat of the insurgent movement.

In addition, to make it worthwhile for the people to resist the subversive movement and join the government effort, a system of rewards can be set up. It is just as important to show that cooperation leads to rewards as it is that noncooperation will lead to punishment. Increases in pay and citations for bravery or patriotism encourage popular response, and periodic offers of individual amnesty are made to induce underground and guerrilla followers to defect to the government cause. On the other hand, severe punishment is administered for withholding information concerning subversive matters.

Strategy, Tactics, and Countermeasures

Civilian Mobilization

In order to prevent the passage of military and police measures which may be detrimental to the community, civic leaders should play a role in determining pacification policy. Civic groups loyal to the government can assist in information programs and in organizing civic demonstrations. With military leadership self-defense units can be organized in cities and villages for guard duty and patrols, and normal police and traffic functions.

The recruitment of civilians for self-defense units serves several purposes. Since the military structure provides a built-in surveillance system, it prevents them from joining the underground, and releases police and army personnel for aggressive action against the insurgents. Families and friends of members of the self-defense units are less likely to provide information to the insurgents if it may endanger the lives of relatives and close friends. Recruiting civilians into territorial and local police units, and into regular army units absorbs much of the slack in employment or displacement caused by underground activities, thereby reducing the number of unemployed and unoccupied persons who might turn to the underground.

(U.S. Army Photograph)

A Special Forces medic conducts sick call for Rhade villagers in Vietnam.

Administrative Measures

The administrative branches of government, in performing their normal public service functions, can be effective in maintaining the support of the people. Visible civic improvements indicate an interest in public welfare, while aid to displaced persons and refugees can prevent them from joining the underground. Relocation measures, when administered in a sympathetic manner, may induce loyalty to the government of the people affected.

The most important aspect of pacification is what the people perceive the intent of the government to be. If it appears that the government is attempting to solve their difficulties and is taking aggressive action to remove the threat of insurrection, they will provide support against the underground. In the Philippines Ramón Magsaysay demonstrated that the administrative forces of government in developing nations could be called upon directly to assist the people, if the people would directly assist the government. He had the people report graft, corruption, and poor administration directly to him and he took prompt action against the violators, thereby building popular confidence in the government. In Malaya the British selected highly qualified Chinese-speaking British civil servants, each of whom was assisted by a Malayan Chinese, to administer the relocation centers. This contributed greatly to the success of their program.

Resistance and revolution create social disorganization and stress in a civilian community. Such things as government controls, relocation, guerrilla and underground threats and reprisals have a significant effect upon the day-to-day behavior of individuals. To provide an effective administrative program during social disorder and stress, it is important to understand human behavior under these conditions and to know that procedures which are effective under normal conditions may not be effective under conditions of stress. Some of the lessons learned through evaluation of U.S. experiences in relocation and administration during World War II will be discussed next.

Individuals Under Stress

Investigations have shown that although people throughout the world may differ profoundly in their beliefs and customs, they display many common behavior patterns under stress. When people are subjected to threats, personal discomfort, loss of their homes, inadequate food, loss of means of subsistence, or restriction of movement due to controls or confinement, they react in similar ways. They display one of three types of behavior: excessive cooperation, apathy or withdrawal, or aggressive action against authority.[38] Stress may lead to violent, inappropriate behavior or it may breed fear which in turn leads to the creation of rumors. The rumors usually take the form of atrocity stories or plots of betrayal, and frequently lead to attacks upon people who have little or nothing to do with the causes of the stress. If these behaviors are recognized as symptoms of aggression arising out of stress, efforts can be made to control it. In dealing with aggressive behavior of this type, it is usually cheaper to provide some form of relief from the stress than to impose punishments which might arouse new aggressions. Outbursts of aggressive behavior are generally followed by a period of calmness and relief and advantage should be taken of this period of cooperativeness to modify conditions.

Strategy, Tactics, and Countermeasures

Administration of Individuals Under Stress

To maintain his authority under conditions of social stress an administrator should be careful to grant simple requests which can be easily accommodated; he should be equally careful to refuse aggressive demands which would encourage destructive behavior. He should make only such rules or threats as he can enforce, lest he foster disrespect for his authority.[39] One way for an administrator to counter the stress created by social disorganization is to use existing leaders to create new organizations. In selecting leaders, members of minority groups or factions should not be chosen simply because they favor the administration. A better criterion for selection of a leader is his ability and capacity to understand the needs of the people and to solve the existing problems.[40]

To relieve tension, the constructive tendencies of people should be encouraged and opportunities should be provided for achieving social stability by creating economic security, providing opportunities to perform work which is valued and lends prestige. Self-government, education, sports, and recreation are useful, constructive ways of relieving stress and tension. However, it is dangerous to assume that when the administrator introduces these innovations, the people will accept and participate in them. They must first feel a need for the new things. Plans for social change should be tried out on a small scale in a part of the community, and be evaluated before they are applied to the entire community. In this way, large errors which might antagonize the community and increase stress are avoided.

Communication is an important tool in the administration of people under stress. By informing people of the necessity of change and the means of accomplishing the change, much of the people's uncertainty can be removed. The communication process should work both ways—it is important to know how people feel about innovations as well as to tell them what the changes are. The use of both formal and informal channels of communications within the community can facilitate the two-way transfer of information and the ultimate acceptance of innovations.[41]

Civic Action and Information Campaigns

Construction of hospitals, schools, sanitation and irrigation facilities, highways, and other public works demonstrate the government's interest in the people. Agricultural assistance from crop planting to harvesting earns the gratitude of the farmers. Emergency distribution of food, clothing, and medical aid is also effective. In all these endeavors, the role of the military and of civilian representatives of government is emphasized. Permitting the military to aid in assistance programs or work in the fields as they did in the Philippines has done much to remove prejudice toward them. Intensive information campaigns are conducted to discredit the insurgents and to convince the citizens that it is in their own personal interest, as well as being their obligation, to support the state. Such campaigns are reinforced by actions designed to nullify the appeals of the underground and to reaffirm the government's desire to help.

Movies, newspapers, radio, loudspeakers, house-to-house visits and other means of mass communication are used to explain the problems of the emergency to the people. People who have lived under stress for prolonged periods of time become apathetic and avoid learning of new situations for fear of personal involvement. There is probably no better way to ensure that each person knowns about the situation than to visit him personally, not only because immediate contact as such is good, but because in many areas of the world there are few radio sets to receive government broadcasts. Crystal sets have been given to people in rural areas, and radio receivers with loudspeakers have been set up in markets and other public places to help the government maintain communication with the public. Discussion groups are organized in which people learn how to combat the subversive movement. In Malaya, as noted previously, the British sent captured guerrillas from village to village as members of the British information campaign to provide firsthand information on the subversive movement. For the civic action and information campaigns the government must continually have information from the people to determine what their fears and interests are and how they are reacting to underground and government propaganda. Opinion polls of select groups within the community can provide this information.

SUMMARY

Undergrounds are difficult to detect and destroy in the early stages of their development. As a movement expands and reaches the militarization stage, certain steps can be taken in the attempt to destroy it. A first step has been to cut off any external supply and sanctuary to which the guerrillas may have access in order to force them to rely upon internal support for their existence. As the guerrillas lose their source of supply, they are forced to attack security forces, thus providing an advantage for the conventional forces, who are better organized and better trained. Secondly, the activities of the underground which are the internal source of supply, intelligence, and recruits for the guerrillas must be curtailed. This is done by denial operations. An intelligence network with a system of informers is an essential factor in identifying members of the underground and infiltrating its ranks. Still other effective means are rewards for information on underground activity and amnesty to any insurgent who surrenders.

To gain the cooperation of the people and to acquire necessary intelligence sources, the security forces must provide the populace with a feeling of security. This can be accomplished through relocation of rural families into defensible areas, an action which also cuts the guerrillas off from a readily accessible source of new recruits and provisions. Militias are organized in order to mobilize a large segment of the populace and programs are initiated

Strategy, Tactics, and Countermeasures

in an effort to show the people that the government is sincerely interested in changing the social, economic, and political conditions of the country.

FOOTNOTES

1. Gene Z. Hanrahan, *Japanese Operations Against Guerrilla Forces*, Techincal Memo ORO-T-268 (Washington: Operations Research Office, March 16, 1954).
2. For a discussion of German occupation policy and adminstration in Europe, see Arnold and Veronica Toynbee (eds.), *Hitler's Europe (Survey of International Affairs, 1939-1946)* (London: Oxford University Press, 1954), pp. 91-125.
3. U.S. Department of the Army Pamphlet 20-243, *German Anti-Guerrilla Operations in the Balkans, 1941-1944* (August 1954), p. 13.
4. Edgar M. Howell, *The Soviet Partisan Movement, 1941-1944* (DA Pamphlet 20-244) (Washington: Department of the Army, August 1956), pp. 107-109; see also Ronald Seth, *The Undaunted: The Story of the Resistance in Western Europe* (New York: Philosophical Library, 1956), pp. 177-187.
5. Department of the Army, *German Anti-Guerrilla Operations*, p. 75.
6. Ibid., p. 17.
7. Howell, *Soviet Partisian Movement*, p. 119.
8. Department of the Army, *German Anti-Guerrilla Operations*, p. 76.
9. Ibid., p. 77.
10. George K. Tanham. "The Belgian Underground Movement, 1940-1944" (unpublished Ph. D. dissertation, Stanford University, 1951), p. 319.
11. Department of the Army, *German Anti-Guerrilla Operations*, pp. 18-19.
12. D. M. Condit, *Case Study in Guerrilla War: Greece during World War II* (Washington: Special Operations Research Office, 1961), p. 19.
13. Hanrahan, *Japanese Operations*.
14. Lorna Hahn, "Algeria: The End of an Era," *"Middle Eastern Affairs. VII* (August 1956), 286-293.
15. Edwin F. Black, "The Problems of Counter-Insurgency," *United States Naval Institute Proceedings* (October 1962), 22-39.
16. Joseph Kraft, *The Struggle for Algeria* (Garden City, N.Y.: Doubleday, 1961), pp. 92-110.
17. Harry Miller, *Menace in Malaya* (London: George Harrap and Company, 1954), p. 139.
18. Ibid., p. 121.
19. Ibid., p. 140.
20. Ibid., p. 151.
21. Ibid., p. 146.
22. Ibid., p. 182.
23. Black, "Problems."
24. Charles T. R. Bohannan, "Unconventional Operations" (Seminar in *Counter-Guerrilla Operations in the Philippines, 1946-1953;* Fort Bragg, N.C., June 15, 1961), p. 62.
25. Miller, *Menace*, pp. 209-210.
26. R. N. Anderson, "Search Operations in Palestine," *The Army Quarterly* (January 1948), 201-208.
27. Ismael D. Lapus, "The Communist Huk Enemy" (Seminar, Fort Bragg), p. 20.
28. Medaro T. Justiniano, "Combat Intelligence" (Seminar, Fort Bragg), p. 43.
29. Frank Kitson, *Gangs and Counter-Gangs* (London: Barrie and Rockliff, 1960), pp. 100-101.
30. Toynbee, *Hitler's Europe*, p. 149; see also Howell, *Soviet Partisan Movement*, p. 120, and Slavko N. Bjelajac, "Case History in Area Operations" *Army*, XII (May 1962), 30-40.

31. E. K. Bramstedt, *Dictatorship and Political Police* (New York: Oxford University Press, 1945), pp. 142–144.

32. Uldarico S. Baclagon, *Lessons From the Huk Campaign in the Philippines* (Manila: M. Colcol and Company, 1960), p. 31.

33. Justiniano, "Combat Intelligence," pp. 44–45.

34. Seth, *The Undaunted*, pp. 36–37.

35. Otto Heilbrunn, *Partisan Warfare* (New York: Frederick A. Praeger, 1962), p. 69; see also Howell, *Soviet Partisan Movement*, p. 119.

36. Kitson, *Gangs*, pp. 74–76, 94.

37. Napolean D. Valeriano, "Military Operations" (Seminar, Fort Bragg), pp. 32–33.

38. Alexander H. Leighton, *The Governing of Men* (Princeton: Princeton University Press, 1945), pp. 252, 263.

39. Ibid., pp. 268, 282, 275.

40. Ibid., pp. 331, 342.

41. Ibid., pp. 358, 363–365.

PART II

CASES OF UNDERGROUNDS IN RESISTANCE AND REVOLUTION

INTRODUCTION

To demonstrate the variety of roles which undergrounds have played in both resistance and revolution, seven widely different geographic areas in which underground movements have operated have been selected for discussion. Some worked in support of guerrilla forces in World War II resistance movements, some in support of conventional forces. Both Communist and non-Communist revolutionary movements are represented.

In each description, an attempt has been made to highlight those aspects which make the movement unique as well as to demonstrate its similarities to other undergrounds. The descriptions are not intended to be complete or representative accounts, but are intended to show the wide scope and range of underground activity. In each account, the historical setting is first presented, as background to the organization, administration and operational functions, and activities.

The administrative categories covered are recruitment, logistics, finance, communications, and security. The operational categories are intelligence, psychological operations, sabotage, and escape and evasion. The countermeasures taken by the security forces are also described. Since it was not possible to obtain comparative information, the extent of coverage varies from study to study.

Each of the accounts highlights a particular aspect of an underground, its activities, or the situation in which it operates.

The role of the various undergrounds in the French resistance (1940–45) is presented to illustrate how the underground coordinates its activities with those of conventional troops.

The story of the Yugoslav resistance (1941–45) illustrates the problems involved when two resistance groups fight each other as well as the occupier. The small Communist Party of Yugoslavia organized a partisan army and civilian National Liberation Committees in liberated areas, thereby gaining control of local and, later, national government agencies. In this way the Communists were able to turn a resistance movement into a revolution. The relationship of the underground to one area guerrilla leader in the Royalist movement is discussed in some detail.

In Malaya (1948–60), after a phase of unsuccessful political activity, the Communist Party organized a fighting force (MRLA) and a civilian support force, the *Min Yuen*. The collapse of the guerrilla fighting force was closely related to the inability of the underground to provide intelligence, supplies, and other support for the guerrilla force in the face of the British countermeasures.

Cases of Undergrounds

Algeria (1954–62) provides an example of a non-Communist revolution in which independence was won primarily through political rather than military means. The interesting counterrevolution of the OAS is treated briefly.

The story of the Communist insurgency in Greece (1945–49) demonstrates the importance of an external sanctuary. Yugoslavia, after being expelled from the Communist bloc, denied the Greek Communists sanctuary within its borders, and their underground was unable to provide a reliable steady source of supplies. Shortly afterwards, the movement failed.

The Communist insurgency in the Philippines (1946–54) is of interest primarily because of the wide range and the effectiveness of the countermeasures used. The capture of the underground leaders in Manila and the effective social, political, and military reforms instituted by the government led to the collapse of the Huk movement.

Palestine (1945–48) demonstrates the effectiveness of combined underground and political activities. The use of cordon and search as a countermeasure against underground activities is another feature of this description.

CHAPTER 6

FRANCE (1940–45)

BACKGROUND

When Germany invaded Poland in September 1939, Britain and France fulfilled their treaty obligations and declared war on Germany. There was little fighting on the Western Front, however, until May of the following year; then it took the German forces only 6 weeks to defeat the French Army. By the terms of the armistice signed on June 22, 1940, by Marshal Henri Pétain, Alsace-Lorraine was annexed to Germany, and two French zones were established. A Northern Zone was occupied and directly administered by German military forces, and an unoccupied Southern Zone was established, to be administered by Pétain and a French Parliament.*

Even before the armistice was signed, Col. Charles De Gaulle escaped to London and founded the Free French movement, and by the end of 1940 local resistance groups were also emerging. At first there was little regional coordination or national cooperation, although the British, largely through their Special Operations Executive (SOE), immediately began to support the resistance effort in order to supplement the Allied war operations.

In September 1941 the French National Committee was created in London under the leadership of General De Gaulle, and came to include spokesmen for the different political parties. In the spring of 1943 the French National Committee in London merged with the North African administration, headed by Gen. Henri Giraud, to form the French Committee of National Liberation. Thus the resistance organizations working outside of France were technically unified under De Gaulle. Efforts to unite the total resistance movement within France resulted in the creation of the National Committee of Resistance (CNR), composed of delegates from all the major underground groups in France. The titular political head was Jean Moulin, while a General Delestraints (Vidal)** was named leader of the Secret Army, the military arm of the underground, known officially as the *Armée Française de L'Intérieur*. The CNR collapsed in the fall of 1943, however, after both Moulin and Vidal were captured by the Gestapo.

Early in 1944 the name French Forces of the Interior (FFI) was adopted for all resistance units; the FFI was led by J. M. P. Koenig. Membership in French underground organizations totaled about 200,000 by 1943, approximately doubling a year later,[1] with perhaps about one-fourth being part of the armed *Maquis*. They operated in an area controlled initially by 500,000 occupation troops, police, and auxiliary forces.[2] These occupation forces were, of course, greatly increased as the threat of an Allied invasion became clear.

---

*After the Allied invasion of North Africa in November 1942, the Germans occupied the whole of France.

**Vidal was General Delestraints' underground cover name.

Cases of Undergrounds

The undergrounds in France contributed to the Allied war effort in several ways. The intelligence gathered by the French resistance gave the Allies detailed knowledge of German activity in France; the members of the French resistance gave invaluable assistance to Allied OSS and SOE agents. Downed Allied airmen and other military personnel were able to escape capture through underground escape and evasion nets. In addition, the French underground organized a secret army and directed preparations which led to a general uprising of the French resistance forces after D-Day.

Following the D-Day invasion the FFI Secret Army fought with the Allied armies against the Germans. The underground forces assisted the Allied invasion by sabotaging railway lines, roadways, and communication services. Additional FFI attacks were made against German troop concentrations and fuel and munition dumps. These actions created considerable confusion in the German rear, and were of value to the Allied Expeditionary Forces.

(UPI Photo)

Gen. Charles De Gaulle and Gen. Henri Giraud (left) in May 1943. The two Free French leaders created and assumed joint presidency of a council to govern liberated French territory and to lead their countrymen against the Axis powers.

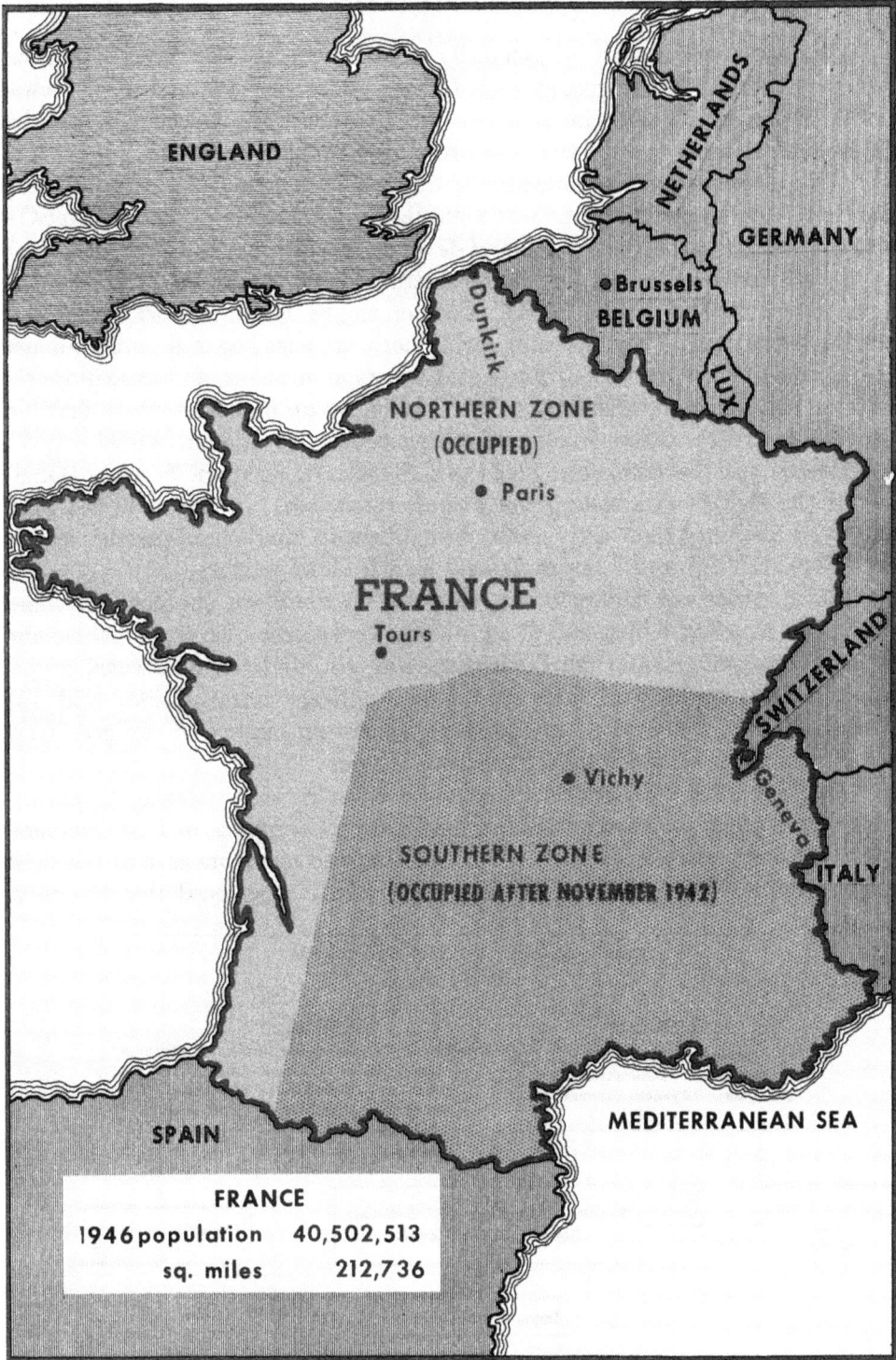

Figure 5. Map of France.

ORGANIZATION

The resistance movement generally reflected the prewar French political and social scene. Many groups were formed under the sponsorship of prewar political parties. Their aim was two-fold: to resist the German occupation forces and to work toward the overthrow of the Vichy regime.

Two agencies were established in London in the fall of 1940 to support the French underground. One was a special French section of the British SOE and the other was the *Bureau Central de Renseignements et d'Action* (BCRA), set up as part of General De Gaulle's general staff headquarters. Although nominally independent, the BCRA relied on the SOE for support in carrying out its operations. The following spring both agencies began to infiltrate men into France to perform intelligence and sabotage missions, and to explore the state of resistance groups already functioning, set up new ones where feasible, provide instructors, and work out escape routes for hidden Allied POW's and airmen. In the summer of 1942 the U.S. Office of Strategic Services (OSS) joined the British in assisting the French resistance. In late 1943 the SOE and OSS combined their activities and established a single headquarters known first as SOE/OSS and later as Special Forces Headquarters.

Three important underground groups in the Southern (unoccupied) Zone were: (1) *Combat*, which had at its disposal organizers, sabotage teams, and other operational groups; (2) *Libération*, with an intelligence network, propaganda office, professional cells, and a paramilitary organization; and (3) *Francs-Tireurs*, the weakest of the three, consisting mainly of refugees from the Northern Zone and intellectuals from Paris.

Early in 1942, BCRA undertook to coordinate the resistance movements in the unoccupied zone and created a Coordination Committee of four members: one representative from London (Jean Moulin) and one from each of the three major groups. On October 2 of that year BCRA announced that the three

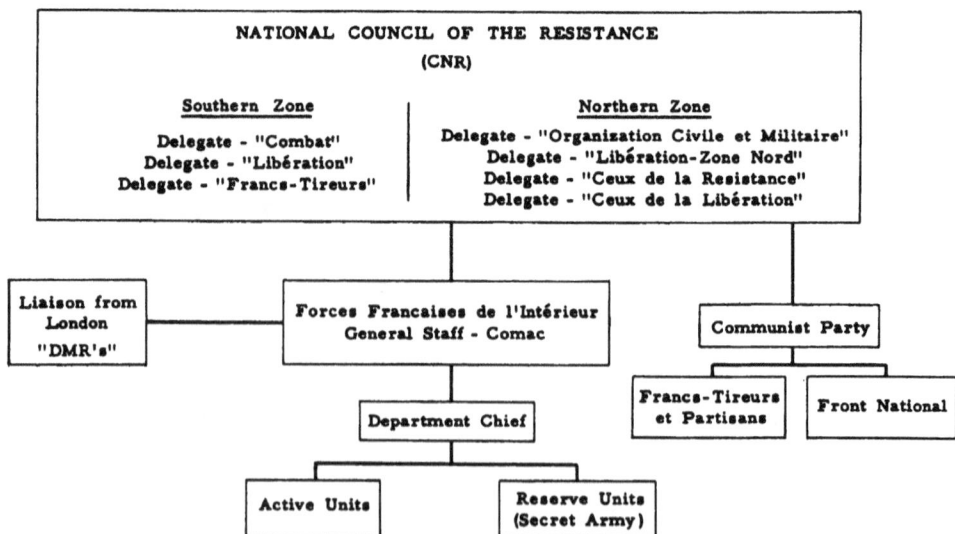

Figure 6. *Organization of Undergrounds in France.*

movements had recognized Moulin as the head of the resistance within France.[3] It further stated that the paramilitary organizations of the three would form a Secret Army acting under orders from De Gaulle. The staff of this army was to be structured along regular military lines, and on December 9, 1942, the command was given to General Vidal.

Before this synthesis, *Combat* had been organized geographically into six areas, *Libération* into seven. A uniform subdivision was now adopted which followed the departmental administrative system of the Vichy Government, under which the zone was divided into six regions. The coordination committee took the name of Committee of Directors, and assumed political control through regional and departmental political commissioners duly selected by the commanders of the region. Eventually, the committee assumed leadership of the Secret Army.[4]

In the Northern Zone, where pressures of the German occupation were felt directly, initial resistance groups immediately began such activities as collection of intelligence data and sabotage.

In February 1943, a *Comité de Coordination* for the resistance groups of the Northern Zone was set up under the guidance of General Vidal. The four largest non-Communist groups, *Organisation Civile et Militaire* (OCM), *Ceux de la Résistance* (CDLR), *Ceux de la Libération* (CDLL), and the *Libération-Zone Nord* (LZN) agreed to pool their forces in the Secret Army.

As in the Southern Zone, the occupied area was divided into military regions concordant with the prewar French administrative system.[5]

After the German occupation of the Southern Zone, the resistance groups in both zones were united. Jean Moulin persuaded representatives from all forces to establish the national Council of the CNR,[6] which was composed of representatives from all major underground groups, as well as advisers from labor unions and political parties. This council became the highest authority on French soil, but it recognized the supreme authority of the French Provisional Government in London. General Vidal was reaffirmed as commander of the Secret Army.

Just as a stable and unified resistance seemed to be taking shape, both General Vidal and Jean Moulin were captured. The various groups then became reluctant to pool their resources, and, in effect, the Secret Army ceased to exist before becoming a functioning unit of operation. In the spring of 1943 the Communists joined with the Secret Army to form what became the French Forces of the Interior (FFI). Initially the Communist *Front National* and its strong paramilitary unit, the *Francs-Tireurs et Partisans Français* (FTPF), withheld cooperation. Not until General Vidal's capture in late February 1943 did the FTPF send a representative to the Committee of Directors.*

The CNR continued to exist formally under the nominal leadership of Georges Bidault. A central committee of the resistance groups was created, with eight members. In February 1944 the central committee decided that

---

*The Communist Party had been the largest and best organized of the French parties before the war; because it had been outlawed in 1939, it had already established channels of communication which could be put to use for underground activity.

the FFI was to be placed under the leadership of the CNR. The general staff of the FFI was called the Committee of Action (ComAc). This organ was also supposed to arbitrate differences and misunderstandings between the military delegates (liaison agents sent from London, who had great influence because they controlled the flow of supplies to the French underground forces) and the local and regional underground organization. In May 1944 the name of the supreme staff was changed to *Comité d'Action Militaire de CNR*.

The organization of the lower-level resistance units is not well known. A regional directive of the Communist FTPF suggested the creation of hierarchical triangular cells, with each leader responsible to the cell chief of the next higher level.[7]

Paramilitary organizations set up by the various movements (in both zones) were organized within a territorial framework corresponding to the prewar French departments (or states), and were divided into active and reserve units. An active unit might be either a "normal unit," a "*Corps Franc*," or a "*Maquis*."* These troops, supervised by departmental chiefs, had as their major tasks general guerrilla warfare and sabotage. The *Corps Franc* executed traitors condemned to death by military tribunals. The *Maquis*, largely made up of youths who were evading the *Service de Travail Obligatoire* (STO) (forced labor in Germany), were usually permanent inhabitants of mountain retreats, and their prime duty was to stay alive until such time as they might be needed as fighters.

The reserve units were composed of older men and others who had not previously enlisted.

The Communist FTPF had its own independent organizational scheme. It included three kinds of paramilitary units. (1) groups of partisans, who continually carried on guerrilla operations against the enemy (2) *francs-tireurs*, grouped or isolated, who had legitimate occupations, but devoted their free time to the underground struggle and (3) "patriotic militia," who carried on sabotage and attempted to demoralize the enemy.

UNDERGROUND ACTIVITIES

ADMINISTRATIVE FUNCTIONS

Communications

A great deal of organizational effort was needed to establish a reliable communications network between the various organizational groups and geographical regions, as well as between the country and the London authorities. Although the courier system was too cumbersome, slow, and exposed to com-

---

*Maquis comes from the Italian word *Macchia* meaning thicket or undergrowth. It originally meant those resistance units in redoubt or mountain areas. Later it was extended to mean all those engaged in clandestine resistance against the Germans.

promise for use in everyday routine communications, couriers—crossing neutral frontiers with forged papers—were dispatched in considerable numbers until the liberation.

In the early days of the war, mail was sent by carrier pigeons and fishing vessels. Spain was considered to be a safe area through which mail could be transmitted to British authorities in Gibraltar. By this route "2622 messages were dispatched from France in April 1944 alone." [8]

The first trained radio operators were sent to France in the spring of 1941. Half of these were caught and executed by the enemy. In theory, these radio contacts with London did not seem to present problems; in practice, however, the establishment of a workable system proved to be quite difficult. Common problems were the lack of replacement parts for equipment, and heavy loss of personnel. By 1943 the efficiency of radio communications and the security of the operators were greatly improved. Such innovations as the so-called "V plan," which allotted operators a variable time and frequency schedule, and the introduction of a system of "blind messages," which were sent at certain hours to certain operators without the latter being required to confirm their reception, confused German listening posts. The Germans could not tell whether the operator heard and located in a certain spot at a specific time was the same one heard later at a different location. [9]

Written messages were often deposited at private homes, coffeehouses, stores, and "letter-boxes" to be picked up by relaying couriers or other agents and forwarded to their destination.

Since the security problems involved in transmitting written information were enormous, secret meetings were held among members of the various levels of organization.

Recruitment

The *Organisation Civile et Militaire*, the most highly organized of the resistance groups, drew members at all levels from the railway service (SNCF), the post, telephone, and telegraph service, and other public utilities. Non-political in structure, its effective membership was estimated to be 40,000. [10]

Ceux de la Résistance, a paramilitary organization with its greatest strength in Paris, recruited members from all classes. Its membership in 1943 was estimated to be 1,000 and its leader stated that it could have 35,000 troops ready by D-Day. [11]

Ceux de la Libération, another paramilitary unit concentrated chiefly in and around Paris, had a maximum strength of perhaps 35,000. [12] Its staff included *Routiers* (members of the French Teamsters Union), and it had agents among the police and the fire brigade.

In view of the lack of records and the loose structure of the organization, the figures for the membership of the French underground after its unification into the FFI may not be accurate. In the Northern Zone there were estimated to be about 182,000 "enlisted underground," 20,000 *Maquis*, 59,000 OCM and CDLR, and 200,000 reserves. Strength in the Southern Zone was estimated at about 47,000 "enlisted underground" and 50,000 reserves. To these figures must be added the ORA (*Organisation de Résistance de l'Armée*), units of the prewar French Army whose strength in September 1943 was estimated to be 30,000 in the Southern Zone and 10,000 in the Northern Zone.[13] As late as June 1944, however, the FFI admitted that it did not possess reliable data concerning the numerical strength of its forces.

No numerical data is available for the strength of the various youth groups, but it is known that one of the largest was the Catholic Association of the French Youth (ACJF). Its activities were suppressed in the Northern Zone, but permitted under Vichy rule.

Although the number of Jews in France did not exceed one percent of the population, they were one of the most active elements in the French resistance movements. They made up 20 percent of the membership of the Southern Zone resistance groups *Combat* and *Libération*, and there were other groups in which Jewish membership amounted to 25 percent of the total. Although their situation became particularly dangerous after the Germans ordered them to wear the Star of David on their clothing (May 1942), and though threatened with complete annihilation, they still played prominent roles in the underground. Of the supreme command of the FFI, two members were Jews. Several small independent Jewish groups arose soon after the armistice. At the end of 1942 several of these groups in the Northern Zone became part of the Communist paramilitary force, FTPF.

A large percentage of other ethnic groups (Poles, Italians, Germans, Austrians, etc.) participating in the French resistance could be attributed to two factors: considerable prewar immigration and the wartime influx of political refugees. The Communists had tried persistently to gain the sympathy of these people through their prewar industrial organization, the Organization of Immigrant Workers. The best-organized national group were the Poles. Approximately 82,000 Polish officers and men had reached France in 1939 and had taken part in the Battle of France. After the surrender in 1940, these, with the help of Free French authorities, formed an extensive organization designed to evacuate their own men as well as Allied airmen stranded in Europe. Prewar Polish immigrants also created the Polish Organization for the Fight of Independence (POWN), which cooperated with the FFI and maintained close contact with Polish authorities in London. It even received its own supplies by Allied parachute drop. In 1944 it became integrated with the FFI with the understanding that control would revert to Polish authorities after the liberation. The thousands of Germans and Austrians who had found political asylum in France after 1933 also formed underground units. Small national groups of Italian deserters and anti-Franco Spaniards were also active in the *Maquis* of the Southern Zone.

Finance

External Sources

Most of the funds for the resistance came from outside France, mainly from London and Algiers. Funds sent or brought to France by French, English, and United States agents consisted of French francs, Free French treasury bonds, and U.S. dollar notes. Before the summer of 1943, only French francs and U.S. dollars were used, but as the francs were in the form of a deposit sent by the Bank of France to England before June 1940, the Germans were able to establish their serial numbers and hence to trace their source of distribution in France. Later, U.S. dollars were sent, as they could be easily and profitably exchanged for francs within France. If the handling of dollars became risky, they were sometimes taken to Switzerland, where they could be exchanged for francs.[14]

Internal Sources

The resistance was able to raise a considerable amount of money in France itself. The sources for this income were—

(1) *Expropriation of public funds.* This operation, ordered by the FFI authorities and carried out by local resistance organizations, was directed mainly against local branches of the Bank of France, tax collectors' offices, and similar public institutions. Many staff members of these institutions cooperated with the underground. At first, receipts were made out by the underground for all expropriated funds, but this was soon stopped, as bandit groups would pose as legitimate agents and collect funds.

(2) *Donations.* Because banks and commercial enterprises found it difficult to conceal their donations from the German authorities, the amounts received from such sources failed to meet the underground's initial expectations. In addition, donations from individuals were not significant.

(3) *Internal loans.* Obtaining loans from wealthy individuals, and occasionally banks, became the responsibility of the *Comité de Financement* (COFI), which was established in February 1944 under the auspices of the CNR.

Logistics

An extensive technical apparatus was needed to assure proper functioning of supply operations carried out clandestinely by parachute drops or airplane landings. Considering the technical details involved, the necessity of using ground crews who could not be trained properly, and the tremendous proportions which these airlift operations assumed toward the end of the war, the results achieved were remarkable.[15]

A primary task was to find suitable ground for parachute drops and landing operations. Grounds for these operations had to be makeshift, for the regular airports were controlled by the enemy. They did, however, have to meet certain minimum requirements. They had to be large enough, not too close to enemy troop encampments, not located in sectors of extremely dense antiaircraft implacements, and accessible to resistance personnel. When such an area had been found, its coordinates and bearings, along with the pertinent conditions of the surrounding territory, were submitted to the authorities in

England for review. It was the responsibility of SOE or BCRA specialists in the field to verify the accuracy of these specifics.

In London, the submitted details were closely scrutinized. The computed risk of losing a plane and its crew was weighed against the benefits to be derived from supplying the particular resistance group. The British Air Ministry had to approve the proposed drop zone. When a site was approved, a ground file was set up, including a map of the area, an operation number, and a BBC message and recognition letter for communication with the recipient group in France.

In France, a Reception Committee, or "R.C.," was then organized. The R.C. leader was given the BBC code phrase and instructed to listen to the proper BBC broadcasts in order to receive notice of the incoming shipment. The resistance group's headquarters would also listen to the broadcast and notify the R.C. leader when the signal was received, in case the R.C. had missed the signal. At the site, pits were dug and camouflaged so that the received containers could be temporarily stored. This enabled the Reception Committee to make a quick withdrawal after the drop, a maneuver often expedient because a circling plane often aroused German suspicions and prompted searches of the area by German patrols. Then, too, temporary storage was often necessary because there was not enough nighttime left to conceal the transportation of the containers. After a site had been used once, its safe use greatly decreased, so resistance groups changed sites frequently.[16]

The "specialists" of the SOE and BCRA were organized into committees called COPA (*Comités des Opérations de Parachutage et d'Atterrissage*). They not only supervised the preparation of parachute and landing grounds and the actual operations themselves, but were also in charge of distributing properly the parachuted supplies. The COPA representative also inspected security measures and caches prepared by the FFI for hiding containers, etc., and contacted the crews designated to assist in the operations. The preparation of aerial operations thus included—

(1) The preparation of suitable terrain in a distance of at least 200 to 400 meters from the nearest buildings.

(2) The assignment of a team of four to five reliable men with a radio transmitter.

(3) Transportation of parachuted supplies.

(4) Designation of temporary hiding place (cache) such as a forest, a thicket, a barn, or a camouflaged excavation.

A minimum of 20 to 30 such grounds had to be prepared for each department; approximately one ground was needed to supply two *trentaines* (groups of 30 men). Instructions were issued as to the technical procedure to be followed during an actual air-supply operation. Such an operation would be announced by a previously agreed-upon code-phrase on the BBC, whereupon ground crews would be alerted, and lamps arranged in a certain pattern would be lighted when the airplane could be heard. If the airplane was supposed to land, passengers to be picked up would wait in readiness at an exactly designated point of the provisional field.

Toward the end of the war, new methods of directing airplanes to their targets were used, such as directional radio beams and ground-to-airplane radiotelephones. However, these new devices were never available in sufficient numbers.

Security

An effective security system, so indispensable to a successful underground organization, was not a strong-point in the French resistance movement. The Communists, who had been forced underground in the summer of 1939 and had developed effective security procedures, were the resistance group best protected against enemy penetration and detection.

An "Instruction from the ComAc," issued in August 1944 and directed to all regions and departments of the FFI, advised on the basis of past mistakes and examples how the underground forces should be deployed:

Do not make the mistake of constituting excessively large units and to give them too rigid an organizational frame.

Except for the national leaders of the underground, members of the resistance should not travel outside their region of action.[17]

The size of the various resistance groups was not, however, dependent on the will of the organizers: there were many factors which caused a confluence of men in certain areas, and they had to be taken care of whether the leaders of the region liked it or not.

The need for security measures increased sharply when the resistance was flooded with young men who were evading forced labor conscription by the Germans.

Throughout 1943, officials from the SOE and BCRA traveled through France. Before leaving England, each of these agents went through detailed preparations, and was given a series of aliases, cover stories, and identity cards. In each place where he was to contact underground officials, he would be provided with a "safe house" in which to stay, and an alternate "safe house" if the first was known to the Germans. Each agent had a pill which he could take if subjected to unendurable torture; many of the officers within France carried these cyanide capsules for security reasons. Agents agreed beforehand that if one did not show up for an appointment, he was presumed to be captured. The arrested agent would wait 2 days before releasing any information to his captors, thus allowing the network time to reorganize itself. Especially important was changing the location of "safe houses."

As the number of parachute drops increased, it became more difficult to provide safe areas for drops and landings. Precision was immensely important, as the landing strip had to be outlined by fires at night. Usually, if an expected airplane did not arrive within 2 hours of the appointed time, the Reception Committee returned home. By 1944 the Air Operations Office in London and the FFI in France had built up a network of 600 landing grounds.[18] Although 150 reception committees stood by each month, an average of only five operations succeeded.

Cases of Undergrounds

Before the Allied invasion of North Africa, the Southern Zone which was still unoccupied by German troops provided a reasonable amount of security for underground workers in the Northern Zone who were in need of sanctuary. Since official permission was needed to cross the demarcation line, forged papers were necessary.

It must be understood that the French underground took for granted the passive, if not active, support of the larger majority of Frenchmen. Many of those considered traitors were sought out and eliminated.

OPERATIONAL FUNCTIONS

Psychological Operations

The FFI used propaganda to raise the morale of the native population and lower that of the enemy. The main effort in this field was the publication of clandestine newspapers. The National Library in Paris lists 1,034 illegal newspapers which were printed at least once during the course of the war. While the great majority of these did not publish consistently, several publications did, and these played significant roles. The major journals and the number of editions published were—

> *L'Humanité*—the official organ of the French Communist Party—(315 editions)
> *Les Volontaires de la Liberté* (99 editions)
> *Libération Nord* (190 editions)
> *Combat* (78 editions in all, including some published under other names)
> *La Voix du Nord* (65 editions)
> *Libération-Sud* (52 editions)
> *Défense de la France* (47 editions)[19]

Information for the newspapers was obtained from the British Broadcasting Corporation and Swiss Radio Geneva. Large supplies of paper were required for resistance psychological operations. *Combat* alone, for example, accounted for three tons per month in 1944.[20] Many persons in the printing trade offered paper when it was needed and some was obtained from paper factories; large amounts were also purchased on the black market. One clandestine editor set up a commercial shop which was sanctioned by the Vichy authorities, and imported paper from Germany.

The problem of printing was largely solved by the cooperation of printers, who offered their services at a great risk. Since it was necessary to shift installations from one location to another, it was ideal for the editor to have his own machine. The editor of *Combat*, for example, set up his office in a cave near Lyon.

The printing methods of *Défense de la France* are probably typical of those of the larger publications. First mimeographed, later printed on hand and mechanical presses, this paper finally acquired a linotype machine and even published illustrated supplements.

Many of the illegal publishers had the cooperation of transportation agencies and government workers in solving the circulation problem. For example, workers of the French National Railways would see that the paper was deposited at the various stops on their trips, and would also allow men with suitcases full of newspapers to travel free on the trains. Post office facilities were also used for circulation, and newspapers were circulated in churches and in the Paris subway.[21]

Most of the illegal newspapers carried the *Croix de Lorraine*, signifying their allegiance to General De Gaulle, and each one had an eye-catching patriotic slogan to appeal to the readers.[22] For example, *Combat* had the slogan "One Chief, De Gaulle; one fight, for our liberty." The newspapers contained three sections. One gave news of Allied military accomplishments gathered from radio news sources in Britain and Switzerland. Another section was concerned with political propaganda; it might include a list of patriots arrested, sabotage missions accomplished, etc. The third, which was sometimes in code, would give directions to guerrillas and other patriots.[23] These newspapers were encouraged by Allied authorities, although no information is available on possible direct financial and material support for this operation from London. Many of these clandestine papers became legal publications after the liberation.

Other propaganda items were the posters given to members of the various youth groups to attach on walls.

A unit responsible for carrying on propaganda against the enemy was created in 1942. Known as the TA (*Travail Allemand*), it was composed largely of groups or individuals who had access to official German military and civil sources. The objectives of the TA were—

> . . . anti-Hitler propaganda, whether written or verbal, and the formation of a national organization on a broad basis for peace inside the Wehrmacht and the German administration.[24]

Pamphlets were distributed clandestinely among the German troops. Some of these may have been printed in France itself, but most of them were printed in Great Britain and dropped by parachute into France.

Minority groups within the enemy ranks constituted a target for underground propaganda. Once a loudspeaker truck was used to broadcast desertion appeals to Russian soldiers with the German forces, and this propaganda may have been a factor in the removal of these troops from the area.

One underground unit, the Communist FTPF, issued instructions in June 1944 on themes to be used in propaganda material directed toward the German troops. This pamphlet, called *Comment Faire La Guerre* (How to Wage War) contained desertion appeals, information about the collapse of the German front in the East, and other war news.[25]

Publicity on behalf of the French cause was directed at Allied and other nations by the Free French authorities in London.

Cases of Undergrounds

Intelligence

One important contribution to the Allied cause by the French undergrounds and individual agents working in France was in the field of intelligence. It was this intelligence which provided the Allied commanders with information which greatly aided them in planning the details of the invasion of June 1944.

Whereas the various resistance groups, with the exception of the Communist National Front, were generally confined to their own regions, the network which specialized in intelligence extended throughout France. Altogether, about 100 different intelligence networks operated at various times, new ones being created from the remnants of the old.

The earlier networks were organized by British intelligence services. They were formed under the direction of British agents who had remained in France after the armistice, or under British or French agents secretly dispatched to France, with contact addresses. As early as November 1940, De Gaulle sent General Roulier to France to organize a French intelligence apparatus in the occupied zone, with branches in the unoccupied zone to provide avenues of withdrawal. This network was later called the *Confraternité Notre-Dame*. Another network organized about this time was the *Alliance*. All major political parties produced their own organizations, a number of which were built up under the direction of the United States OSS. The French section of SOE also parachuted into France 366 French, English, and Canadian agents, 80 of whom died in landing, and 15 of whom were killed in battle. It has been estimated that about 100,000 persons, including 35,000 women, participated in these "special services." [26]

The significance of the OSS and SOE intelligence operations was that Allied conventional forces used a net of underground workers to support their intelligence operations. In many cases, it was difficult for the underground to provide specific pieces of intelligence since they were not always informed of Allied strategic plans. Therefore Allied agents parachuted into occupied territory and contacted the underground. The underground provided food and shelter and assisted the agent in obtaining the necessary information. Once the information was obtained, the agent would be smuggled out of the country through an escape and evasion net bringing with him any necessary maps and documents.

Sabotage

An important task of the underground forces in France—and one of the main reasons for their existence, as far as the Allies were concerned—was sabotage and armed diversion behind enemy lines. To assist in this objective, a series of contingency plans were drawn up. These "color plans" were developed to provide targets for a program of widespread sabotage against German war materiel. Comprehensive lists of targets were dropped into France in colored containers, each color indicating the type of target to be attacked. The BBC notified the underground by code the time of the initial landings on

D-Day and the color plan to be used, thus signifying the time and type of attacks to be carried out. Although seven were originally called for, only three of the plans were actually carried out; those against the railway system, against roadways, and against railway terminals.

Training programs were set up in Great Britain to brief agents on sabotage techniques for use in attacking such targets as railroads, locomotives, highways, power lines, canals, etc. Also manuals on techniques were published and distributed to underground personnel.

Major sabotage operations began after the invasion of June 6, 1944. It was the responsibility of FFI saboteurs to interrupt the flow of German troops and equipment to the coastal areas where Allied troops had landed. To this end the Free French authorities in London, in close cooperation with the Allied General Staff, assigned two main tasks to the French resistance: to sabotage the German war effort and protect, from destruction by the Germans, objects and installations in France which would be valuable for future Allied operations. Both sabotage and countersabotage were to be carried out by the underground. The French authorities in London felt that two considerations had to be borne in mind in selecting targets for sabotage: wherever it could be avoided, facilities that could contribute to France's postwar economic revival should not be destroyed, and unnecessary losses of the population had to be kept to a minimum in order to maintain the morale of the resistance.

In certain cases an understanding was reached between the local underground leaders and patriotic managers of plants and public services as to where sabotage should be carried out. To encourage this cooperation, the British gave assurances to the French authorities in England that plants blown up to further the Allied war effort would be given priority when reconstruction began after the war.[27] Managers of plants and chiefs of various public services (e.g., cable lines, roads, and bridges) were asked to indicate the most vulnerable targets and to provide the resistance with maps and plans of facilities under their charge. Several plant managers who heretofore had collaborated with the Germans now cooperated with the underground in order to establish for themselves a record of patriotism.

Escape and Evasion

The large number of British troops abandoned at Dunkirk and the thousands of Jewish *émigrés* attempting to evade the oncoming Germans constituted a serious responsibility for the French underground. Others who required the use of escape routes were Allied personnel, escaped POW's, deserters from forced labor camps, and French underground officials who desired to confer with Allied personnel in London. Under the chaotic conditions which existed, the first escape routes developed spontaneously. Refugees were usually put into family homes, and then placed in the hands of a *passeur* who would conduct them to Spain, Switzerland, or directly by boat to Gibraltar or England. Some measure of coordination was achieved between the escape nets of Belgium and France. There were two main escape routes out of Belgium. One went from Brussels through Châlons-sur-Marne, Lyon, Grenoble, and

Toulouse to the Spanish border. The fugitives were given false papers and dressed in clothing similar to that of the natives. The other route went through Paris, where the escapees changed their false Belgian papers for false French ones.[28] In time, and with experience, the underground devised means of providing false papers, "safe homes," and security measures to ensure the relative safety of the refugees.

COUNTERMEASURES

As already noted the Germans divided France into two zones. In addition, Alsace-Lorraine was annexed to Germany. The Northern Zone, being of extreme strategic importance, was placed under the direct authority of German occupation officials, while the Southern Zone, unoccupied by German troops, was administered by officials approved by Marshal Pétain and the newly created "Vichy Government."

Activities aimed at detecting and annihilating the underground resistance groups in Northern France were directed by the *Geheime Staatspolizei* (Gestapo). Besides the intelligence work of their own agents, the Gestapo received the cooperation of many French Fascist youth and sports organizations, as well as of collaborating individuals.

Because of the laxities of security measures of the FFI, Gestapo agents were successful in infiltrating many of the underground networks. Interception of written messages and breaking of codes often provided agents with access to clandestine channels. Once they had infiltrated the system, they often learned underground procedures and met personnel involved in the resistance effort. In the Southern Zone, several small groups of people who had tended to be anti-British before the war or who sincerely hoped for a French-German understanding actively collaborated with the Germans. The greatest danger for the French resistance in this zone came from such militant organizations as the *Milice Française*, which became an auxiliary police force supplied with arms by the Germans. Composed mainly of ambitious pro-German sympathizers, *Milice Française* also had a youth and women's auxiliary. Other pro-Fascist groups were the *Parti Populaire Français*, the *Garde Mobile Républicaine* (GMR), and the *Légion Française des Combattants*.

The Germans also used considerable numbers of Frenchmen as undercover agents. Most were attracted by immediate prospects of material gain. Often this would be something such as gasoline, which they would then sell on the black market.

The Vichy Government, sanctioned by the Germans but administered totally by French officials, provided legal justification for many to cooperate with the government. Moral justification was given by Marshal Pétain, who exerted tremendous influence and authority during the first years of the war.

It was because of such individuals that time after time various groups of French underground lost their leaders, and that radio transmitters, supply dumps, or *Maquis* redoubts were betrayed to the Gestapo and underwent seizure and liquidation. It was largely because of them that the Gestapo almost succeeded in preventing the unification of the resistance and actually delayed it for a year when, with the help of an informer, they captured Jean Moulin and General Vidal, the civilian and military representatives of General De Gaulle. Nevertheless, it is estimated that no more than 3 percent of the population ever actively collaborated with the enemy.[29]

FOOTNOTES

1. Marcel Vigneras, *Rearming the French (U.S. Army in World War II)*, Office of the Chief of Military History, Department of the Army (Washington: Government Printing Office, 1957), pp. 304–305.
2. F. O. Miksche, *Secret Forces* (London: Faber and Faber Limited, 1950), p. 73.
3. War Department Special Staff, Historical Division (Historical Manuscript File) *French Forces of the Interior—1944* (General Services Administration, Federal Records Center, Military Records Branch, Washington, 1944), pp. 106–108.
4. Ibid., pp. 25–30.
5. A map containing delimitations of military regions of French resistance as prepared by the *Commission Militaire de la Résistance* for the *Bureau Historique de l'Armée* in Paris can be found in the French Resistance Collection in the Hoover Library, Stanford University, California.
6. French Resistance Collection (Hoover Library, Stanford University, California), Folder 24, No. 7.
7. War Department, *French Forces*, p. 320.
8. H. Michel, *Histoire de la Résistance* (Paris: Presse Universitaire), p. 77.
9. War Department, *French Forces*, p. 419.
10. Ibid., pp. 28–29.
11. Ibid., p. 28.
12. Ibid., p. 25.
13. Ibid., p. 29; see also French Resistance Collection, Folder 5, Nos. 4 and 5.
14. E. Reval, *Sixième Colonne, Un Grand Peuple Lutte pour sa Libération* (Paris: Thonon, S. E. S., 1945), p. 382.
15. Ibid., p. 44.
16. Ibid., pp. 379–380; see also War Department, *French Forces*, p. 917.
17. French Resistance Collection, Folder 5, No. 13.
18. War Department, *French Forces*, p. 44.
19. Raoul Aglion, *The Fighting French* (New York: H. Holt & Co., 1943), p. 271.
20. Mme. Granet, "La Presse Clandestine en France" in *European Resistance Movements, 1939–1945* (New York: Pergamon Press, 1960), p. 184.
21. Ibid., pp. 187–189.
22. Ibid., p. 189.
23. Ibid.
24. Free German Movement in Great Britain (ed.), *Free Germans in the French Maquis: The Story of the Committee "Free Germany" in the West* (London: I.N.G. Publications, 1945).
25. D. Knout, *Contribution à l'Histoire de la Résistance Juive en France, 1940–1944* (Paris: Editions du Centre, 1947), p. 102.
26. J. Joubert, *La Libération de la France; Comment la France Fut Occupée; Comment la France Fut Liberée, 1940–1945* (Paris: Payot, 1951), pp. 69 ff.
27. WD–FFI, War Department, *French Forces*, pp. 1011, 1326–1327.
28. George Tanham, "The Belgian Underground Movement 1940–1944" (Unpublished Ph. D. dissertation, Stanford University, California, 1951).
29. U.S. Office of Strategic Services, "French Pro-Fascist Groups" (U.S. OSS Research and Analysis No. 1694, Washington), p. 2.

CHAPTER 7

YUGOSLAVIA (1941–45)

BACKGROUND

In preparation for the invasion of the Soviet Union, Hitler secured a neutrality treaty with the Kingdom of Yugoslavia on March 25, 1941. Reaction within the ranks of the Yugoslav Army was quick. Two days later, high ranking officers successfully overthrew Prime Minister Dragisa Cvetkovic and revised the neutrality pact. This coup d'etat forced Hitler to delay his previous plan to invade the Soviet Union: he now had to send troops into the Balkans in order to secure his southern flank. German troops attacked Yugoslavia on April 6, 1941, and within 2 weeks, the Royal Army collapsed and the government went into exile. Within several months the Axis powers had dismembered the country. Parts of Slovenia were annexed to the German *Reich;* Bulgaria incorporated Macedonia; Hungary annexed the fertile northwest province. A Croatian dream was realized when an "independent" Croation state * was created under the leadership of Ante Pavelic. He was permitted to form a regular army, the *Domobran,* and his armed Croat *Ustashi* ** became the terrorist group which was to eliminate over one-half million Serbs, their traditional rivals, and almost all Jews by the end of the war. Although German military occupation authorities were in control in Serbia, a puppet government was formed under the leadership of Milan Nedic, who had at his disposal a small military force called the Serbian State Guard, and some legal *"Chetnik"* *** units.

Military control was maintained throughout strategic points of the country by the presence of about 200,000 Axis troops. The German occupation was complicated by three conflicting policies. The Italians, who were aware of the Croatian nationalism and desire for the return of Dalmatia, which the Italians occupied, used Serbian *Chetnik* bands against Pavelic's Croatian raiders and later against the Communist-led Partisans. The Italians were not opposed to the Serbs in particular, but they did oppose the Croatian territorial claims. The Nazi leaders supported the Croats since they suspected the Serbians of being responsible for the 1941 coup, and objected to the Italian aid to the Serbs. The local German military commanders favored neither the Serbs nor Croats and were not opposed to supporting legal Serb *Chetnik* units to fight the Communist-led Partisans.

---

*Which immediately declared war against the United States.

**The *Ustashi,* a Fascist underground movement, was formed by Pavelic during Alexander's reign and after the German occupation became a military force comparable to the German SS troops.

***Cheta is Serbian for an armed band and has been used to describe those small guerrilla bands which have been important throughout Serbian history, especially in fighting guerrilla war against the Turks. Later, the *Chetniks* became an official organization which trained its members in guerrilla warfare in order to use them as a paramilitary organization. In 1941 Kosta Pecanac, leader of the *Chetnik* organization, went over to the Germans. Those who followed him were called "legal" *Chetniks* and were later incorporated into Nedic's armed forces. Those who refused to follow Pecanac and joined Mihailovic were termed "illegal" *Chetniks.*

Cases of Undergrounds

Marshal Tito and his staff at Partisan headquarters in the Yugoslav mountains. Left to right, front row: Colonel Filipovich, Edvard Kardelj, and Marshal Tito. Back row: Radonja, Tito's secretary; Chalakovich, Secretary of the National Anti-Fascist Council; Koebeck, Minister of Education; and Lt. Gen. Sreten Zujevich.

Gen. Draja Mihailovic (second from right) confers with his guerrilla commanders at his headquarters in Yugoslavia.

222

Organized resistance began in the spring of 1941, when Col. Dragoljub (Draja) Mihailovic retreated to the Bosnian mountains with a staff of Royal Army officers and some regular army units, which eventually became known as the Royal Army of the Fatherland. Not until June of that year after Germany attacked the U.S.S.R. did the Communist Party of Yugoslavia, (CPY), led by Josep Broz (Tito), begin resistance activities, which were to be the first stage of a revolutionary movement designed to win control of the government. Between 1941 and 1945, Tito and his Partisans, who were to number over 200,000 by the end of the war,[1] survived seven major military campaigns directed against them by the Germans.

The British, who maintained diplomatic recognition of the Royal Yugoslav Government-in-exile in London, approved of General Mihailovic being named Defense Minister in January 1942. Plans were made to assist his resistance movement, consisting largely of Royal Yugoslav soldiers and illegal *Chetnik* bands totalling at that time 10,000 to 15,000 men, supported by perhaps 50,000 active sympathizers.[2] At first the resisters were not encouraged to carry out large-scale military operations against the enemy. This was a reflection of the personal views of Mihailovic, who felt that too much armed action would provoke the Germans to reprisals which would be disastrous to the movement as a whole. The initial plan was to hold off any mass uprising until the Allied invasion of the Balkans.

In November 1942 the high command of the Communist-led Partisans announced the creation of the Anti-Fascist Council of National Liberation. To expand the movement they organized rural soviets in the "liberated areas." People in these areas were trained for work in various aspects of guerrilla and underground warfare. In the summer of 1943 Tito announced that his "council" was to be considered a "provisional government," thus hoping to present the Allies at the Teheran conference with a *fait accompli*. At this conference the Allies dropped their plan to invade the Balkans and began to increase their aid to Tito, while slowly abandoning support of Mihailovic. Allied support of the Partisans ultimately contributed heavily to the collapse of the non-Communist resistance movement.

By the summer of 1943 civil war replaced resistance as the primary activity of these two clandestine armed forces. Conflict or collaboration with the Axis became merely a tactical maneuver within a broader atmosphere of civil strife. In addition, the forces of Mihailovic and the Serbian population of western Yugoslavia were constantly threatened by the Axis-armed Croat *Ustashi*.

The entry of Russian troops into Yugoslavia in 1944 enabled Tito's forces to gain a foothold in Serbia for the first time since the outbreak of the war. Tito and his Partisans entered Belgrade with the Red Army. It was only a few months later that he announced the creation of a "united front" government, which was soon transformed into a Communist dictatorship.

Figure 7. Map of Yugoslavia.

ORGANIZATION OF THE NATIONAL LIBERATION MOVEMENT (PARTISANS)

The origin of the Partisan movement can be traced back to the clandestine activities of the Communist Party of Yugoslavia (CPY), which was outlawed in 1921 and forced to go underground.

In 1923 a school for Yugoslav Communists was opened in Moscow. Most of the original students were non-Serb. By the time of King Alexander's death in 1934, these men had established regional committees of the CPY in all parts of the country.

The first test of party organization and efficiency came during the Spanish Civil War. The Communists clandestinely recruited volunteers who were then channeled to the ranks of the Republican Army in Spain. Once there, party members received practical experience as political commissars and military tacticians.[3]

In 1937 the CPY had 1,500 members; in 1939 the membership was 3,000; in 1940, 6,000; and by the beginning of 1941, 12,000.[4] When the CPY began resistance activities against the Germans, following Hitler's invasion of the Soviet Union, the major party leaders (including party secretary Tito)—all of whom had studied in Moscow—were appointed to positions of responsibility. The most important of these leaders were Edvard Kardelj, who was assigned to direct actions in Slovenia; Milovan Djilas, who was sent to Montenegro; and Svetozvar Vukmanovic, who stayed with Tito in Bosnia.

The major task of these men was to direct the "liberation" of territory nominally under German occupation, and convert it into "liberated areas" under CPY control. The first step in this procedure was generally for the party to organize groups of citizens—who were usually sympathetic to, although not members of, the Partisans—as "terrain workers." These individuals were responsible for collecting information on enemy troop movements, and anti-Communist individuals and their activities.

As the "national liberation movement" expanded, the Communists were able to lay the groundwork for control of the civil affairs of each of the "liberated areas." The basic administrative unit became the National Liberation Committee.* "Liberation committees" were soon found on all levels: local, district, and regional. Although not all of the members were Communists, party members controlled the actions of the committees.

In the liberated areas, the task of civil administration had become urgent. The national liberation committees filled this vacuum by providing schools, courts, and local governments. They also recruited men for the Partisans, set up primitive industries, and gathered food and supplies for the guerrillas. In theory, these committees were elected. However, in practice it was common to find that the only candidates were Partisans and the electors were usually

*National Liberation Committee (NLC) was a term used on both the national and local levels. The National Liberation Committee, created at Jajce in November 1943, was in effect the cabinet of the Communist provisional government. The local National Liberation Committees set up in "liberated areas" behaved in a manner similar to a Russian soviet.

Cases of Undergrounds

limited to Partisan supporters. The committees did, however, have some degree of spontaneity and in many cases recruited efficient local people who eventually won local support by their effective jobs.[5]

The Communist Party also created national liberation committees in many occupied areas for the purpose of fundraising and recruiting for the Partisans in the mountains. They were instrumental in smuggling food and medical supplies to the guerrillas, and providing them with intelligence information, as well as performing sabotage against railroads. Since food was extremely scarce in the Partisan-held regions of Bosnia-Hercegovina, Dalmatia, and Montenegro, the underground had to smuggle supplies from the peasants in the fertile plains of the Sava Valley to the guerrillas in the mountains.[6] When an area was under control, youth leagues, women's anti-Fascist councils, cultural centers, and Communist-armed militia were established.*

As the war progressed, provisions were made to establish more and more local soviets in towns which were "liberated" by the Partisan forces. Even if they could only hold a town temporarily, the administrative machinery was set up for eventual control by party personnel.

Largely because it its long experience in clandestine operations, the Communist Party was able to organize an effective underground to aid the Par-

---

*The term "Communist" was not often used in connection with these newly created organizations, as it suited the purpose of the party leaders not to associate themselves with the Soviet Union too quickly. When some overanxious party leaders in Montenegro proclaimed that province to be a Communist "Soviet Republic," they were immediately reprimanded.

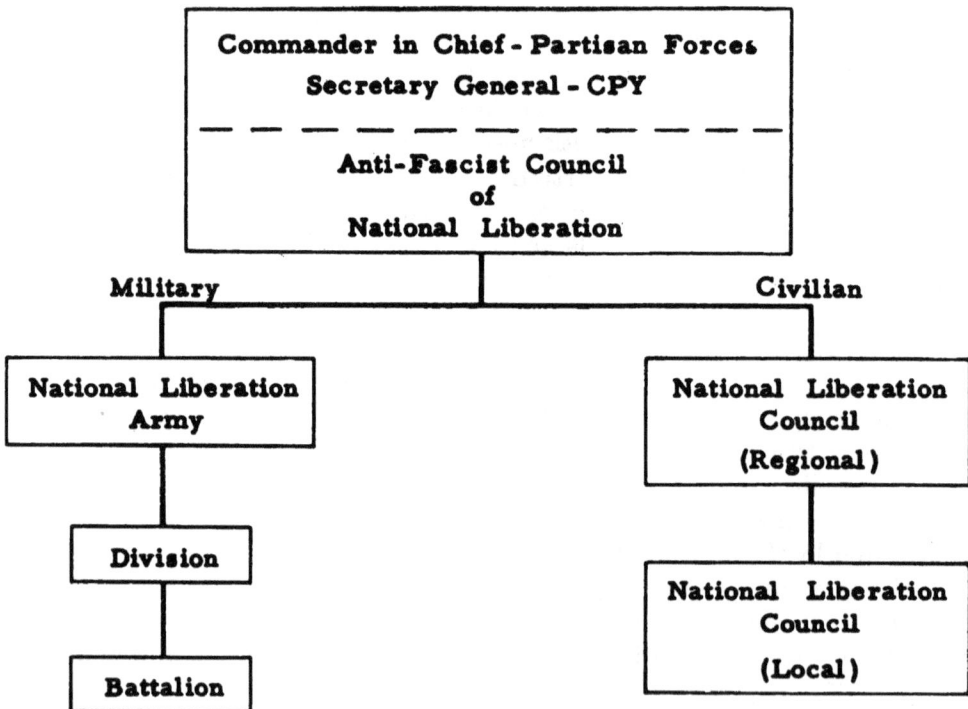

Figure 8. Organization of National Liberation Movement.

tisans. Specific targets for party infiltration were labor unions, the Royal Army, the government bureaucracy, student groups, and other political parties. The illegal existence of the CPY had allowed Tito, by 1941, to head the only political party in Yugoslavia trained in the techniques of underground existence, and which had a professional following in all areas and many institutions of the country.

The conditions of the war allowed the Communist Party to broaden its range of activities. The most significant of these was to include recruitment of non-Communists in a nationwide resistance movement, aimed initially at harassing the Axis occupation forces, and ultimately at bringing about a successful revolution.

The Communist desire to enlist non-Communist support led to the development of the "united front." In presenting the united front as a joint effort to oppose the Axis occupation forces, and hence to serve the population as a whole, the Communists wished to show the difference between their organization and that of General Mihailovic, which they characterized as a movement designed to further the interests of Serbian nationalists. The Communist Party succeeded in rallying elements of various political and cultural groups to the National Liberation Front (NLF); however, it was the only political organization which entered the NLF as a whole.

In November 1942, the first meeting of the Anti-Fascist Council of National Liberation (AVNOJ) took place at Bihac, in Bosnia. Delegates representing the various local liberation committees attended this meeting from all over Yugoslavia, and here Tito proclaimed a "popular front against fascism" and a National Liberation Army. This could not yet be called a representative governing body. Its purpose was to unite already existing NLC's and to form new ones. Since party members organized these front groups, they always assumed the leadership positions.[7]

One year later, on November 29, 1943, Tito was appointed Supreme Marshal of the National Liberation Army, and President of the National Committee of Liberation, with all the powers of a provisional government. Tito officially excluded the officials of the exiled Royal Government in London from participation in activities of this movement.

Tito's military forces were initially organized into regular "Partisan" units and "proletarian" units, usually brigades. The former was made up largely of party members, veterans of the Spanish Civil War, and other patriots who could not, or did not want to, join the forces of Mihailovic. The latter were made up primarily of students from the Universities of Belgrade, Zagreb, and Ljubljana. By 1943 there were about 60,000 armed men under Partisan command.[8]

After the Allies began to supply Tito and his resistance organizations on a large scale, participation in the Partisan movement increased significantly. A reorganization of the military took place at this time.. The Partisan forces were drawn up into 11 "Lenin Brigades." Later, the military command system included armies and corps. By the end of the war in 1945, Tito claimed to have a force of 200,000 men.

Cases of Undergrounds

The CPY dominated an actual *de facto* government which ruled large portions of Yugoslavia. The entrance of Russian forces into Belgrade in September 1944 enabled Tito to assert full control. Although only a small fraction of either the Partisan fighters or the members of the local governing committees were actual Communists, and although many people on the lower level of both the underground and the armed forces were not Party members, the Communists were able to impose their will on the people by obtaining command of key posts at all levels. By providing people with such necessities as food, schools, and security, as well as carrying out active resistance against the Germans, the Communists eroded the population's will and ability to resist the Partisan organization.

The Partisan strongholds were located in the central mountain areas. Outside of this area they carried on considerable political and military activity in Slovenia, Croatia, and Macedonia. In Slovenia they formed a liberation front which was led by Communists and supported by various liberal groups. From this front group, units were formed and maintained close contact with Tito's mountain headquarters. In Croatia there had been a Communist underground organization from the beginning of the war, but no large-scale units were formed until the summer of 1942. The Yugoslav and Bulgarian Communist Parties both claimed the right to lead the Macedonian resistance. The Communist underground organization exploited the promise of an independent Macedonia, and by the summer of 1943, under Tito's leadership, Macedonian partisans were operating in this area.[9]

The initial steps of attaining power had been taken during the war. From 1941 until the end of 1944 the CPY consolidated its control over all agencies of the National Liberation Front. Through the various local liberation committees, they had gained political control of the country. By the end of 1944 any potential opposition had been eliminated or neutralized through the "people's fronts." It was physically impossible for any other political force to challenge the armed Communists.

In the larger cities and plains villages of Serbia, where Partisans had been relatively inactive, the National Liberation Council appointed local administrative officials in the name of the provisional government. Since the CPY was in control of the Police and Interior Ministries, plus the now conventional army, there could be no opposition to these appointments. Once the larger metropolitan areas were under firm control, suppression of any non-Communist activity was carried out without compromise. Political rallies were broken up, newspapers were taken off the newsstands and burned. The final step of consolidation of Communist power was achieved by "legal means." National, regional, and local elections were held on November 11, 1945. There was only one candidate for each position, each one being approved by the Communist Party.

ACTIVITIES OF THE NATIONAL LIBERATION MOVEMENT

ADMINISTRATIVE FUNCTIONS

Communications

The Yugoslav Communists were in direct communication with Moscow during the war years by radio transmitter and courier. The large amount of correspondence indicates that the system functioned fairly well.[10] When instructions or statements were received from "grandpapa" (code name for the Moscow authorities) they were mimeographed and distributed throughout the country. Couriers bearing the messages of the central committee would hide these materials in suitcases, or tie them under such objects as the stretchers of Red Cross personnel. The large attendance at the party meetings at Bihac and Jajce suggest that communications within the country also functioned fairly efficiently.

Communication with the Allies was maintained through the radio transmitters operated by British liaison officers assigned to Partisan units. In addition, information was relayed by code in open broadcasts of the BBC.

Recruitment

Before 1940 recruiting was done through the channels of the party. In seeking out possible converts, trained cadres usually concentrated on labor unions and student groups. Potential recruits were observed carefully by the Communists before being selected as members.[11]

Although attempts to organize their own unions failed, the Communists always sought to guide and manipulate. In this way, the party was often able to exert an influence far beyond its numerical strength, and to guide the activities of many supposedly non-Communist sport societies, cultural clubs, reading seminars, and singing groups. At the outbreak of the war individual party members and leftwing sympathizers took steps to mobilize youth organizations. Consisting largely of students from the Yugoslav universities, these individuals promoted the idea of the necessity for a "united front" against the enemy. Guided by the "liberation front," this movement was to disassociate itself from the traditional channels of Yugoslav bureaucracy. Since many non-Serbs saw this movement as one which could break the Serbian hold on national politics and administration, the initial enthusiasm for "united front" participation was encouraging.

It was under such conditions that the Communists were able to gain control of resistance activities in the area. Once a local National Liberation Committee was set up, recruitment took the form of forced mobilization. There was no real attempt to increase the membership of the Communist Party, but if possible all available manpower was transformed into labor brigades, intelligence nets, and courier systems through the united front groups.

Cases of Undergrounds

In regions where popular support for the Partisan movement was not great, the Communists forced the population into a position where participation in underground activities was unavoidable. Blackmail, extortion, assassination, and threats of informing the Germans of "treason" were not uncommon tactics used to achieve cooperation.

Some guerrillas who had been fighting with units of the Royal Army joined the Partisans when they concluded, largely from Allied broadcasts, that the latter were more active in resisting the German enemy. Other opportunists joined when it became clear that Tito was going to emerge as the "strong man" at the conclusion of the war.

The great majority of wartime recruits never became active members of the CPY. At the conclusion of the war many of them returned to their native villages to find the Communist bureaucracy in full control.

Finance

Before the war monthly dues were assessed each party member, and funds were solicited from sympathetic leftwing groups. In addition, the party owned clothing stores, conducted black-market operations, and printed and sold lottery tickets. It can be assumed that leaders received funds from Moscow during their frequent visits to the Soviet Union.

During the war taxes were collected in "liberated areas." One of the Communist municipal administrators had this responsibility. They were especially eager to tax the wealthy landowners, because the receipt of payment would serve as a tool for blackmail. If a wealthy person refused to pay after the first time, it would not be difficult for the Partisans to let the German occupation officials know that "there were people engaging in illegal activities." Sometimes this was done as a matter of course to eliminate any possible threat to Communist authority.

The measure of control that the National Liberation Committees exerted over the financial resources of certain areas is indicated by the following statement made by Tito at the Fifth Conference of the CPY:

> The Slovene National Liberation Committee, which has rallied all patriotic Slovenes, representatives of various political groups, had floated a 20 million lire liberty loan. The Committee has also declared a compulsory people's tax to be paid by every Slovene receiving a regular income.[12]

Logistics

Although the Partisans for a time operated an arms factory at Uzice,[13] Tito's men were in need of every type of logistical support until supplies were dropped regularly by the Allies after the spring of 1944. Tito repeatedly requested material aid from the Soviet Union. Communications with Moscow indicate that he did not, however, receive it.[14]

At the beginning of the war Partisan groups made periodic attacks on police posts in order to obtain arms. These raids served as training supplies from army depots which now were in the hands of the occupation forces, while students in Belgrade persuaded officials in the War Ministry to give the resistance access to military equipment.

After the collapse of Italy in the spring of 1943, all the Yugoslav resistance forces rushed to seize the large quantities of arms, equipment, and stores that the Italians had stored on the Adriatic Coast. At this time the Allies instructed Italian commanders in Yugoslavia to hand over all their facilities to Partisan units. Before the Germans could begin effective countermeasures, the Communists had removed most of the supplies from large Italian dumps.

One month after the first allied liaison officer was introduced to Tito, a Partisan delegation went to Cairo to draw up plans for large-scale Allied support of the guerrillas. This included dropping airborne supplies to Partisan units in Yugoslavia, training new recruits in southern Italy, and caring for the wounded Partisans.[15]

Security

Members of non-Communist resistance organizations were accepted into the Partisan organization as individuals, in order to create the impression of a united front, but never as a group with any authority. Thus, the Communists were able to isolate these elements from former affiliations.

The Communists had a secret police called the Department for the Defense of the People (O.Z.Na.) which was styled along the lines of the Soviet model. This unit was charged with supplying intelligence and liquidating individuals who were disloyal to the Partisans.[16]

OPERATIONAL FUNCTIONS

Psychological Operations

In the interim between the signing of the Soviet-German nonaggression pact and the German attack on the Soviet Union, Communist propaganda in Yugoslavia simply condemned the conflict as a war between "Fascist imperialists," while praising the principles of Marx and Lenin. From the German invasion of Yugoslavia in April 1941 until June of that year, the CPY concentrated its attacks on the Yugoslav Government, making a particular effort to discredit the remnants of the Royal Yugoslav Army. After expanding activities of their own revolutionary movement, the Communists attempted to discredit all other resistance groups then springing up in various part of the country, and to label the King and his entourage in London as traitors. Propaganda directed at the general population stressed the theme that the *Chetniks* were tools of the Serbian monarchy collaborating with the Germans and Italians to defeat the "democratic" partisans.

At the same time, the Communists made exaggerated claims as to their actual strength and support, such as telling the inhabitants of an isolated mountain village that the whole country was in sympathy with the partisan movement under Tito, which was the only real national resistance organization. Being ignorant of the political or military situation, the villagers found it difficult to believe otherwise.

Throughout Yugoslavia there was a deep sympathy for Russia and for Slav brotherhood, which was fully exploited by the Communists in building their underground organization.[17] The Communists also stressed the theme that the Russian revolutionaries personified the sort of behavior the Yugoslav Partisans were striving for, and emphasized the ties between Russians and Yugoslavs.

As the war progressed, the Communists tried to picture the Royal Army and *Chetniks* as collaborationists, while claiming credit for themselves for resistance operations that, in fact, were achieved by the Royal Army and/or *Chetniks*. They claimed to be responsible for keeping 16 German divisions pinned down in Yugoslavia, giving no credit to other resistance groups. They claimed responsibility for destruction of the Vishegrad bridge, which in fact was destroyed by units of Mihailovic.[18]

While the bulk of their propaganda effort was concentrated within Yugoslavia, the Communists also maintained liaison agents in London charged with the task of urging the Allies to abandon Mihailovic and support Tito. They stressed the military value of the Partisan forces to the Allied cause. They pointed to the "united" nature of the movement, playing down or trying to deny the fact that the Partisans were controlled by the Communists. Mihailovic was pictured not only as ineffective, but as a German collaborator.

Propaganda was under the direction of the Department of Agitation and Propaganda (Agitprop), run by the Central Committee of the Communist Party. Agitprop agents infiltrated many social, cultural, and athletic organizations throughout the nation. At meetings they condemned anyone who challenged the righteousness of the Soviet Union, Tito, and the Partisans. When a town was seized by Tito's forces, Agitprop agents immediately asserted the authority of the National Liberation Movement, and urged the population to participate in the resistance effort. When a National Liberation Committee established itself in control, agents distributed propaganda whenever possible.

In the "liberated areas," schools were opened, and compulsory attendance was required at lectures on the history of the Bolshevik movement. Youth groups sang Russian and other Communist songs, while theories of communism and the history of the Russian movement were taught to prospective members of the National Army of Liberation.

The official newspaper of the CPY, *Borba*, continued publication during the war. An illegal press in Belgrade printed additional propaganda materials, and it can be assumed that the Communists took over all press facilities in each town they "liberated." They also printed and circulated fake *Chetnik* and German documents purporting to show collaboration between the two groups. Communist efforts were made easier after the Allies abandoned Mihailovic, and United States and British planes dropped leaflets urging all Yugoslavs to cooperate with the Partisans.

There were no clandestine radio broadcasts originating in Yugoslavia. However, a powerful transmitter did operate from Tiflis, in the U.S.S.R., calling itself "Radio Free Yugoslavia." [19] In addition, the British Broadcasting Corporation maintained a consistent schedule of broadcasts to Yugoslavia in support of the Partisan movement.

Since the Communists knew that a general atmosphere of uncertainty and fear made it easier for them to maintain security of their forces and at the same time obtain supplies, terrorism was utilized. Terrorism took the form of *discriminate* terror directed at eliminating those influential persons among the population who were active or potential rivals. This was usually carried out in "liberated areas" by officials of NLC's. *Indiscriminate* acts of terror were usually carried out with the purpose of blaming the act on the Royal Army, thus driving the people to support the Partisans. A method of gaining immediate response in this way was to disguise Partisan soldiers as *Chetniks* and raid a Croatian village. Such raids served to inflame the Croat population against the Serbian-dominated movement led by Mihailovic.

Intelligence

Since the National Liberation Army was a mobile fighting force, living for the most part in the hills and being constantly on the move, the inhabitants of the villages were relied upon to gather information concerning German troop movements. In villages not under their control, Communist cadres had small networks of informers who reported pertinent information; this information was then relayed—usually by courier—to Partisan headquarters in the field or in a "liberated area."

Within the "liberated areas," the National Liberation Committee engaged the services of the entire population in the gathering of intelligence. Teenagers, for example, were used as scouts; women going to market were asked to report daily gossip. Information was sought on three general categories: occupation and Royal Army troop movements; political activities pertinent to the security of the village or liberated areas; the political situation in neighboring territory, which might indicate whether or not the area could be "liberated." Trained party personnel evaluated the information brought in and relayed it to responsible persons. Of great importance to the Partisans were the long range Allied war plans, and the attitude of the Allied High Commands toward the Partisans. Because of the work accomplished by Communist agents in London, Tito was evidently briefed on many Allied decisions without proper authority.[20]

ORGANIZATION OF THE ARMY OF THE FATHERLAND

When the Yugoslav Army surrendered to the German forces on June 15, 1941, many Royal officers refused to lay down their arms. Initially setting up headquarters in Montenegro, Colonel Mihailovic, soon after named Defense

233

Cases of Undergrounds

Minister by the government-in-exile, began efforts to coordinate a movement to carry on resistance against the occupying powers.

One of the first groups to join Mihailovic was the *Chetnik*—a Serbian military veterans organization—whose name was generally given by outside observers to the entire organization led by Mihailovic.

Mihailovic had the support of several political parties which had been influential before the war, most of which were Serbian-dominated.* The party leaders formed a National Committee in the summer of 1941, but the Committee did not meet in a joint session until January 1944 at the Congress of Ba.[21]

The organization expanded into regions of Yugoslavia that were of immediate strategic importance to the Germans, and in areas inhabited predominantly by Serbians. The hinterland of almost all regions of the country, and especially Serbia, was at some time during the war under the domination of the forces of General Mihailovic.

Mihailovic's forces consisted of operational mobile guerrilla units composed of full-time fighters who had temporary bases in rural areas, and "home guards" or part-time guerrilla battalions composed of older men who assisted the regional commander in maintaining general security. The guerrilla units were commanded by Mihailovic and his High Command. Regional commanders set the policies and strategy for propaganda, intelligence, sabotage, and escape and evasion operations in the areas under their jurisdiction. The regional commanders, or "corps commanders," as they were sometimes called, usually were assigned to their home provinces where they acted independently and largely on their own initiative. They confined their activities to their own provinces. A lack of cooperation between commanders of adjoining districts is suggested by one observer who noted that "moving into an area of another commander was not popular, especially if you had the enemy at your heels at the time."[22]

Within each region of operations was a civilian underground committee responsible to the regional military commander. Underground groups were located in towns, ports, industrial centers, enemy garrisons, and areas close to these, where guerrilla organizations could not operate. They were organized in order to reduce the military, economic, psychological, and political potential of the enemy, and as a support organization for the guerrillas. They cooperated in such efforts as escape and evasion, propaganda, sabotage, logistics, and recruitment for the guerrillas. They were also used to take enemy pressure off the guerrillas. If, for example, the enemy forces were attempting an offensive against the guerrillas, the underground would intensify its activities. At the same time, a successful military engagement by the guerrillas seemed to raise the morale of the underground and the local populace.

A major activity of the underground was the collection of intelligence. Separate intelligence nets were formed to work with the intelligence divisions of the regional commanders, the Home Guard, and the local underground head-

---

*These parties were: the Radical Party, the Socialist Party, the Republican Party, the Independent Democratic Party, the Democratic Party, the Serbian Agrarian Party, and the National Party.

quarters themselves. There were also underground agents infiltrated within the civil administration, student groups, and shipping and railway agencies. In addition, a "secret army" was called for. This force was to be composed of men who were to keep weapons hidden in preparation for a general uprising at the time of an Allied invasion.

In villages and towns where the population was largely loyal to Mihailovic, "civilian comittees," or local governments, were established to carry out local administration, care for refugees, and indoctrinate the population on ways to deal with enemy troops and spies.

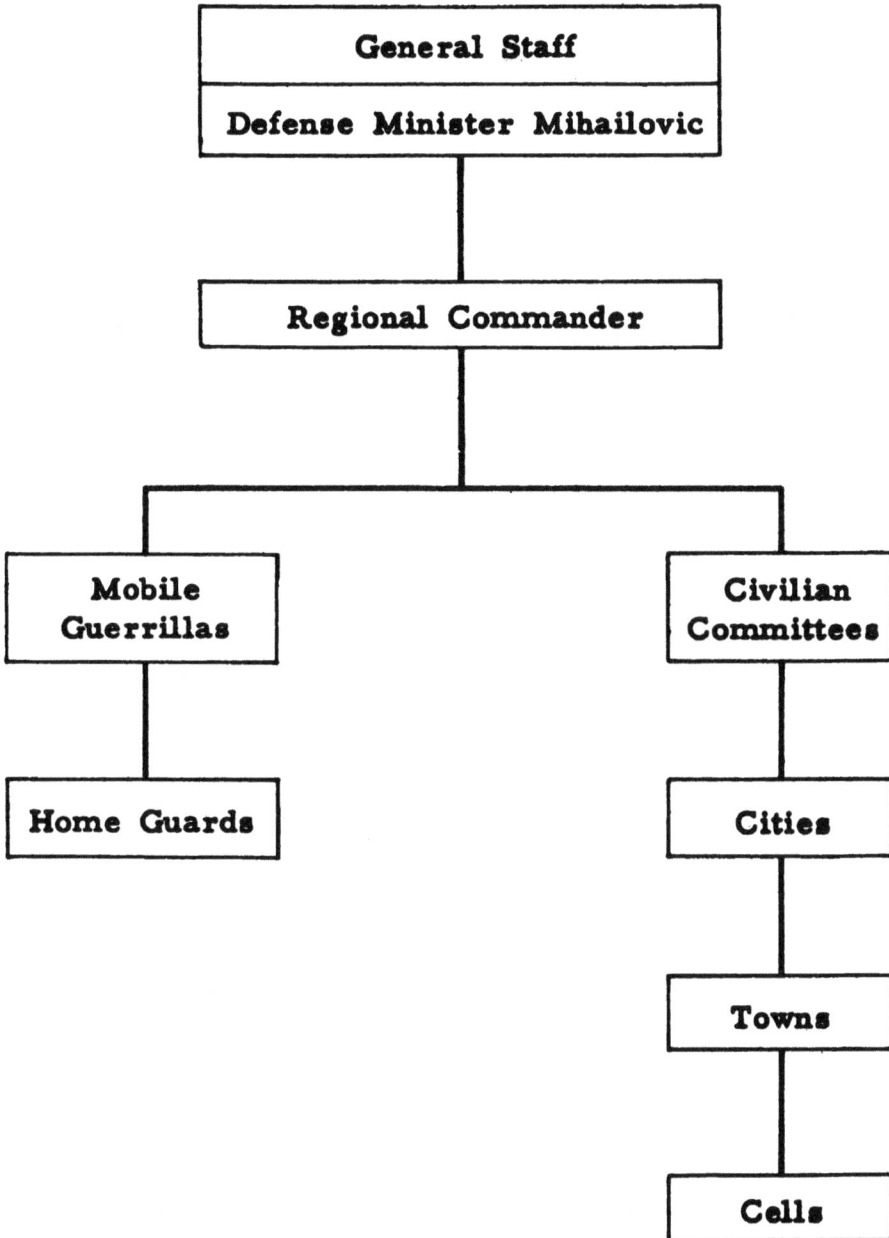

Figure 9. Organization of the Army of the Fatherland.

ACTIVITIES OF THE ARMY OF THE FATHERLAND

ADMINISTRATIVE FUNCTIONS

Communications

According to a British observer, Mihailovic had a remarkable wireless intercommunications system which helped him maintain some coordination in a loosely knit resistance effort.[23] It was through the wireless, in fact, that the Allies first heard of Mihailovic's clandestine activities. When British liaison officers arrived, they followed the general policy of controlling resistance communications with governments-in-exile, and forced Mihailovic to send and receive all such messages through British Special Operations Executive (SOE) channels in Cairo and London, and to use British codes. They supplied him with additional radio transmitters for use in internal communications, however.

When messages were sent to agents outside the country, the BBC would confirm via Switzerland and Rome their successful transmission by a code phrase in a regular broadcast.

Regional commanders, residing with the armed guerrillas in the mobile bases of the hinterland, contacted their subordinate underground agents by couriers. Communication channels led from urban areas into guerrilla strongholds, from town to town, as well as out of the country for contact with Yugoslav officials in Cairo and London. A courier network also kept Mihailovic informed of activities in Belgrade.[24]

Recruitment

The non-Communist resistance movement relied on two main groups for support. One included members of the Royal Yugoslav Army and of various *Chetnik* groups who regarded their resistance activities under Mihailovic as a legitimate exercise of military orders. The other consisted of individual Yugoslavs who were anxious to serve the cause of their country. The Serbian landowning peasants, a politically conservative group which distrusted the Communists and viewed Mihailovic as the leader of a cause which sought the return of the monarchy, formed the backbone of the underground and intelligence networks, and when civil war began between the Partisans and the Royal Army, their allegiance lay with the latter.

With each command region, the initial step in recruitment was to obtain the support of influential people, since their actions could be expected to stimulate the cooperation of others. To gain support on as board a base as possible, the movement sought to select leaders on a nonpolitical basis.

Selective recruitment was directed toward individuals who had access to utilities and industries operated by the enemy. Such people as railroad engineers, ship captains, port authorities, and civil servants were requested to use their working facilities in ways helpful to the underground. The fact that the Serbian Orthodox Church supported the movement was often a help to Mihailovic in obtaining recruits. The majority of the priests in some provinces actually lived with resistance groups and supervised recruiting ceremonies.[25]

Finance

No national organization existed to coordinate the collection of resources for the resistance. In the local regions, the civilian committees enforced some type of tax system. Influential individuals (such a bankers) who joined the underground offered the use of their private financial resources and the resources of their business enterprises. In addition, their employees were urged to make financial contributions. Officials of the Nedic government gave money to Mihailovic from the reserves of the Serbian National Bank,[26] and it can be assumed that officials of the Serbian Orthodox Church were called upon from time to time to place financial reserves of the Church at his disposal.

Logistics

In 1941 the Germans were in such a hurry they had no time to take thorough measures to disarm the Yugoslav Army and since many followers of Mihailovic were military men, supplies were acquired from many of the regular army depots. The acquisition of additional arms and food was one responsibility of the local undergrounds, while the Home Guard was responsible for protecting the caches where these supplies were stored. Since there were many mountain villages which were never occupied by enemy troops, no special techniques were required by local underground members in order to obtain the necessities for daily life. It also seems likely that officials of the Church could be counted upon to provide food, clothing, and sanctuary (especially in isolated monasteries) to underground personnel and their associates who were in need of assistance. Mihailovic never received material aid from the Allies beyond some medical supplies and radio transmitters; thus the supply problem was always acute.

When printing and publishing supplies were needed by the guerrillas in order to maintain their illegal presses at their bases in the woods, requisitions were made through individual printers who cooperated with the underground effort.

Individual underground agents also secured the cooperation of coastal ship captains when their vessels were required by the underground.

In Montenegro, some *Chetnik* units made accommodations with Italian troops when threatened by Partisan forces, because they speculated that the Italians could be used as a source of supply after their expected surrender. The fact that this expectation was never fulfilled was a prime factor in the ultimate collapse of the loyal resistance.

Security

The resistance high command and its agents kept few written documents. This, and the fact that compartmentation of the movement kept any one individual from being informed of total resistance activities, prevented the movement from becoming compromised if a member was captured.

The cellular structure of the underground, which demanded only indirect contact between the resistance leader and the lower echelons of the movement, prevented the high command from knowing the actual numerical strength of the clandestine organizations. The cell structure was, however, a useful security measure in that it also served to limit the information any single underground worker might have.

As a result of accommodations made with the Italians, there were never any *Ustashi* incursions in areas where Italian troops were stationed. Considering the large number of Serbs assassinated by this Croat terror group, Italian preventive measures were of prime importance to Mihailovic.

In the military organization, special agents responsible to the regional commander checked the activities of subordinate units. The underground and Home Guard networks often had security agents who reported breaches of security to the local or regional command.

OPERATIONAL FUNCTIONS

Psychological Operations

Mihailovic tried to present his movement as one composed of democratic elements acting in the interest of all Yugoslavs as opposed to the Partisans, who were dominated by the monolithic Communist Party. He also emphasized the fact that the movement was serving the strategic policy of the "Allied cause." Serbian nationalists appealed to their compatriots in an effort to reassert Serbian influence in the political activity of the country, while officials of the Serbian Orthodox Church requested allegiance to the cause of Mihailovic on religious principles.

Urging that Partisan politics be avoided in regions protected by his officers, Mihailovic asked that persons of all religions and shades of political opinion unite behind a policy of liberation from the enemy. There were Croatian and Muslim officials in Mihailovic's headquarters, in guerrilla units, undergrounds, and support organizations. A province commander had a battalion composed of Croatians and an underground support group. Potential political friction was to be avoided by placing a military man, the regional commander, in charge of underground resistance activities. Specific operations within a village or town, however, were planned by the local civilian committees. Their targets were not only the enemy occupation forces but also Yugoslav groups acting against the legal Royal Government (i.e., the Partisans and the Croatian *Ustashi*).

Propaganda was aimed at publicizing the purpose and accomplishments of the underground and the guerrillas, in order to bolster the national spirit of the local population and lower the morale of the occupying soldiers. Both the guerrillas and the underground workers printed and regularly distributed newspapers in several languages, along with numerous pamphlets and leaflets. Where radios were available, the population was encouraged to listen to the daily broadcasts of the BBC which gave accounts of resistance.

Often the guerrillas would launch an attack against occupying forces when the primary objective was not to inflict a military defeat per se, but to give a psychological boost to the underground workers.

Intelligence

The guerrilla force, the Home Guard, and the underground in each province had their own separate intelligence nets. The Home Guard (into which the nets of village organizations were integrated) and underground organizations were obliged to relay pertinent information to the regional commander. The individual guerrilla units had intelligence agents mainly concerned with their immediate target, but useful intelligence from this source was also forwarded to the regional commander. Where intelligence depended upon the observations of peasants who were normally unconcerned or unacquainted with military matters, the information received was often misleading and "wildly exaggerated." [27]

The administration of the Italian and German police was infiltrated, usually by individuals who were not members of regular intelligence networks but of the underground. Women, who could find employment as interpreters and secretaries with the occupation forces, proved to be particularly effective in carrying out intelligence and counterintelligence missions.

Sabotage

To keep reprisals at a minimum, Mihailovic and his forces carried out most sabotage missions in such a manner as to deflect the blame from the inhabitants of the immediate area. Sabotage of a coal car on a railroad train was planned so that the explosion took place hundreds of miles away. A similar technique was applied to barges carrying goods down the Danube River.

One major sabotage mission was carried out with the cooperation of Allied liaison agents. In September 1943 resistance forces battled with Axis troops and destroyed a critical bridge along the Belgrade-Zagreb rail line. [28]

Escape and Evasion

No special network was organized for escape and evasion activities. When the necessity for assistance presented itself, aid to an escapee was expected from any and all persons involved in the resistance. This included providing shelter, security passes, and means of transportation.

Persons in need of escape and evasion facilities included army officers fleeing from the enemy, civilian deportees liberated by the underground, suspected members of the resistance, escaped prisoners of war, resistance officials being sent on missions to Allied countries, and Allied personnel. Perhaps the most consistent escape and evasion operation carried out under Mihailovic was the aiding of Allied airmen shot down over Yugoslavia while they were en route to, or returning from, bombing missions over Rumania. Mihailovic issued "safe conduct" passes assuring them of Yugoslav Army and *Chetnik* aid in areas through which they were passing. Even after the Allies had withdrawn their military missions, his troops continued to supervise the safe evacuation of these stranded fliers. The major facility provided for the rescue of these men was a landing strip where small aircraft could land and pick up the fliers, who were then flown to Italy.[29] Some were also transferred to the Dalmatian Coast where boats took them across the Adriatic.

COUNTERMEASURES

The initial step of the occupying powers was to break up the existing state of Yugoslavia. In addition to the regular German and Italian Army units, elite SS troops, Bulgarian and Hungarian military divisions, and Croat *Ustashi* were used as occupation troops. Police measures were enforced by the Gestapo, aided by quisling police and authorized armed home guard units.

The major concern of the Germans was to protect the communication and transportation lines running south from Serbia to Greece and east to the Black Sea. The presence of Mihailovic supporters in Serbia forced them to maintain about 200,000 German and satellite troops in the region.[30] Not until the Italian collapse in 1943 did the Germans consider Tito to be a more formidable foe than Mihailovic.

In dealing with these resistance forces, the Germans also tried to exploit the national rivalry between the Serbs and Croats, and authorized armed militia units and puppet governments in German-occupied Serbia and the newly created "independent" state of Croatia. The Croat *Ustashi*, which was violently anti-Serb in sentiment, was given a virtual free hand in raiding Serbian villages. According to estimates, over half a million Serbs and almost all Jews were killed. Independent Serbian armed bands did not hesitate to retaliate against Croats and Muslim *Ustashi* and the Germans made no attempt to interfere.

Since they were confronted with two rival resistance groups whose enmity toward each other sometimes surpassed their animosity for the Axis occupiers, the Germans often had to take into account the existing situation between the partisans and the loyalists before planning operations against either or both. Thus, the Germans and Italians were not unwilling to give material support to one or the other resistance organization when it suited their purposes.[31]

Although the Germans and Italians were allies, their occupation policies often conflicted. Considering Yugoslavia to be their sphere of influence, the Italians often felt that the Germans were interfering in their affairs. This was especially true in Croatia, where the situation was further complicated by friction between the German Army, the Gestapo, and Foreign Office officials.[32] The Italians, having successfully negotiated arrangements between themselves and certain Serbian *Chetnik* groups, were able to restrict violence in their areas of occupation. At the same time Germans were often carrying out full-scale counterguerrilla operations against forces under Mihailovic.

The Germans offered rewards for the capture of top resistance leaders. Both Tito and Mihailovic had a price of $50,000 on their heads. The Germans also instituted a program of harsh reprisals on the population for all subversive activities. A military order stated that for each German soldier killed, 50 to 200 Yugoslavs would die. The net effect, however, was to increase the population's hostility rather than instill fear. These tactics contributed to the decline of the population of Yugoslavia, which decreased by over 10 percent during World War II.

The Germans were successful in deciphering the code used by Tito for his internal communications and were often aware of planned Partisan action long before it took place. The Germans were able to infiltrate Tito's High Command in hopes of capturing the Partisan leader. It was only by a narrow escape that Tito evaded arrest by German paratroopers who were dropped at his hideout in Dvar, Bosnia.[33]

FOOTNOTES

1. Otto Heilbrunn, *Partisan Warfare* (New York: Frederick A. Praeger, 1962), p. 182.
2. David Martin, *Ally Betrayed* (New York: Prentice-Hall, 1961), p. 45.
3. U.S. Senate, Committee of the Judiciary, *Yugoslav Communism—A Critical Study* (Washington: Government Printing Office, 1961), p. 52.
4. Ibid., p. 50.
5. Hugh Seton-Watson, *The East European Revolution* (New York: Frederick A. Praeger, 1951), p. 219.
6. D. A. Tomasic, *National Communism and Soviet Strategy* (Washington: Public Affairs Press, 1957), pp. 66; see also Seton-Watson, *Revolution*, p. 124.
7. U.S. Senate, *Yugoslav Communism*, p. 97; see also Milovan Djilas, *The New Class* (New York: Frederick A. Praeger, 1957), p. 73.
8. Martin, *Ally Betrayed*, p. 45.
9. Seton-Watson, *Revolution*, pp. 122–123.
10. See Moshe Pijade, *La Fable de l'Aide Soviétique à l'Insurrection Nationale Yugoslave* (Paris: Le Livre Yugoslav, 1950), passim.
11. Vladimir Dedijer, *Tito* (New York: Simon and Schuster, 1953) p. 117.
12. See U.S. Senate, *Yugoslav Communism*, p. 93.
13. Seton-Watson, *Revolution*, p. 125.
14. See Pijade, *La Fable*, passim.
15. Martin, *Ally Betrayed*, pp. 234–235.
16. See U.S. Senate, *Yugoslav Communism*, p. 124.
17. Seton-Watson, *Revolution*, p. 119.
18. Martin, *Ally Betrayed*, pp. 41–42.
19. Ibid., p. 33.
20. U.S. Senate, *Yugoslav Communism*, pp. 102–103.
21. Martin, *Ally Betrayed*, p. 190.
22. Jasper Rootham, *Miss Fire* (London: Chatto and Windus, 1946), p. 28.
23. Ibid., p. 29.

Cases of Undergrounds

24. Wilhelm Hoettl, *The Secret Front* (New York: Frederick A. Praeger, 1954), p. 154.
25. Martin, *Ally Betrayed*, photo between pp. 172–173.
26. Hoettl, *The Secret Front*, p. 154.
27. Rootham, *Miss Fire*, p. 63.
28. Martin, *Ally Betrayed*, p. 41.
29. Ibid., p. 249.
30. Heilbrunn, *Partisan Warfare*, p. 182.
31. Hoettl, *The Secret Front*, pp. 154, 165.
32. Seton-Watson, *Revolution*, pp. 79–80.
33. Martin, *Ally Betrayed*, pp. 238–239.

CHAPTER 8

MALAYA (1948–60)

BACKGROUND

Ever since the first Chinese Communists began organizational work in Malaya in the 1920's, the membership of the Malayan Communist Party (MCP) has been drawn largely from the Chinese community, and the party itself influenced by China.[1] Before World War II it enrolled some 5,000 members and controlled a mass base of 100,000 people in front organizations.[2] The party expanded greatly after the Japanese invasion of Southeast Asia in 1941. During the war its leaders, helped by the British 101 Special Training School, built up in the jungles an effective guerrilla force called the Malayan People's Anti-Japanese Army (MPAJA)*, which led the resistance against the invaders, and a civilian support army called the Malayan People's Anti-Japanese Union (MPAJU). They obtained arms from the British in return for a promise to follow British military orders.[3] Despite this agreement, the party platform of February 1943, like that of 1940, announced the party's intentions of eventually obtaining independence for the Malay states from Great Britain.[4]

During 1944–45, the MCP made plans for developing the MPAJA into a permanent armed force through which it could seize power after the defeat of Japan. In April 1945 it ordered the formation of secret MPAJA units which would remain incognito and stay in the jungle, while those MPAJA members who had worked with the British would remain in "open" units. In addition, arms were to be cached for use, if necessary, in a war against the British. The party also, between September 1945 and May 1946, adopted a program for advocating liberal reforms, while saying nothing about communism or its intention of establishing Communist rule.[5]

The Japanese surrendered on August 15, 1945, and in early September the MCP demanded "self-rule" pending independence. British Commander Mountbatten refused, however, to discuss politics, and in mid-September British troops established effective control throughout Malaya. In the interim between the Japanese surrender and the British reentry, however, the MCP and its affiliated MPAJA–MPAJU set up many local administrative organs called people's councils or committees, and took the opportunity to eliminate "collaborators" and any potential rivals. In certain areas the Malays, in reprisal, attacked non-MCP Chinese.[6]

After the British returned, only a part of the MPAJA—the "open" section—demobilized, and these maintained contact through veterans' associations. Approximately 4,000 men in its "secret" section kept their arms.[7]

From the end of 1945 the party concentrated on gaining control of the

*The MPAJA was established in March 1942.

245

Cases of Undergrounds

labor movement in Malaya, as a prelude to overthrowing the government. Mainly through its ability to apply physical coercion, it managed to dominate the new trade union movement. At the same time it created the satellite New Democratic Youth League and began to infiltrate such groups as the Malay National Party and the Malayan Democratic Union. Subsequent efforts to form a nationwide Communist-controlled united front failed, however.[8]

This failure coincided with the change in line of the international Communist movement, signalled by the reorganization in September 1947 of a new Communist organization known as the Cominform.

In February 1948 two international Communist conferences held at Calcutta were attended by Malayan Communists, who, along with other delegates from Southeast Asia, were ordered to adopt a policy of armed uprising in the name of "national liberation." In March the central committee of the MCP called for a "mass struggle and armed revolution" against British "imperialism." The following May, the MCP ordered the greater mass of the party to go underground; militant action was taken to disrupt the economy through strikes and terrorism; and agents were sent to remobilize the MPAJA and reconstitute it as the Malayan People's Anti-British Army (MPABA).[9]

The MCP began immediately to implement this policy and organized strikes during April and May. They were accompanied by terrorist actions: workers were intimidated, rubber factories were burned, and contractors and managers were murdered. Most of the terrorism was carried out by the Mobile Corps, formed from the secret section of the MPAJA. Plans were also crystallized for mobilizing the "open section" of the MPAJA as soon as the government outlawed the party—a move which the latter expected.

In June the government issued drastic emergency regulations and outlawed the party, whereupon party leaders returned to the jungle and began a full-scale insurrection. Their two major forces were the guerrilla army—the newly constituted Malayan People's Anti-British Army subsequently renamed the Malayan Races Liberation Army (MRLA)—and the civilian support group known as the *Min Chung Yuen Thong (Min Yuen)*. The total strength of the MRLA never exceeded 6,000 men; the *Min Yuen's* strength is estimated to have ranged between 10,000 and 100,000 active participants.[10]

The Malayan Communists attempted to pursue the general strategy of Mao Tse-tung and follow a protracted three-phase struggle which would pass from guerrilla warfare to the eventual establishment of large 'liberated areas." They produced such texts as *Strategic Problems of the Malayan Revolutionary War*, obviously inspired by and modeled after Mao's *Strategic Problems of China's Revolutionary War*, published in 1936.[11] However, the MCP movement was handicapped by two factors: Mao's strategy for mobilizing the masses was designed for rural, agricultural areas and was ineffective for jungle-based fighters, and the Chinese character of the movement limited its appeal to less than half of the total Malayan population. These two factors, coupled with an effective countermeasures program which concentrated on resettling the jungle-edge "squatters" who had aided the guerrillas, were among the major reasons for the eventual failure of the insurrection. However, the "Emergency," as the insurrection came to be called, lasted for about 12 years. To put it down, the government used 49,000 British and 250,000 native troops and police.[12]

In 1951, concluding that their violent tactics of indiscriminate terror would not succeed, the Communists turned to a strategy of discriminate terror, attacking only British military forces and installations. On the political front, the MCP openly acknowledged its errors and attempted to rid itself of the reputation of being a Chinese movement by trying to attract Malays and Indians; it also sought to organize a united front with other groups.[13] These tactics brought few results, however, and in 1955 party leaders offered to make a negotiated peace with the British. The latter insisted instead upon the unconditional surrender of the guerrillas, and although they offered to pardon individual rebels, they refused to grant amnesty to the party as a whole. The renewal of the British offer of independence in 1957—this time on terms which completely suited the Malays [14]—cut the remaining ground from under the Communists, and individual surrenders increased until only a few hundred guerrillas remained active by the end of the following year. By 1960 the government removed the emergency restrictions imposed 12 years before. In the middle of 1962 it was estimated that there were 500 to 550 MCP guerrilla members located in the rugged terrain of the Malay-Thai border.

ORGANIZATION

Prior to launching the "Liberation War," the Malayan Communist Party had passed through three phases of growth: the prewar conspiratorial phase, which lasted until the start of the Sino-Japanese War in 1937, the wartime resistance phase, and the postwar "united front" phase. In each period the party organization was revised to fit current policies.

Upon entering the fourth phase, that of "armed struggle," the party reorganized once again along lines which its leaders hoped would enable it to conduct a guerrilla campaign patterned after the Chinese Communist experience. Careful theoretical planning and a highly disciplined organization, however, could not cope with the special problems of jungle warfare and the skillful countermeasures program instituted by the government. "As a result, the experience of the party has been one of replacing the large and the impressive with the small and the barely functional; big organizational units had to give way to smaller and more scattered ones, minimum objectives became only ambitious plans, and audacious actions degenerated into ineffective plots." [15]

Theoretically, the central committee of the Malayan Communist Party determined and controlled the operations of the rebellion. The members of this committee were in turn senior officials of one of the 11 regional committees. These regional committees, supported by local cells and cadres, were in charge of the two major units of the Communist movement, the Malayan Races Liberation Army and the *Min Yuen*.

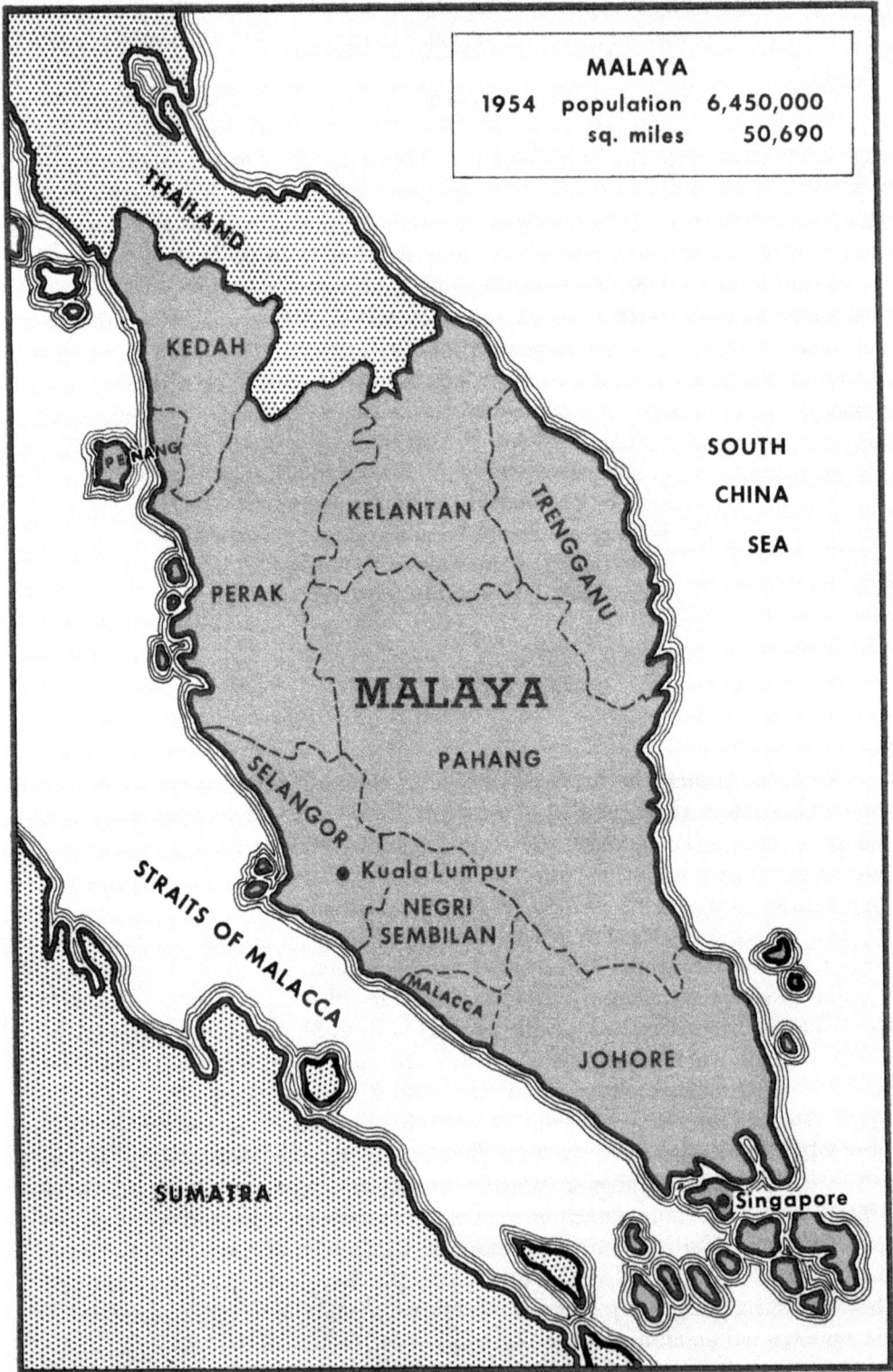

Figure 10. Map of Malaya.

The MRLA, originally composed of 3,000 to 4,000 men,[16] and never exceeding 5,000 to 6,000 men,[17] was intended to be a revival of the old MPAJA. Actually only about 60 percent of those who had fought in the wartime resistance army responded to the call of the MCP in 1948.[18] Their leaders later admitted that they had neither enough men nor enough supplies. And "without supplies they were forced to live off the country, which meant that they had alienated the sympathy of the very people which they needed for their mass movement." [19]

Each guerrilla unit, usually composed of 40 to 50 men, had a political liaison officer appointed by the regional committee. Although he attempted to coordinate the units' activities with the policies set down by the MCP, poor communications made unified action impossible; each unit's operations were determined by the immediate circumstances and the will of the political commander.[20] The lack of coordination of operations severely handicapped the rebellion.

A terrorist unit, known as "The Blood and Steel Corps," also existed. This corps stole, punished, and intimidated for the party. It was the activities of this group which led the government to label the MCP members "bandits," a term later withdrawn in favor of "Communist terrorists." [21]

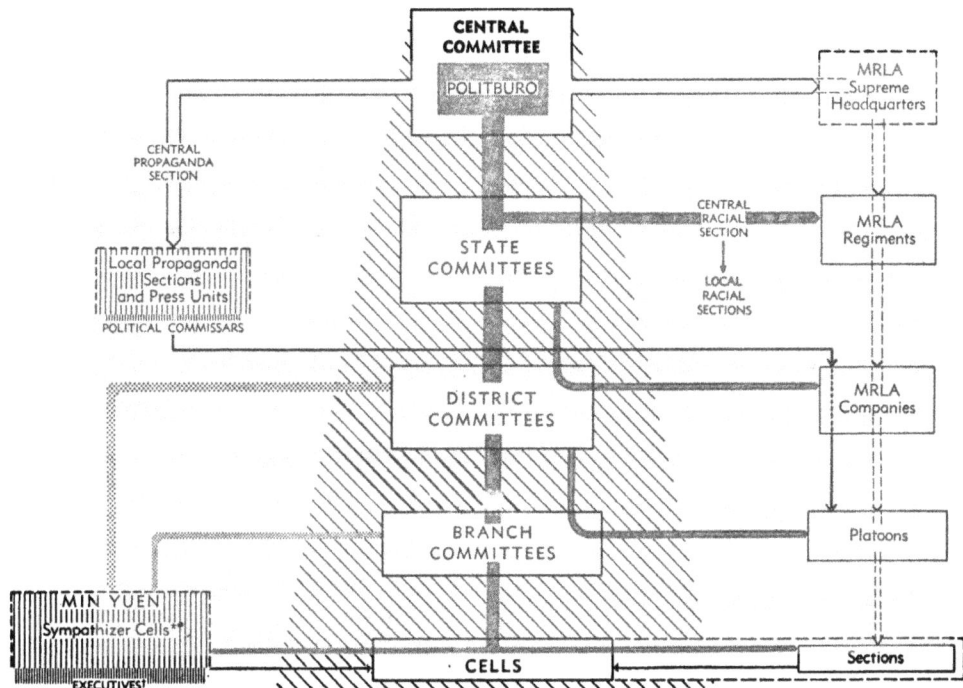

*Self-Protection Corps, Cultivation Corps, Anti-British Alliance Society, Anti-Traitor's Corps, Student's Union, Women's Union, etc.
†Individual underground workers

Reprinted from *Guerrilla Communism in Malaya* by Lucian W. Pye with permission of the publishers, Princeton University Press. © 1956.

Figure 11. Malayan Communist Party Organization During the Emergency.

Cases of Undergrounds

The *Min Yuen*, or Popular Movement, was the civilian arm of the MCP. Its two principal duties were to supply the MRLA with logistical support, and to serve as an intelligence network to keep the guerrillas informed of government troop movements. Initially most of the members—estimated at between 10,000 and 100,000—were Chinese squatters who lived at the edge of the jungle.[22] Many of these people had settled in these areas during the war, hoping to escape from the Japanese occupation officials who were known to be much harsher on Chinese than on Malays. The *Min Yuen* also was charged with organizing new, or dominating already established, sport, youth, and women's groups.

Directed by the Communist Party, but not organized into any recognized unit, were individual underground workers who performed duties within the labor movement. Some of these party members were union leaders who avoided any activity reputed to be Communist-dominated; others were rank and file members who agitated openly for "reform," "wage increases," etc. Their activities permitted the MCP leaders to manipulate union activities to conform with Communist strategy. Thus they could say that their changes were in line "with popular labor demand."

The absence of efficient interregional communication systems, the inability to maintain single unit "regiments" in the jungle, and the success of colonial efforts at relocating rural elements of the population brought about a general separation between the disintegration of the various Communist political and military commands. District commanders found themselves forced to draw up their own strategy, without the knowledge or approval of the high command. At the same time, decisions made at the central committee reached the lower echelons too late to be effective.[23] As typical guerrilla activity failed to secure the MCP's objective, it was succeeded by large-scale and indiscriminate terror, which weakened their claims to be the champions of the Malayan people.[24]

UNDERGROUND ACTIVITIES

ADMINISTRATIVE FUNCTIONS

Communications

As the British pressed their counterattack, the Communists were divided into smaller, isolated units with little intergroup accessibility, and communications within the ranks of the insurgent movement became increasingly difficult. Centralized direction was slow to be achieved and even slower to be transmitted. The Politburo could not meet often; it had to draw up plans on a yearly basis, and assign these on a quota system to subordinate units for execution. Change, rapid response to new situations, surprise, mobility—all the tactical desiderata of guerrilla warfare—could not be utilized because of the difficulty of transmitting orders from higher to lower echelons with the necessary authority and speed. Radio transmitters were almost completely lacking;[25] the party had to rely upon couriers, and major decisions were often a year in reaching their destination. (Such was the case in 1951, when the central committee decided to abandon terrorist activities in order to develop united front tactics and concentrate on political warfare.) Couriers were recruited from the ranks of the *Min Yuen;* often this work was a prelude to their becoming full-scale guerrilla fighters. The *Min Yuen* ran two courier systems, one for its own internal communications and one for the MRLA in the jungle. Government security forces were successful in breaking communications between the various levels of committees.

One other factor appears to have weakened the MCP communications system. Interviews with enemy personnel who surrendered to the British revealed attitudes of distrust of all but face-to-face communication. These persons presumably tested the reliability of the information by their estimate of the informant. Such attitudes place incalculable strains on any communication system, particularly one being gradually worn down and worn out.

Finally, the MCP probably faced the additional complication of maintaining communications across national frontiers. Like all Communist parties except the Soviet and the Chinese, it would be likely to look to some outside authority for approval of its current policies. Following the meetings in Calcutta in 1948, and especially after 1949, overall strategic guidance for Communist activity in Southeast Asia came increasingly from China. Thus a communications channel probably existed between China and the MCP high command, possibly through the Permanent Liaison Bureau of the World Federation of Trade Unions after 1949. Early in 1950, according to one authority,[26] Chinese Communist cadres were introduced on a regular basis, infiltrating by way of Thailand or by sea. After the British effectively sealed off the border, however, sustaining such communication became increasingly difficult for the duration of the Emergency.

Recruitment

The outcome of the recruitment effort of the insurgent movement was shaped by several factors. Probably most significant was the fact that the rebellion was the creature of the Malayan Communist Party, which was traditionally associated with the Chinese community in the plural, divided society of Malaya. Communist Party leaders assumed that Chinese elements in Malaya would be readily attracted to the movement, as indeed many were. However, the realization that 95 percent of the underground insurgents were Chinese only served to alienate the other ethnic groups. Throughout, the insurgency remained a predominantly Chinese movement.

On the other hand, the MCP was helped by the fact that through its wartime resistance it had acquired the reputation of being the only organization which was serving the interests of Malaya. When the rebellion began, many people, including peasants and workers, still felt that the MCP was serving their interests—particularly after the British, who represented "colonialism," denounced the guerrillas as "bandits," an epithet which failed to elicit a truly anti-Communist response and later was discarded in favor of "Communist terrorists," a more accurate descriptive term. Others, including Chinese who had been too young to fight during the war, were anxious to fight during the Emergency, especially after the Communist victory in China. The glories of Mao's Peking regime aroused a patriotic response in the overseas Chinese communities. China's intervention against the U.N. forces in Korea played no small part in this attraction. Recruitment for both guerrilla and underground work increased at several points during the Emergency in reaction to events outside Malaya such as the Chinese Communist victory on the China mainland and the French reverses in Indochina.

Although the MCP could safely count on the vast majority of overseas Chinese for support, the character of the rebellion required the party to select recruits carefully. Persons of two sorts were needed: those who could withstand the rigors of the jungle and those who could provide food and supplies from urban areas.

Of broader importance in making the rebellion a primarily Chinese movement were the social and economic conditions in postwar Malaya. Two circumstances helped to convince the Chinese that they were the "have not" segment of the population. For one thing, technological advances eliminated the need for many of the Chinese unskilled laborers. Also, the Chinese were not allowed to participate, at least not on an equal basis with the Malays, in the civil administration of the country. This was a field primarily reserved for Malays.[27] The Malay Federation citizenship rules, which discriminated heavily against the Chinese, served further to alienate them. In June 1949 only one-sixth of the Chinese population had citizenship rights.[28]

To the Chinese, who felt a sense of isolation, being asked to join a group was important; furthermore, many of them felt that the MCP was the one element in Malayan society concerned with the individual. Hence many were eager to join the insurgents.

Communist sympathizers were impressed with the results achieved by the MCP, a thoroughly disciplined organization. Such recruits were not unaware that the Communists controlled various front groups, but rather than being shocked at the Communist manipulation of these organizations, they were impressed by the MCP strength and shrewdness. Even if an individual did not always agree with the goals of communism, he could admire the political skill of the MCP.[29]

While front members were undoubtedly aware that their leaders were carrying out secret activities for Communist goals, they considered this a legitimate act of authority. The recruit felt some thrill at belonging to a group which did some illegal things. The shift from participation in a front group to membership in the party was a very selective process, and it was common to serve in a front group before becoming a full-fledged party member. The conferring of party membership was viewed by the individual as a promotion. Friends of a party member or sympathizer often joined the Communist movement, but usually out of friendship rather than dedication to a cause.

The MCP also had other agencies, less overtly political, which were used to acquire adherents. Youth and sporting groups provided stimulating leisure-time activity. Also, with employment hard to find, it was sometimes necessary to join a Communist union in order to secure a job.

Once the 500,000 Chinese in the rural population were impounded in guarded villages, recruitment for the MRLA was hampered. Nevertheless, the Communists would seek out individuals in fields adjacent to the jungle and demand that they secretly deposit food and supplies for the MRLA at pre-arranged points. Once an individual had helped the rebel cause it was almost impossible for him to avoid continued service for fear of being reported to the government officials, which the Communists frequently threatened to do.

Finance

Immediately after the Japanese surrender, the British authorities learned that the MCP had amassed a large sum of Japanese script to finance Communist activity. When the British repudiated this script it hurt some loyal merchants, but it was especially painful to the Communists.[30] However, the MCP succeeded in securing funds from a variety of sources. A major one, especially prior to the June 1948 enactment of protective legislation, was the treasuries of the unions which the Communists organized and otherwise dominated. Though some local unions and associations of workers (chiefly Chinese) ante-dated World War II, the MCP organized the first trade union federation in 1941. It achieved some status, not only because it served the workers but also because it became a target for the Japanese. After the war the MCP again turned to the labor movement and again was successful in organizing and gaining domination over the first postwar unions, more or less associated in the General Labor Unions (GLU). In February 1946 these unions and others were reconstituted as the Pan-Malayan Federation of Trade Unions (PMFTU), affiliated with the international Communist World Federation of Trade Unions.[31]

The GLU and PMFTU collected or controlled the funds of the local unions. They rebated to the locals only what they determined to be necessary. It is estimated that at least 30 percent of all funds collected went to these central offices. Between 1945 and the passage of the 1948 legislation that put an end to such practices, their success was sufficiently alarming to cause the British Government to send trained union officials from England to assist in countering their efforts. With the passage of the restrictive legislation aimed at requir-ing a "proper system of registration" and with the opening of the Emergency—the armed rebellion—the MCP and its trade union leadership absconded with union funds. Subsequently the Malayan trade union movement was slowly rebuilt on non-Communist bases.[32]

When union treasuries could no longer be tapped by the Communists, the *Min Yuen* became largely responsible (together with the Blood and Steel Corps) for the collection of funds. They extorted large amounts from land-owners, mine operators, and transport companies, and smaller sums were

Cases of Undergrounds

"taxed" from the workers on the rubber plantations and tin mines. An effective technique to obtain money was to persuade the mine workers to strike; after they had succeeded in obtaining higher wages, the *Min Yuen* would demand payment for the support given the strikers during the time they had not worked. One observer noted that "the laborer had to pledge a percentage of his earnings to the party."[33] A branch of the MCP, the Communist Protection Corps, took the responsibility of eliminating those who refused to pay these subscriptions.

Notwithstanding the ingenious fund-collecting schemes of the Communists, the effective countermeasures of the British created a desperate financial crisis within the MCP. By 1951 the terrorists relied on raids on villages, where they obtained money and supplies by threat. This hampered the *Min Yuen* in its collection of voluntary contributions. It was forced to blackmail the richer citizens of the towns to get money for the rebellion. The urban *Min Yuen* groups operated successfully for a time in Penang, Kuala Lumpur, and Johore.

Selling stolen rubber and tin on the black market proved profitable. The Communists are said to have collected 1,500,000 English pounds from the sale of stolen rubber.[34]

Although the British were able to prevent supplies from China from being delivered to the MRLA by boat, it is suspected that funds or other material support reached the rebels through the facilities of one or another of the Chinese Communist Party's vehicles.[35] Several Chinese millionaire businessmen and other overseas Chinese living in Singapore were also suspected of having given financial aid to the MCP.

Logistics

From the beginning of the Emergency, MCP officials complained that the Chinese tactics which they followed did not meet their needs: Mao Tse-tung's system of establishing "rural retreats" in order to guarantee the guerrillas a steady supply of food was of little value in the dense jungles of Malaya.

Although the *Min Yuen* was able at first to obtain supplies from "lost drops" made by the British during the war, this source was soon exhausted. At an early stage the rebels were forced to make terrorist raids on village police stations to procure firearms and other supplies. Trucking firms in Singapore were known to have "lost" their cargoes of foodstuffs and other goods on trips from the port city to points inland, and there is also a possibility that arms were arriving from China through neutral Thailand. The border area between Thailand and Malaya on the China seacoast is still a locus of MCP activity.

Food was supplied directly by the Chinese squatters living on the edges of the jungle. After many of these people were relocated by the British Director of Operations, Lieutenant-General Sir Harold Briggs,* and placed

---

*Briggs was author of the plan which eventually proved successful in combatting the insurgency, although he ruined his health and died shortly after he left Malaya in late 1951. See Victor Purcell, *Malaya: Communist or Free?* (Stanford: Stanford University Press, 1954), pp. 65ff.

under police guard, the *Min Yuen* found its job of supplying the rebels much more difficult. In these circumstances a local inhabitant who managed to steal supplies and elude the guards was forced to join the guerrillas, as he would be unable to return home. Thus his usefulness as a source of food was ended, and he now became only another mouth to feed. The guerrillas tried clearing jungle areas and growing crops, but this only made them vulnerable to the searching eyes of the British reconnaissance pilots who had been introduced into the Briggs Plan for the first time. By 1952 the Communists had to concentrate most of their efforts on obtaining enough food to stay alive.

The food problem was not simply one of acquiring sufficient supplies, but of getting them to where they were needed, and a situation developed where excessive stockpiles existed in some areas while isolated units were near starvation.[36] Their inability to obtain food forced the MRLA and *Min Yuen* to reorganize into small, mobile gangs. In time their logstics operations broke down completely.

Security

When the British clamped down on Communist activity in labor unions, the Communist organizers fled the urban centers to join their comrades in the jungle. Cadres were infiltrated into unions, however, to provoke general agitation. Their ultimate goal was to dominate the Singapore non-Communist Trade Union Confederation in order to eliminate it as a potential competitor. This long range plan met with little success, however.

When open revolt began in 1948, the major security problem concerned the protection of the MRLA bases in the jungle. As the intelligence and counterintelligence system often proved inadequate, however, it was difficult to devise successful security measures. Since most Chinese in Malaya were looked upon by the British as suspect, security measures for those in the *Min Yuen* and the MRLA involved not secrecy of identity but secrecy of purpose and function. Thus a Chinese squatter would be a farmer by day and a soldier by night.

A significant breakdown in the MCP intelligence-security system became noticeable as the British expanded their Chinese-squatter relocation program. The Communists were unable to adapt their insurgency to the loss of almost half a million persons.[37]

OPERATIONAL FUNCTIONS

Psychological Operations

From the end of World War II to the declaration of the Emergency in 1948, the MCP and its affiliates tried to present themselves as organizations which had defeated the enemy in the interest of Malaya. When its initial attempts to be recognized as a legitimate political party by the British failed, the MCP returned to clandestine means of spreading propaganda among the the population. Its primary target was the urban Chinese community. Because the Chinese had traditionally been denied participation in Malayan politics, the MCP saw the Chinese community as a likely source of support. It urged the Chinese to demand broader participation in the affairs of the country, and tried to convince them that it was the British who were denying them their proper place in Malaya. Two newspapers, closed down by the British in 1948, repeated this accusation. The *Minh Sheng* (Voice of the People) and the *Sin Min Chu* (New Democracy) made statements to the effect that the British were more vicious than the Japanese.

Segregation from the prewar colonial administration had led the MCP to establish its own schools, clubs, etc., which made it possible for the MCP to approach the Chinese community through already existing organizations. These provided places where people could participate in political discussion. Apparently the prospective Communists were impressed with knowledge shared as well as the emotional intellectual stimulation.[38] At the discussion periods, party members and sympathizers would distribute Communist literature.

Personal informants rather than impersonal sources were preferred, as many Chinese tended to be suspicious of the printed word. A common attitude was expressed by one surrendered Communist:

> When people talk to you, you can tell whether they are sincere or not and thus you can tell whether the information is reliable. You can't do this with a newspaper.[39]

Although the MCP took great pains to try to present itself as a respectable political organization which desired to represent all ethnic groups in Malaya, the fact that it directed its major propaganda efforts toward the Chinese community only served to alienate the other nationalities. The name of the guerrilla army was eventually changed to Malayan Races Liberation Army in order to attract recruits from outside the Chinese community, and the All Races General Labor Union, created in 1950, hoped to achieve the same goal. Neither of these measures, however, met with much success. Furthermore, the relative economic prosperity of the tin and rubber industries, coupled with the fact that the British were preparing the Malayans for independence, soon rendered ineffective the MCP charges that the country was being exploited by the colonials.

Concomitantly, the failure of this propaganda program to stimulate any significant support for the insurgency outside of the party and the Chinese community resulted in a definite change in propaganda policies in late 1948. The new psychological operations included attempts to create an atmosphere of terror throughout the country. This policy was designed to bring about two conditions favorable to the Communist cause: disruption of the Malayan economy and forced cooperation of the indigenous population.

Terrorist measures included both murder and sabotage. Colonial officials, native administrators, and government sympathizers were murdered. Sabotage was directed primarily against rubber plantations and tin mines, but also against transportation facilities (railroads, buses) and communication lines.

These missions were carried out either by a terror squad (the Blood and Steel Corps), units of the MRLA based in the jungle, or *Min Yuen* "Protection Corps" which were under control of the local committee. The terrorist campaign boomeranged, however, and in 1951 the Politburo of the MCP directed all agencies connected with the insurgent movement to cease terrorist activities against the indigenous population. The *Min Yuen* leaders had complained that the indiscriminate terror of the MRLA was making it impossible for them to carry out even routine work among the populace and they feared it would drive the masses to greater support of the government.[40] Thus the Communists were forced to return to a policy of encouraging sympathetic support by nonviolent means. References were now made to the Communist success in China and the "War of Liberation" in Korea. While this made some impression on the Chinese population, it failed to alter the general attitude of Malays and Indians toward the movement.[41] In brief, after the first flush of postwar success, and especially after 1950–51, the MCP failed to find the tools or themes for conducting successful psychological warfare. Its movement ran downhill, declining continually in membership.

Intelligence

After the revolt broke out in 1948, the primary intelligence needed by the MCP was information concerning government troop movements and police operations. The *Min Yuen*, composed largely of Chinese settlers living on the edge of the jungle, was responsible for gathering most of this information. The Sakai aborigines of the Malayan jungle were used as an intelligence network in order to secure the safety of the guerrilla bases.[42] Rebel intelligence operations were fairly effective until the British relocated the Chinese squatters into new villages, thus depriving the rebels of their major source of information.

Although the British made little effort to single out individual Chinese who might be suspected of being members of the *Min Yuen*, the activities of this organization were severely restricted because a large proportion of the rural Chinese population was forcibly moved into "new villages." It is known, however, that these resettled Chinese, while at work in the fields, were approached by guerrilla soldiers, who demanded information or supplies. Although the new villages separated the guerrillas from their source of supply and intelligence, they did not automatically eliminate Communist Party activity within the new village compounds.[43]

The Emergency continued and the rebellion had increasing difficulty maintaining its effectiveness. Threat and blackmail replaced persuasion, as the population could no longer be relied upon to give adequate security and intelligence to the rebels on their own initiative.

Sabotage and Terrorism

In the initial stages of the revolution, the MCP sought to convince the population that the MRLA was a force strong enough to act where and when it desired, hoping in this way to attract people who otherwise would hesitate to join an illegal movement.

By slashing rubber trees and blowing up tin mines, the MCP hoped to disrupt the economy of Malaya by frightening the owners away and terrifying the workers into cooperating with them. Sabotage against rubber plantations and tin mines ceased, however, when it became obvious that destruction of these raw materials was hurting the insurgents and not seriously harming the growing prosperity of these industries.

Initially in the revolt it seemed as if the MCP used terror and assassination on a selective basis. The killing of several wealthy plantation owners made quite an impression on the Chinese labor force, who still hoped that the MCP would serve the interests of the working class. However, as their tasks became more difficult and as straightforward guerrilla activity failed to achieve the expected objective, the MCP resorted to indiscriminate, violent, and brutal forms of terrorism.[44]

The colonial authorities became aware of the seriousness of the movement when High Commissioner Sir Henry Gurley was assassinated on October 7, 1951. Raids against rubber and tin estates deprived Chinese and Indian workers of a livelihood. Robbery, killings, and sabotage hurt innocent people. The Malayans demanded action from London. They got it at the beginning of 1952 in the form of Sir Gerald Templer, who came as both military and civilian commander of Malaya with a new directive [45] announcing the government's policy of self-government and a united Malayan (i.e., Chinese and Malays) nation. Templer became a controversial figure who nonetheless carried out the Briggs Plan.

In the interior the *Min Yuen* soon complained that terrorism was making difficult their activities of collecting voluntary contributions of money and food from the villages. By 1951 the British learned that the central committee had instructed all insurgents to refrain from indiscriminate acts of terrorism and to concentrate on military forces and installations. The rebels were then at a point where they needed all the sympathy and support they could get simply in order to survive; terrorism was only further alienating the population. However, the decrease in terrorism led to a decline in Communist influence over the general public, since this influence had been based upon fear rather than sympathy.

From 1952 until the end of the Emergency, whenever terrorism increased, as it did from time to time, the British understood that the *Min Yuen* and the MRLA were in desperate need of food and supplies and were taking extreme measures merely to stay alive; they were, in other words, no longer using terrorism because they felt it would help the movement. And correlatively the British supplied increased protection for the affected areas and people.

COUNTERMEASURES

After the first outbreak of terrorism in 1948, the British declared a state of emergency, and emphasized that the government would take any and all steps necessary to maintain law and order. Ordinances were passed to permit detention of suspects for up to 2 years. Individuals involved in subversive activities were banished from the country. Wishing to give the impression that this was not a political struggle between British rule and "communism," which as an ideology seemed to be gaining in popularity throughout Southeast Asia, the British originally did not refer to the insurgents as Communists, but as "bandits," although they had no doubt as to the identity of the enemy. However, this policy did create some confusion within the ranks of Emergency personnel. There was, for example, no clear delineation of the roles which the military, the police, and other agencies should play. Also, the label "bandits" proved to have grave disadvantages, as it had been employed by the Japanese against resistance members during World War II, and by the Nationalist Chinese in referring to the forces of Mao Tse-tung. During the Korean conflict, the British reversed their previous definition of the insurgents and referred to them as "Communist terrorists" (CT's). This legalistic definition of the problem limited the Communists' efforts to dramatize their cause in order to win popular support. At the same time, it permitted the government, through the use of ordinary administrative channels, to mobilize effectively a large amount of men and resources.[46]

Under these conditions, the British began to regard the situation as a problem necessitating four fairly distinct types of action, each of which implied a somewhat different logic of policy choice. These activities were—

(1) Strategic measures designed to destroy the MCP and destroy its appeal.
(2) Responses to the tactical moves of the Communists to prevent this strength from growing.
(3) Continuing the normal functions of government which would be pursued had there not been a Communist challenge.
(4) Government policies concerned with bringing about social and economic changes in Malaya, in order to win positive popular support for the government.[47]

The measures taken to destroy the power of the MCP were twofold. Legal steps were taken to expel known Communists from labor unions. At the same time, the British authorities, in consultation with British labor leaders, stimulated the growth of non-Communist unions in Kuala Lumpur and Singapore.

A broader program of strengthening public morale was carried out by trying to obtain from the people a greater degree of commitment in the struggle against the Communists. Here the British generally found the Malays, who feared the strength of the Chinese, to be willing allies.[48] This policy, however, obligated the government to inform the public more completely concerning its

Cases of Undergrounds

operations against the enemy. (To a degree this contradicts conventional methods of a police force fighting criminals, as it is accepted traditionally that the public does not have the right to demand information in police files.) This meant the government would also have to compensate for the insecurity brought about by increased popular participation. The authorities urged the community to join legal organizations as a means of expressing group indignation against Communist activities.

The result was the founding of the non-Communist Malayan Chinese Association (through which the British attempted to show that they did not consider the total Chinese population unreliable), the Independence of Malaya Party, and the Malaya Labor Party. The British realized that the masses supporting these political organizations would probably oppose British rule in the future, and that this policy involved some amount of risk, especially since many of the effective political leaders had been eliminated by the Japanese during the war. It was, however, considered preferable to allowing the public to sympathize with the enemy.

To many observers the key to success in Malaya was the Briggs Plan, drawn up by Lieutenant-General Sir Harold Briggs after the Communists resorted to active warfare in 1950. The aim of the plan was to cut off the

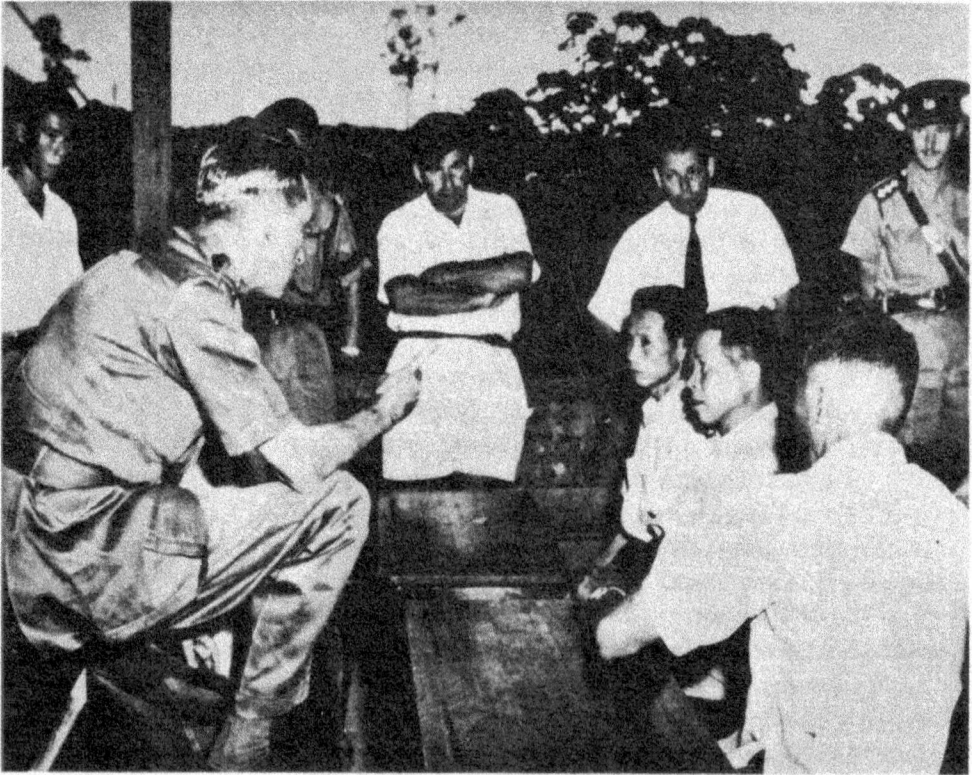

(UPI Photo)

Gen. Sir Gerald Templer talks to the elders of a village at Kampong Pelawan in Lower Perak during a tour of the Malay Federation.

flow of supplies to terrorists in the jungle and to sever their communications with the Communist cells among the populace. It also called for unified control of operations against the terrorists. This is what in fact occurred when General Sir Gerald Templer was placed in supreme command of both civilian and military operations by being appointed High Commissioner (civil) and Director of Operations (military). Full and unified authority was provided for the overall direction of the total civil, military, police, and economic effort. War Emergency Committees were established for federal, state, and local districts.

The object here was to bring local political figures into a direct working relationship with the British Armed Forces and the Malay security police, and thereby make the populace feel that it was playing a role in forming government policy. This allowed for a high degree of day-to-day coordination between local civil and military authorities. Templer combined the special Emergency Information Service with the regular information service. The new Director General of the Department of Information worked directly with top policy planners, and was empowered to use information and propaganda as he saw fit.

In the effort to isolate the guerrillas from the villagers who aided them, the Briggs Plan launched a novel and highly successful resettlement program whereby about 500,000 Chinese squatters were moved into "new villages."

(Courtesy of the Natural Rubber Bureau)

The High Commissioner unlocks information boxes in King's house in the presence of six community leaders from Tanjong Malim.

This cut off a main source of food and information for the terrorists. In these new villages, strict security measures were established, including the creation of "home guards" for protection; food was rationed to restrict the amount that could possibly be passed on to the guerrillas, and modern communication channels which could inform government forces of any immediate danger of Communist activity in the area were established.[49] But whether an event—such as an ambush—took place near a new village or in an established town, Sir Gerald never hesitated to impose on the situation the full force of his authority. The ambush at Tanjong Malim in March 1952 and the subsequent measures of punishment and defense which were imposed there served as a harsh but helpful example of the new countermeasures.[50]

The British recruited and trained 25,000 regular police and 50,000 special constables by 1953. Approximately 250,000 home guards were recruited to protect the villages and new settlements. An additional 35,000 men were enlisted into the regular forces. A total of 360,000 security forces were opposing 5,000 guerrillas and an estimated 12,000 underground fighters, a ratio of approximately 24 to 1.[51]

The Malay-Thai border was effectively closed and British gunboats patrolled the coast, thus cutting off the possibility of outside sanctuary and external supply. Military actions were carried out by combat patrols of squad and platoon size.

Intelligence activities were centralized, however, and the British operated on the principle that the unit commander closest to the tactical front could make the decision and act immediately to exploit any intelligence information.

Malaya was included in an overall plan to make many of the colonies of Great Britain equal members of the Commonwealth of Nations. Thus one of the major themes, that of "imperialism," became ineffective when the British offered Malaya its independence separate from Singapore. Independence was achieved in 1957. At this time the responsibility for dealing with the Emergency was placed in the hands of indigenous Malayans. Under their administration the Emergency was brought to an end in June 1960.

FOOTNOTES

1. See Gene Z. Hanrahan, *The Communist Struggle in Malaya* (New York: Institute of Pacific Relations, 1954), p. 7. Interest in the Communist Party developed largely from the Chinese propensity to form secret societies. These societies, brought by the first Chinese immigrants into Malaya, were all considered offshoots of the *Thian Ti Hui* (Heaven and Earth League), subsequently known as the *Hung* (Food) League or Triad Society. While these associations were originally religious or self-help groups, they assumed a political character during the end of the Manchu Dynasty rule and then degenerated largely into criminal associations, although several of them did useful welfare work among the Chinese population. Because of the illegal nature of many of their activities, many of these societies were outlawed in 1889. See Victor Purcell, *The Chinese in Modern Malaya;* Background to Malaya Series No. 9 (Singa-

pore: Donald Moore, 1956), pp. 4–6. Legal or illegal, open or clandestine, these societies continued to meet and to interest themselves in politics, running the spectrum from far left to center, and right. For details see Leon Comber, "Chinese Secret Societies," Appendix A, Chapter XI, in N. S. Ginsburg, (ed.), *Malaya* (New Haven: HRAF, 1955).

2. J. H. Brimmell, *Communism in South East Asia* (New York: Oxford University Press, 1959), p. 148.

3. See F. Spencer Chapman, *The Jungle is Neutral* (London: Chatto and Windus, 1949), pp. 16–17, for comment on the agreement between the MCP and the Force 101 Special Training School; see p. 248 for a reiteration of the agreement by which later cooperation was agreed upon. By this time, Force 101 had been transformed into Force 136; see p. 232.

4. J. H. Brimmell, *A Short History of the Malayan Communist Party* (Singapore: Donald Moore, 1956), pp. 14–15.

5. Hanrahan, *Communist Struggle*, pp. 51–54.

6. F. S. V. Donnison, *British Military Administration in the Far East* (London: HMSO, 1956), p. 385.

7. Ibid., p. 17; see also A. Doak Barnett, *Communist China and Asia: Challenge to American Policy* (New York: Vintage Books, 1961), p. 486.

8. Brimmell, *A Short History*, p. 18; see also Virginia Thompson and Richard Adloff, *The Left Wing in Southeast Asia* (New York: Sloan, 1950), pp. 131–148, for a more detailed summary of these events.

9. Brimmell, *A Short History*, pp. 19–20.

10. U.S. Department of State, *World Strength of the Communist Party Organizations* (Washington: Bureau of Intelligence and Research, January 1959), p. 6.

11. Hanrahan, *Communist Struggle*, p. 63. These documents are reprinted, pp. 101–130.

12. For costs of the Emergency, see Hanrahan, *Communist Struggle*, p. 79.

13. Ibid., The party document here cited appears in part, pp. 130–133.

14. Russell H. Fifield, *The Diplomacy of Southeast Asia 1945–1958* (New York: Harper, 1958), pp. 399–408.

15. Lucien Pye, *Guerrilla Communism in Malaya* (Princeton: Princeton University Press, 1956), p. 87.

16. Barnett, *Communist China*, p. 486.

17. Hanrahan, *Communist Struggle*, p. 68; see also Pye, *Guerrilla Communism*, p. 98.

18. Hanrahan, *Communist Struggle*, p. 67.

19. Vernon Bartlett, *Report From Malaya* (New York: Criterion Books, 1955), p. 39.

20. For the importance of political officers in military units, see Harry Miller, *Menace in Malaya* (London: George Harrap and Co., 1954), p. 103.

21. Pye, *Guerrilla Communism*, p. 88.

22. *World Strength*, p. 6.

23. See Pye, *Guerrilla Communism*, pp. 107–108. Even though the decision to cease terrorism was made in 1951, violence continued for several years until the December 1955 meeting between the Communist leader Chen Peng and Tengku Abdul Rahman. This failed, however, to bring about the surrender of the Communists.

24. Purcell, *Modern Malaya*, p. 46.

25. Pye, *Guerrilla Communism*, pp. 99–102.

26. Hanrahan, *Communist Struggle*, pp. 67–68.

27. Ibid., p. 81.

28. Purcell, *Modern Malaya*, p. 41.

29. Pye, *Guerrilla Communism*, pp. 218–220.

30. Hanrahan, *Communist Struggle*, p. 55.

31. Ibid., pp. 55–58, and Alex Josey, *Trade Unionism in Modern Malaya* (Singapore: Donald Moore, 1956), pp. 14–22.

32. U.S. Department of Labor, *Summary of Labor Statistics in Malaya* (May 1958), p. 9.

33. Miller, *Menace*, pp. 105–107.

34. Ibid., p. 215.

Cases of Undergrounds

35. Hanrahan. *Communist Struggle*, p. 80.
36. Lucien Pye, *Lessons From the Malayan Struggle Against Communism* (CENIS M.I.T., 1957), p. 50.
37. Purcell, *Modern Malaya*, p. 46.
38. Pye, *Guerrilla Communism*, p. 184.
39. Ibid., p. 178.
40. Ibid., p. 105.
41. Hanrahan, *Communist Struggle*, p. 81.
42. Ibid., p. 75.
43. Miller, *Menace*, p. 216.
44. Josey, *Trade Unionism*, p. 2; see also Hanrahan, *Communist Struggle*, p. 67.
45. Miller, *Menace*, p. 203.
46. Pye, *Lessons*, p. 15. Cf. with Briggs Plan and Templers Instructions; see also Miller, *Menace*, p. 139 and p. 203, respectively.
47. Pye, *Lessons*, p. 18.
48. When the British Labour Government had offered Malaya independence or self-determination in 1946–47, the Malays rejected the offer, since it would have combined Singapore with the Malay States and given the Chinese a majority. The ultimate solution separated Malaya from Singapore thus assuring the Malays a majority in the states of the Federation; see also Fifield, *The Diplomacy*, pp. 399–400.
49. See S. N. Bjelajac, "Malaya: Case History in Area Operations," *Army* (May 1962), 30–40.
50. Miller, *Menace*, pp. 206–212.
51. Bjelajac, "Malaya."

CHAPTER 9

ALGERIA (1954–62)

BACKGROUND

After World War II Algeria, although legally part of France, was made up of two distinct communities: approximately eight million Muslims and one million European *colons*. The latter not only controlled most of the land and wealth of the country, but also determined the political representation of Algeria. Most Muslims in Algeria were poor peasants, while most of the 400,000 in metropolitan France were employed as unskilled workers.

Wartime French promises of reform failed to satisfy most of the Muslim political leaders, and a nationalist revolt broke out at Setif in May 1945. This uprising was put down by the French, and severe reprisals were exacted. In 1946 Ferhat Abbas, who previously had sought full assimilation for his people, still hoped that concessions could be gained from France through peaceful persuasion. He founded the *Union Démocratique du Manifeste Algérien* (UDMA), whose aim was internal autonomy. This party, however, decreased in popularity each year. After some time, a group led by Messali Hadj, formed in 1947 as the *Mouvement pour le Triomphe des Libertés Démocratiques* (MTLD) and aiming at full independence, steadily increased its enrollment, due mainly to popular disappointment with French postwar political changes. From the MTLD eventually developed the *Front de Libération Nationale* (FLN), which, on November 1, 1954, launched a war for independence.

Although the FLN began with only a few hundred guerrillas and a few thousand underground members, it was able by 1956 to lead a force of 8,050 armed guerrillas supported by 21,000 underground auxiliaries.[1] The total force was later expanded to an estimated 60,000 men. Its goals were achieved primarily through a prolonged military, political, and diplomatic campaign. By preventing the French forces from winning any big military engagements, it sapped the strength of the army and the French treasury, while also wearing down the will to fight of the French Government and people. Through the skillful use of propaganda, terrorism, and organizational techniques, it was able to gain the support of the great majority of Algeria's Muslims and become the spokesman for the Algerian people. By obtaining the help of Tunisia and Morocco, as well as other African, Arab, and Communist countries, and by playing upon Western colonialism, it was able to put diplomatic pressures on France to negotiate a settlement. The French Government elected on January 2, 1956, on a platform of democratic reforms and peace by negotiation in Algeria, promised reforms based on free elections with a single electoral college which would have assured Mohammedan self-determination. But Premier Guy Mollet was unwilling to carry through his own reforms. This situation caused Ferhat Abbas to join the National Liberation Front, brought about the resignation of Mendès-France from the Cabinet, and further strengthened the revolutionaries. Soon the FLN dominated most of the rural, mountainous areas, while carrying on demonstrations and terrorist attacks within the cities. In 1958 the Algerian situation led to the fall of the Fourth Republic.

Between 1957 and 1960 the French built heavily fortified lines along the Tunisian and Moroccan borders to stop the flow of supplies to the FLN. The French regrouped thousands of villagers into "protected areas" and a Special Administration Service group carried out social-political pacification measures. By 1960 the French had increased their forces from 50,000 to 800,000,[2] using 50,000 mobile reserves, made up of legionnaires and paratroops, with helicopters and close-support aircraft and artillery. However, most of the troops were tied down with garrison duty and the mobile forces were not sufficient to force a military decision.

In September 1959 President De Gaulle offered self-determination to Algeria, and in March 1962 the FLN obtained a cease-fire agreement with France which permitted the Algerians to hold a self-determination referendum. The result of the referendum, held July 1, was independence for Algeria, with some economic and cultural ties with France. Preparations for the referendum—and for Algerian-French cooperation generally—were jeopardized, however, by the formation of the Secret Army Organization (OAS). Composed of French Army officers and Algerian-European civilians, the OAS sought to "keep Algeria French." It waged a terrorist campaign against the Muslims and in metropolitan France in an effort to continue French control over Algeria and perhaps overturn the French Government which had negotiated peace with the FLN. The movement was not large and since the capture of its leaders in April 1962, the OAS seems to have lost its effectiveness.

ORGANIZATION OF THE NATIONAL LIBERATION FRONT (FLN)

The precursor of the FLN, *L'Organisation Secrète* (OS), was founded in 1947 by several men, most of whom were to remain at the forefront of the revolutionary leadership: Mohammed Ben Bella and Mohammed Khider from Oran; Hussein Ait Ahmed, Belkacem Krim, Amar Oumrane, and Rabah Bitat from Kabylia; and Mohammed Boudiaf, Mustapha Ben Boulaid, Larbi Ben M'Hidi, Lakhdar Ben Tobbal, and Yussef Zirout from Constantine. Working clandestinely, they began to recruit members, train people in techniques of guerrilla warfare, and perform occasional acts of violence. (In 1949 the group engineered an assassination attempt against Governor General Naegelein.) In 1950 the French authorities learned of the existence of the OS and arrested Ben Bella, Ben Boulaid, and Zirout. Ait Ahmed and Khider thereupon fled to Cairo, while other members hid. The captured men soon escaped, Ben Bella going to Cairo, the others to the hills. Loss of its leaders

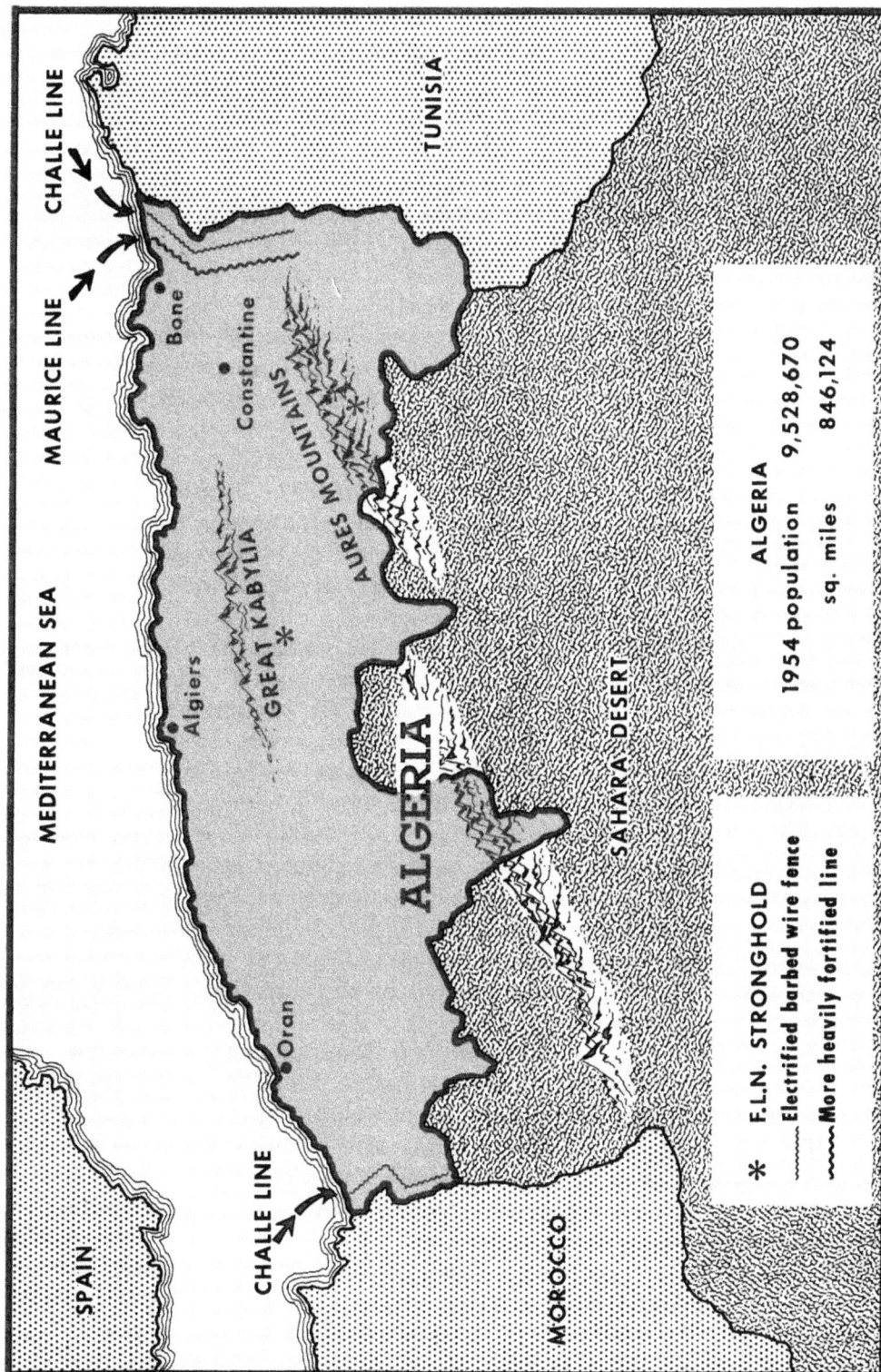

Figure 12. Map of Algeria.

SPAIN

MEDITERRANEAN SEA

CHALLE LINE

MAURICE LINE

CHALLE LINE

TUNISIA

Bone

Constantine

Algiers

GREAT KABYLIA

AURES MOUNTAINS

ALGERIA

Oran

MOROCCO

SAHARA DESERT

ALGERIA

1954 population   9,528,670

sq. miles   846,124

*   F.L.N. STRONGHOLD

———   Electrified barbed wire fence

∿∿∿   More heavily fortified line

severely crippled the OS; the French authorities believed it was destroyed completely.[3]

In July 1954 the main conspirators who had been continuing the organizational activities reorganized formally as the *Comité Révolutionaire d'Unité et d'Action* (CRUA) and laid final plans for revolt. Boudiaf, Ait Ahmed, and Ben Bella comprised the "external" leadership, while Boulaid in The Aures, Didouche in Constantine, Bitat in Algiers, M'Hidi in Oran, and Krim in the Kabylia mountain region constituted the "internal" regional leadership. With the outbreak of the revolt on November 1, 1954, the CRUA merged with other groups in the MTLD who favored armed revolution, and named themselves the *Front de Libération Nationale* (FLN).[4]

In planning its basic strategy, the FLN operated on the assumption that if there was indeed no Algerian nation, as the French claimed, then it would have to create one.[5] The fundamental task, then, was to make a cleavage between Frenchmen and Algerians which would leave the latter no real choice but to support the rebel cause. Thus all its propaganda and military activities were geared to show that it was acting in behalf of the entire Algerian people, who were an entity separate and distinct from the citizens of France. It recognized the crucial importance of convincing not only the Algerians themselves, but also the rest of the world—including Arab neighbors—that Algeria was not, as the French claimed, part of France.

After establishing a base near the Tunisian border, the FLN was soon established as the *de facto* government of about one-third of Algeria, levying "taxes," administering "justice," etc. In many cases, they established governments in areas where local governments had never before existed. Because even Tunisia and Morocco were loath to antagonize France by recognizing a separate Algerian regime, the FLN delayed formally organizing a "provisional government" until September 1958. Once it did so, however, it was promptly recognized by the Arab States, the Communist bloc (although the Soviet Union withheld recognition until 1960), and many African states. By establishing a provisional government, the FLN not only enhanced its position among both the Algerians and outside supporters, but also set itself up as the official spokesman of the Algerian people, the single entity with which the French would be able to negotiate peace.

While the FLN assumed that the United States and the countries of Western Europe would probably not alienate their NATO ally by granting it open recognition, it still sought support among liberal groups in these countries, and tried to persuade politicians to press France to negotiate with it directly. The annual U.N. debates on Algeria were also valuable to the FLN in publicizing its cause and making France's allies apply some pressure on Paris to try to end the war.

Revolutionary strategy within Algeria was based largely on lessons learned from Mao Tse-tung and Vo Nguyen Giap. However, the basic plan of spreading control was the "grease-spot" technique which had been employed by Marshal Lyautey in pacifying Morocco 40 years before. Thus the FLN guerrilla fighters, or *fellaghas*, planned to take control of a few isolated spots, and then spread their influence from these spots over increasingly larger areas.[6]

Important military and political decisions were made at the "Congress" called by leaders of the "internal" forces in the Soummam Valley in August 1956. The rebel army was formally baptized the *Armée de Libération Nationale* (ALN), and a regular command structure was established. Operational theaters, identical with the earlier civil divisions, were divided into six *wilayas*, each headed by a colonel. (The nationalists did not want to name anyone a general because they feared the growth of personal power.)[7] *Wilayas* were in turn subdivided into zones, regions, and sectors, in which operated battalions of 350 men, companies of 110 men, sections of 35 men, and groups of 11 men.[8] With but a few minor changes—addition of another *wilaya* comprising the East Base, or Algerian-Tunisian border area, and the enlargement of battalion units to 600 men and companies to 150—this structure remained intact throughout the war.

Over this hierarchy was placed a five-man general staff, the *Comité de Coordination et d'Exécution* (CCE), originally consisting of Ramdane Abane, Krim, Zirout, Benyoussef Ben Khedda, and Ben M'Hidi. A year later, following a decision to merge the internal and external forces and make the former equal in authority with the latter, the membership of the committee was

(UPI Radiotelephoto)

In the center is Algerian Rebel Provisional Government Premier, Ferhat Abbas. To his right is Foreign Minister Belkacem Krim. Interior Minister Lakhdar Ben Tobbal is at the left.

enlarged to 14. Several political leaders were appointed members and the CCE was given broad executive responsibilities.

The major political decision made at Soummam was the creation of a *Conseil National de la Révolution Algérienne* (CNRA), consisting of 17 full members and 17 associates. Representing all factions within the FLN, it included the "nine historic chiefs" of the CRUA, three leaders of the "interior"— Zorout, Abane, and Oumrane—and three "political" leaders—Ferhat Abbas for the UDMA, Benyoussef Ben Khedda and Mohammed Yazid for the centralists of the former MTLD. It was given authority over the CCE by a provision stating that there should be "priority for the political over the military organization." A year later, however, the interior or military was given the same importance as the external or political, and CNRA was enlarged to include 54 full members.[9]

On September 19, 1958, the FLN formed the *Gouvernement Provisionel de la République Algérienne* (GPRA). Cabinet posts included a Premier (originally Abbas, later Ben Khedda), several Vice Premiers, including Krim, who also held the post of Minister of Armed Forces, and Ministries of Interior, Communications and Liaison, Arms and Supplies, Finance, North African Affairs, Foreign Affairs, Information, Social Affairs, and Cultural Affairs. Posts were distributed among leaders of both the internal and external groups. Decisions made by the GPRA affected all members of the FLN, inside or outside Algeria. When peace negotiations with France were begun, the Provisional Government took steps to tighten its control over the military forces to make sure that political decisions, once reached, could be carried out in Algeria.

Orders from the CNRA which required implementation by the urban undergrounds were generally transmitted by courier to the heads of networks in the key cities. The most important network, that of Algiers, was organized in 1956–57.

Following the Soummam Congress, political-military operations in the "autonomous zone of Algiers" were conducted not by one man, as had been the case previously, but by a council consisting of a political-military chief and three deputies charged respectively with political, military, and supply and liaison activities. Each of the three deputies had under him three men responsible for carrying out his respective activity in each of the three regions into which Algiers was divided. (Region 1 included the two largest sections of the Casbah, Region 2 included the remainder of the Casbah and Western Algiers and its suburbs, Region 3 included Eastern Algiers and its suburbs.) Each of these men, in turn, had subordinates in each of the three sectors into which the regions were divided, and each sector chief had subordinates in each of the three districts into which the sector was subdivided. In theory, a council composed of the heads of each of the three activities was to be formed, but in practice the same man often performed two or perhaps all three duties.

The military branch of each region consisted of three groups, each of which included 11 men: a chief, his lieutenant, and three cells of three men each. Including the regional chief and his deputy, there were thus 35 armed

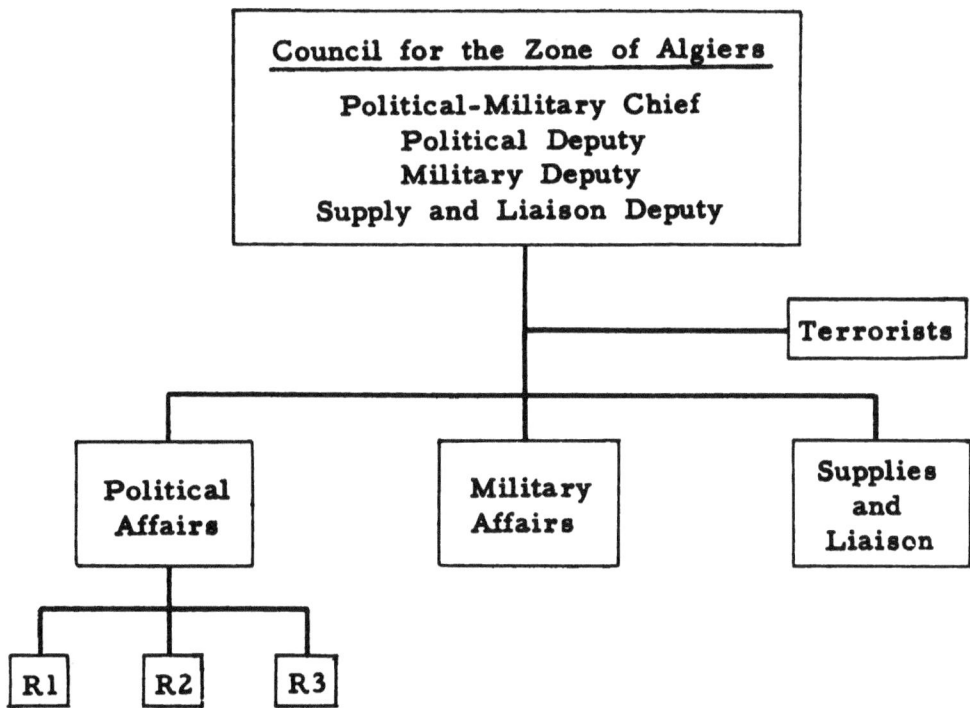

Figure 13. *National Liberation Front (FLN) Political-Military Organization in Algiers.*

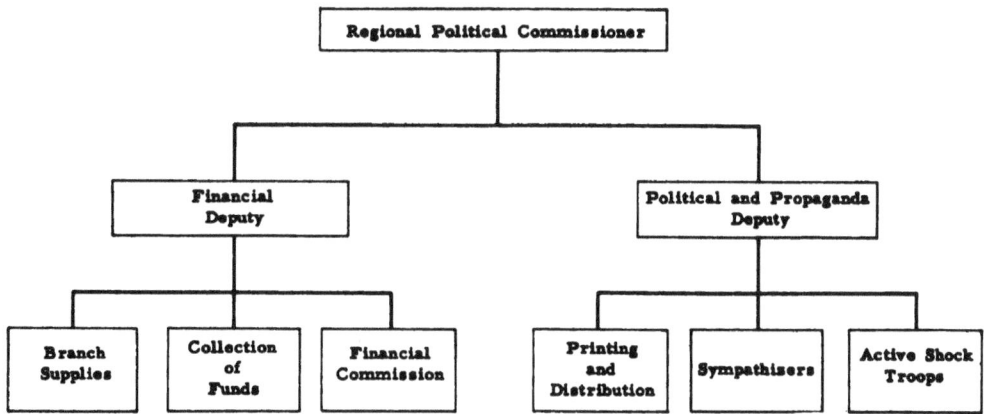

Figure 14. *National Liberation Front (FLN) Political-Financial Organization in Algiers.*

men per region, 105 in all Algiers. In addition to these "military" persons charged with protecting FLN members and their activities were the 50 to 150 hard-core terrorists. They, in turn, often used known gangsters or unemployed persons in terrorist activities.

Although the French authorities destroyed most of this underground apparatus at the end of 1957, it was later reconstituted along similar lines, but on a much larger scale.

Cases of Undergrounds

The political-financial organization is shown in figure 14 but an elaborate system directed by one of the political heads or business leaders of each region was organized to supply the bulk of funds.[10]

ACTIVITIES OF THE NATIONAL LIBERATION FRONT

ADMINISTRATIVE FUNCTIONS

Communications

Communications constituted a problem for the ALN, and facilities were often inadequate. Radio sets were captured or purchased abroad from many countries and were often secondhand. After 1959 a few German Telefunken radios were used both by rebel headquarters in Tunis and by the *wilayas* operating in Algeria, but below the *wilaya* or battalion units most communication was by runner.[11]

In the urban centers, such as the major cities of Algeria, and Paris and Lyon in France, communication was largely by word of mouth. A courier would receive a message from the CNRA or the provisional government, and deliver it to the head of the local network who in turn relayed it to his lieutenants. In Algiers, communication in the Casbah was relatively easy because of the heavy concentration of Muslims in the one area. The close proximity of dwellings, the overcrowding of most homes, and the fact that three generations of a family often lived together greatly facilitated the task of informing large numbers of people within a very short time. In relaying messages, such as orders for a general strike, use was made of educated or influential people who enjoyed respect in the community, or of such people as shopkeepers whose places of business were frequented by many people, and who could, in the course of a day, speak to many people without arousing suspicion.

Recruitment

The original organizers of *Organisation Secrete* first sought support solely among members of the MTLD and among other people already known to favor complete independence from France. Some members of the organization were known to the French authorities as militant nationalists, and were thus under surveillance or, at times, arrest. Others, however, appeared to be cooperating with the legal authorities and worked not only in open, legal jobs but even in respectable posts within the administration. Abderrahmen Kiouane, for example, served as Deputy Mayor of Algiers before leaving the city to join the "external" branch of the FLN.

While the recruiting of political cadres was taking place, mainly in the cities, such men as Khider and Ait Ahmed were recruiting people in the hills and rural villages for the guerrilla forces. Sometimes they would enlist a local chieftain in their cause, and he in turn would get members from his tribal group. Otherwise, individuals who showed ability and dedication were selected and they in turn recruited their friends. To avoid detection much of the training took place at night, with the future guerrilla fighters returning to their normal activities in the village during the day. Some mountain areas were so remote from the French administrative sway, however, that guerrilla training could take place virtually out in the open. Early recruiters had to be careful to recruit people who could be trusted not to reveal, consciously or unconsciously, their activities to the authorities or to anyone who might inform the authorities. At the same time guerrilla forces were being recruited and trained in Algeria, Boudiaf began organizing an underground of Algerian nationalists in Paris and training them for later acts of terrorism and propaganda actions.

With the launching of the revolt and the official formulation of the FLN, the rebels started their drive to gain support from all Algerians. Their name, National Liberation Front, was chosen to indicate that they were fighting not in the name of any one political, religious, or social group, but in the name of all Algerians. They had immediate success in gaining the adherence of the "centralist" faction of the MTLD, several members of which had been cooperating with the OS just before the start of the revolt. As the war continued and rebel successes mounted, and the FLN made it increasingly clear that it was determined to fight until France made major concessions, other organizations and persons were put under increasing pressure to join the FLN or be classed as traitors. Messali refused to join, forming his own group, the *Mouvement National Algérien* (MNA), which also participated in guerrilla operations and some political activities. By 1956 the UDMA, the Ulemas (an influential society of religious scholars), and members of the "Committee of Sixty-One" (composed of Muslin members of the Algerian Assembly) joined the FLN.

The guerrilla army, which began with only several hundred full-time fighters, grew rapidly. Volunteers were induced to join not only by propaganda appeals, but by the offer of pay for what was made to appear very respectable, if not exciting, work. Thousands of unemployed or marginally employed men were easily induced to join. To illustrate that they envisaged a social as well as a political revolution—and simply because all kinds of help was needed—the FLN also organized corps of women. In areas where soldiers were needed but not enough volunteers were forthcoming, terrorism was used to force men to join the guerrillas; a village or a family would be threatened with reprisals if it did not send a specified number of "volunteers." The ALN also used many part-time guerrillas—i.e., people who would participate in a rebel operation at night, but return to their normal activities during the day. Bandits or freebooters were also used, but were seldom considered regular members. Training centers were set up in the hills and soon a large training camp was established in Tunisia. The total strength of the ALN was perhaps 50,000 to 60,000 men at peak strength (although French sources placed the maximum number at 30,000).[12]

Cases of Undergrounds

Individuals were also recruited for underground work in the cities. Kiouane and Ben Khedda worked in Algiers for some time organizing people for intelligence, propaganda, and supply operations. In the early stages, known gangsters whose word was obeyed by Muslims were enrolled to help ensure that the FLN orders would be obeyed. The organization of Muslims in the Casbah was significant not only for underground activities per se, but also for the psychological weapon it gave the FLN when it wished to show that it really spoke for the entire Algerian people, and when it could organize a mass demonstration of Muslims from the Casbah to prove it. Ben Khedda also organized an all-Algerian labor union, the *Union Générale des Travailleurs Algériens*, in 1956, drawing workers away from the Communist-dominated French *Confédération Générale des Travailleurs*.[13] Various student groups and other auxiliary organizations were also formed within France, the FLN organized a large network, part of which operated underground, collecting intelligence, obtaining money and some supplies, etc.

In the early stages of the war, the FLN concentrated not only on getting adherents as such, but on winning over the most respected Algerians in order to "prove" the worth of its cause. By the end of 1956 it had largely succeeded in this; very few Algerian Muslims would consent even to hold a minor office within the French administration. People from all backgrounds and all organizations—including Communists—were accepted into the FLN with one proviso: leaders had to dissolve the organizations which they had headed, and individuals had to break completely with any organization they had belonged to. Former Communists who were discovered to have retained allegiance to the Communist Party were liquidated.

Finance

A great deal of money was raised within Algeria itself. In the hills, the FLN imposed "taxes" on inhabitants of the areas it controlled. Threats, and often acts, of severe reprisals forced compliance when persuasion was inadequate. In the major cities, the FLN underground also collected "taxes" or assessments from people, again aided by force or the threat of force. In France, and even in Belgium, money was also collected from workers and students. The *Union Générale des Travailleurs Algériens* (UGTA), operating in the cities of both Algeria and France, obtained sizable amounts of money from workers. Rivalry, particularly in France, between the FLN and the *Mouvement National Algérien* (MNA)* and their respective unions caused many workers to contribute money to both movements.[14] Acts of terrorism, including murder, were frequent here.

Despite the large amounts collected from Algerians, the FLN would not have been so successful without outside financial help. Tunisia and Morocco, and the Arab League countries, especially Egypt, contributed large amounts. Several independent African countries made at least token contributions.

---

*MNA was founded by Messali Hadj. Although he was the original leader of the MTLD, from which the FLN developed, Messali refused to join the FLN.

Yugoslavia supplied large amounts of money—and also helped in many other ways, such as supplying care for FLN wounded. Communist China, from the start, supplied money to the Algerians based in Cairo. The Soviet Union gave little direct help.

Logistics

Most of the food used by the rebels—principally unleavened bread, peppers, coffee, *cous cous*, mutton, rice, and goat's milk—was obtained from villages. In areas where the rebel forces held *de facto* control, arrangements would be made for each village to supply specified amounts of certain foods. Where villagers were reluctant or unwilling to comply, threats of reprisals obtained cooperation.[15]

Although the *fellaghas* at first had no uniforms, by 1956, when the ALN was formally organized, soldiers had regular uniforms. Most of them were captured from French supply depots, others were bought from French or other sources or donated by Communist China. As Tunisia became more involved in the war, the Tunisian Government, working through Tunisian Merchants, obtained uniforms, coats, shoes, etc., for the FLN. Medical supplies were also captured, purchased abroad by Algerian representatives, or bought in Tunisia or sometimes Morocco.

Arms were acquired from a wide variety of sources. A large amount were captured from French depots, or bought from renegade French and other European sources. The Egyptians sent arms in fairly large quantities, the other Arab countries in small amounts. The wide variety of weapons and equipment included German automatic weapons, Bren guns, French semiautomatic rifles, Lee Enfields and Mausers sent from Egypt, U.S. cartridge belts, British water bottles, etc. In the way of heavy weapons, the FLN had German 81-mm. mortars and antitank mines, 20-mm. Bofors cannon, bazookas, and recoilless rifles, bangalore torpedoes from Egypt, and large numbers of plastic bombs and hand grenades.[16] Sometimes *fellaghas* tricked the French into giving them supplies. On several occasions Algerians went to the French authorities and claimed that they wanted to fight against the FLN. The French gave them supplies and arms, and after getting all they felt they could, the *fellaghas* then returned to the FLN.

Although China and some bloc countries offered arms, the Algerians accepted only small quantities, including some Czech weapons. Their reluctance to accept Communist aid stemmed both from the desire not to become involved with them and from the difficulties in getting replacement parts and ammunition for such weapons.[17]

Supplies entered Algeria in several ways. Those procured in Morocco and Tunisia were slipped over the border; those sent from Egypt went south through Libya and then into Algeria; those purchased abroad were sent by ship usually to ports in Egypt, Morocco, or Tunisia. Several such shipments, such as one sent from Yugoslavia, were captured by the French. Procurement of materials was directed by the Minister of Arms and Supplies in the Provisional Government.

Cases of Undergrounds

Security

Within the army, the security of each unit was generally the responsibility of the Deputy for Political Affairs. Any indications of disobedience, desertion, or disloyalty meant death.

Within the urban underground networks, strong discipline was enforced to prevent betrayals. Any "traitor" was punished by death if caught. If any member of the three-man military cells in Algiers was caught, the other two members of his cell were sent to the mountains to fight, so they could not be questioned by French authorities.

OPERATIONAL FUNCTIONS

Psychological Operations

Propaganda was of crucial importance to the FLN in every phase of activity. With the ALN, each unit commander was assisted by a Deputy for Political Affairs who was responsible for indoctrination and "ideological solidarity" within the unit. The army also received indoctrination through radio broadcasts from the Provisional Government in Tunis.

Great stress was placed on spreading propaganda abroad. For not only did the FLN face the problem of proving to the world at large that its war was justified; it had to prove, first, that Algeria was not part of France, and second, that it ought to be completely independent. In addition to printing up large amounts of propaganda literature in Cairo—and then Tunis—for distribution throughout the world, the FLN also opened several offices in important cities abroad. Members who worked in these offices then established contacts with the local press, delivered speeches wherever possible, and contacted government officials who might be sympathetic. It was hoped that these people would seek to have passed parliamentary resolutions or acts favoring independence for Algeria—or at least nonsupport for France—or possibly take steps to pressure France into recognizing the FLN as the spokesman of the Algerian people and negotiate directly with it. The United States was a major target of FLN propaganda, as Washington had on many occasions helped France diplomatically and materially and was presumed to exert strong influence on her. The U.N. debates also provided an excellent forum, providing press coverage the FLN might not have otherwise received.

As a means of getting direct help from governments or important individuals, the FLN tried to show that the government or person in question would receive benefits from an independent Algeria, be it bases, economic rights, etc. Sometimes individuals were given a direct stake in the outcome of the war. Several oil companies, such as the Italian concern headed by Enrico Mattei, allegedly gave money to help the Algerians in return for promises of oil concessions.

The FLN Radio Cairo and Radio Damascus were able to maintain close contact with the Algerian masses. The constant distribution of leaflets and the weekly distribution of the rebel newspaper *El Moujahid* (The Fighter), which was printed in Tunis clandestinely, provided further propaganda coverage. Display of the green and white FLN flag for Algeria and the singing of patriotic songs encouraged nationalist feelings, while even the mere formal organization and use of uniforms helped keep army morale high. Within Algiers, each political-military unit had a typewriter and a mimeograph machine to print propaganda.

In spreading propaganda to the public at large, the FLN used a wide variety of appeals and techniques. To the people with some education and some general understanding of politics, it distributed vast quantities of literature—printed in Cairo, Tunis, or Algerian cities—explaining why Algerians could no longer live under French rule, why fighting was necessary to achieve independence, and that by cooperating with the FLN they would at last have a chance to do something to serve themselves and their "country." The FLN also stressed that it wished to effect an economic and social, as well as political, revolution, including such basic reforms as equality for women. To religious conservatives and to less educated people went appeals that this was a war for the defense of Islam and Arab culture. *El Moujahid* and other tracts were distributed by the underground in the cities and in France, and radio broadcasts supplemented this written propaganda.

Propaganda was also aimed at the people of France. Frenchmen were told that the war waged by France was unjust, that the FLN was justified in fighting for independence, that the very principles invoked by the FLN were learned from the French Revolution, etc. Some French journalists and scholars accepted money from the FLN to write books and pamphlets supporting its cause. Some existentialists, such as Jean-Paul Sartre, said that the FLN represented progress, and wrote tracts supporting the Algerians. Many church groups either openly supported Algerian independence or indirectly helped the FLN by stating that some of the counterguerrilla measures employed by the French Army, such as the use of torture to obtain information, were un-Christian. The FLN was able to call strikes and boycotts, while its underground organized Muslims in the major cities for demonstrations. This was of great importance, for example, during De Gaulle's tour of Algeria in 1960, when thousands of Muslims from the Algiers Casbah came forth shouting in favor of both De Gaulle and the FLN.

Terrorism was used in both the hills and the cities of Algeria and France. Uncooperative villagers, or those who showed a preference for the rival MNA, suffered reprisals. Often, the threat of such reprisals sufficed to obtain cooperation. Muslims in cities were also killed, or threatened with death by the military groups or terrorists if they did not cooperate.

In the cities, a number of bombings took place, mainly for the purpose of creating general disorder and weakening France's claim to being able to keep order. Bombs were usually either stolen from the French or, more often, made by the underground itself from materials which could be bought or stolen. Acts of sabotage were committed in both Algeria and France, and directed by leaders of the underground apparatuses.

Cases of Undergrounds

Sometimes the mere threat of violence brought large psychological advantages. For example, in 1957 the FLN announced that certain tourist ships or planes leaving France would contain bombs, and warned foreign tourists to boycott French carriers. Although no such bombs actually exploded, fear caused a large drop in French profits from the tourist trade.

All Muslims who held, or who seemed interested in holding, offices under the French administration were threatened with assassination and a few were killed. An example was thus made of Aly Chakal, well-known collaborator who held a seat in the French Assembly, who was killed by the FLN in Paris in 1957. While organizing the Algerian undergrounds, FLN leaders directed assassinations of French officials to indicate the impotence of French authority and the strength of the FLN.[18]

Intelligence

Within the army, each unit had a liaison and intelligence officer who usually directed the gathering of intelligence, as well as supplies, from neighboring villages. He maintained contact with the huge number of civilian auxiliaries who served as "human radar," scouts, intelligence agents, and guides. These *moussebellines* infiltrated French-held villages, prowled the terrain ahead of regular FLN columns, and provided a steady stream of fresh information.

Additional information was supplied by deserters from the French Army, both Muslim and (occasionally) French, and by Muslims who worked in the administration and then defected to the rebel cause.

In the major cities the FLN instructed people to report on the daily activities of French police and armed forces. In Algiers, it also enrolled large numbers of people who had been employed by the French as intelligence agents or spies, and used these double agents to obtain information of French administration measures and troop movements.[19]

COUNTERMEASURES AGAINST THE NATIONAL LIBERATION FRONT

At first the French Government—and most army officials—felt that the uprising in Algeria was simply the work of a "handful of ambitious terrorists," egged on by Cairo, who could be fairly easily subdued. They did not recognize that this might be a well-planned revolt led by people determined to win independence. Military actions were thus piecemeal and ineffective, although their announced aim was to stamp out the revolt completely. Coupled with this underestimation of the actual and potential strength of the FLN was the refusal to conduct any negotiations with the rebel organization.[20]

In order to integrate Algeria more fully with France, a serious psychological, as well as military, campaign was begun to combat the rebels. In 1956 "pacification," or the campaign to "win over the populace," was instituted. It was directed initially by General Salan.[21]

The pacification program, both its civilian and military aspects, was controlled completely by a Resident General in Algiers. Proceeding on the assumption that the French Government would have to promise, and begin to implement, reforms in those areas of mass discontent upon which the rebels could capitalize, a large-scale reform program was begun in the countryside and in the cities. To this end, over one thousand army Special Service units were sent into rural areas.

Each unit was headed by a French Army officer of company grade and staffed with an assistant and a secretary, the latter often the officer's wife. The officer generally opened his residence and headquarters in the midst of an Arab settlement or village, and prepared to administer to the needs of the people, as well as to organize, arm, and train a local self-protection force. The main effort centered on improving the living conditions of the native population. French Army volunteers organized and taught school classes; helped build homes, sanitary facilities, and water supplies; demonstrated improved agricultural and health practices; provided medical services and evacuated the ill and injured to hospitals for surgery. Women physicians and civilian employees taught the Muslim women how to care for their babies.

In an effort to persuade the people that combatting the rebels was a matter of direct personal interest to them, they were asked to assist the pacification effort. Village home defense and raider units were organized, and participants were rewarded for deeds well done. All such operations were accompanied by wide-scale propaganda stressing the greatness of France, France's respect for Algerian Muslim institutions, France's genuine desire to help the people, etc. Propaganda was spread by word of mouth, as well as by leaflets and radio. Each French soldier was taught to be an agent of pacification, and to master political and psychological activities as well as military operations.[22]

In the major cities there was little need for self-defense units, but there was great need for organizing people to make sure they obeyed the laws, and did not cooperate with the rebels by supplying them with intelligence, or by carrying out sabotage operations. In addition to undertaking a large propaganda compaign, including not only the above-mentioned instruments but also rallies and meetings, the French organized a hierarchical control system in which one member of each family was responsible for knowing the whereabouts of the family. The responsible member reported to a floor chief, and, in turn, the floor chief reported to the building chief. All building chiefs reported to a block chief. And so on, up to the highest level.[23]

Pacification won over the population in some areas. In many others, however, it was either ineffective, or actually drove the people over to the cause of the FLN, which promised better reforms without any of the restrictions imposed by the French, and which claimed credit for forcing the French to adopt whatever reform measures where instituted.

Even after pacification had begun, Premier Mollet, in mid-1956, tried to negotiate secretly with the rebels in order to arrange a cease-fire on a basis of granting internal autonomy to Algeria. Some of his subordinates, however, sabotaged these efforts by capturing the airplane containing the five Algerian negotiators. After this, the rebel line hardened and military operations were stepped up. Subsequent peace feelers under the Fourth Republic produced no results, and the official French policy continued to be nonrecognition of the FLN as a spokesman of the Algerian people.

In September 1959 President De Gaulle offered the rebels self-determination and the right to opt for independence. The Algerians subsequently accepted this arrangement in principle, and there followed 2 years of peace feelers and three meetings between Algerians and Frenchmen climaxed by the signing of the Evian Agreements in March 1962. Complete sovereignty over Algeria, including the entire Sahara, was to be granted if the people voted for independence in the referendum, later scheduled for July 1, 1962. The situation was complicated, however, by the uncertain attitude of the French Army, many of whose officers chafed at the idea of being driven from Algeria, and because the Secret Army Organization (OAS), operating in Algeria and composed of French Army and settler elements, seemed determined to launch a civil war, if necessary, to "keep Algeria French."

(UPI Photo)

The four ex-generals responsible for the revolt of army elements and the Europeans in Algeria against the French Government are (left to right): Andre Zeller, Edmond Jouhaud, Raoul Salan, and Maurice Challe.

ORGANIZATION OF THE SECRET ARMY ORGANIZATION (OAS)

Early in 1958 a group of high-ranking French military officials in charge of operations against the FLN in Algeria began to demand more effective government support in France for the campaign. The hero of the Second World War, Charles De Gaulle, reappeared upon the political scene and assured the military that he would, with the proper constitutional authority, work to settle the "Algerian problem" in an "honorable" way. The General declared that Algeria was "organically French." He was installed as President through a virtual coup d'etat, and given very broad powers.

Within a short time, however, the President of the new Fifth Republic began to make peace bids to the rebels and finally offered Algeria "self-determination." The men who were responsible for putting him into office felt De Gaulle had betrayed them. After several unsuccessful attempts to gain more authority, the military junta organized under General Raoul Salan and supported by elements of the army and Foreign Legion attempted a military coup termed the "revolt of the generals," in January 1961. After De Gaulle put down this putsch and disbanded the disloyal army and Foreign Legion elements, remnants from these groups, coupled with European Algerians, formed the nucleus of the Secret Army Organization.

At the time it was organized, the OAS declared that its goal was "to keep Algeria French." After the signing of the Evian Agreements, a faction led by the Algerian student leader, Jean-Jacques Susini—who had been considered the theorist of the organization—indicated it would be satisfied if it merely obtained increased privileges for Europeans within an independent Algeria. At the same time, a few army officers, working with such metropolitan French leaders as Georges Bidault, a member of the French Parliament and former Premier, indicated that their basic aim was to overthrow De Gaulle.

Whatever may have been the actual goals of its leaders, several unusual factors shaped the strategy of the OAS. It could not hope to win the support of the majority of the people of the country in which it carried out major operations. At most, it could gain the allegiance of only one-ninth of the total populace—the Europeans plus a few Muslims. Within France, its strength was confined to a minute fraction of the general populace. Furthermore, whether operating in Algeria or in France itself, the OAS had to draw support from, and operate in, not the rural or mountainous areas, but the few major cities. With its potential numerical strength small and its area of operations sharply limited, the OAS therefore concentrated on one major activity: terrorism.

Its organization was intended to be military in character. It was headed by a supreme council which controlled five staff divisions: the "Organization of the Masses," the "Military Structure," the "Organization of Information and Operations," "Political and Psychological Action," and the "Finance Department." The leaders were all veterans of French military campaigns and sev-

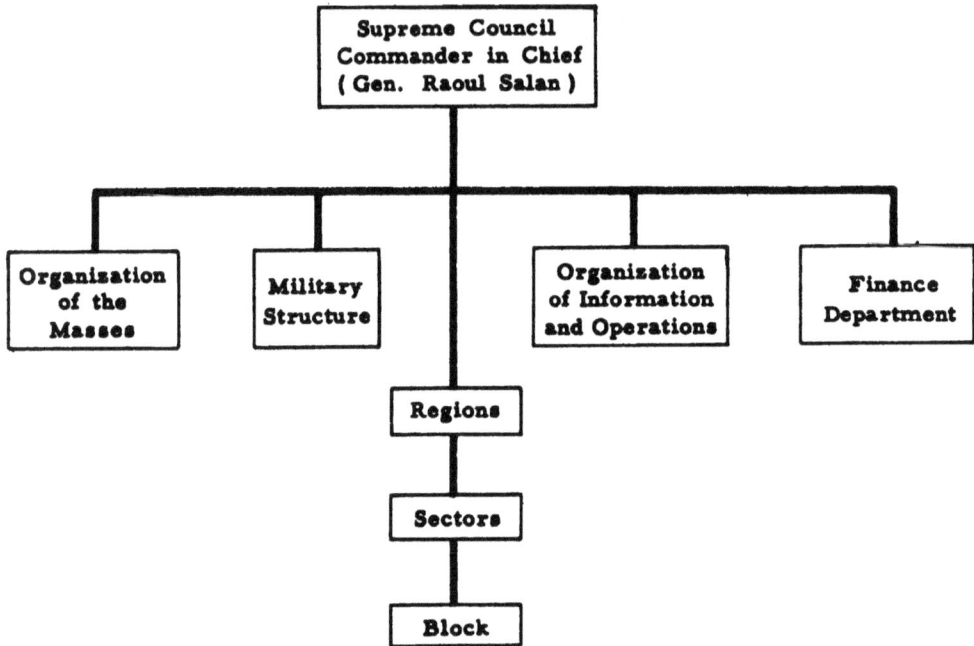

Figure 15. Organization of Secret Army Organization (OAS).

eral of them were experienced in psychological warfare operations. The OAS maintained geographical areas similar to those used by the French military staff. Each geographic region (with headquarters in Algiers, Oran, and Constantine) was divided into sectors, quarters, blocks, and, when necessary, single buildings. In Algiers, with a population of approximately one million, the region was divided into six sectors containing approximately 150,000 people, with a colonel in charge of the region, and captains in charge of the sections. The sector commander had absolute authority and no operations could be carried out in his sector without his permission.[24]

In January 1962 the high command of the OAS had about 25 members. There were possibly 2,000 terrorists and saboteurs in addition to 20,000 block leaders, spies, fundraisers, and agitators.[25] Besides having the passive support of perhaps 85 percent of the million Europeans in Algeria, the OAS hoped to recruit the active participation of 100,000 disbanded militiamen.

Geographical areas were also set up within France. Because of government countermeasures and general lack of popular support, operations in the homeland were not nearly so coordinated or tightly knit. The OAS early in 1962 claimed to have 7,000 members, including 500 plastic-bomb experts, working for the movement in metropolitan France.[26]

Although it was apparently well organized at first, rifts in the OAS began to appear soon after the capture of Generals Jouhaud (the second in command) and Salan in April 1962. There was sharp disagreement, for example, on whether to negotiate with the FLN for further guarantees for European rights, and thus give up the original goal of "keep Algeria French," or whether to continue an all-out fight against Algerian independence.

ACTIVITIES OF THE SECRET ARMY ORGANIZATION

ADMINISTRATIVE FUNCTIONS

Communications

A courier network operated in the urban centers to distribute instructions to the various branches of the OAS. The Secret Army claimed that no assassination was carried out without direct instructions from the area chief. Radio transmitters were probably used also by insurgent commands to communicate with each other, if telephone service was unavailable, or not feasible, for security reasons.

The communication network between agents in North Africa and France probably ran through Spain, since OAS officials resided in Madrid.

Information and instructions for the general population (e.g., the calling of a general strike) were transmitted by leaflet, poster, and radio and television broadcast. In "secure areas" news was easily spread by word of mouth. On several occasions the OAS "captured" local radio stations and made open broadcasts to the people, telling them what was expected of them. They used jamming devices to stop De Gaulle's broadcasts and messages to the people of Algiers.

Recruitment

The OAS drew its active support from three major groups: (1) the *colons*, many of whom were eager to maintain their privileged position in Algerian economic life, and were particularly active in the Operations Organization in Algiers, Oran, and Constantine; (2) career army officers, many of whom had great personal admiration for the leaders of the insurgent high command and felt a sense of frustration aroused by the unsuccessful counterguerrilla campaigns in Indochina and North Africa; and (3) veterans of the disbanded French Foreign Legion (none of whom were French, except some of the high officers). The last-named were generally anti-Arab and professional soldiers of fortune; they had taken part in the "revolt of the generals."

OAS leaders claimed to be in contact with important officials in the French Government and Army; they also received some support from the 60,000 Algerian-born Europeans in the French Army, many of whom were transferred out of North Africa.

The Organization of the Masses was charged with mobilizing the general (often passive) support of the *colons*, particularly members of the local police, firemen, and prison officials. This allowed the active personnel of the Secret Army to move openly in downtown areas of the three major cities of Algeria.

The OAS had hoped that the majority of soldiers in the French Army would not attack their fellow countrymen participating in the movement. However, soldiers eventually did fire on civilians attempting actions against Muslims.

Cases of Undergrounds

Training of new members involved giving the recruits "relatively simple tasks, such as carrying messages, driving automobiles or providing hideouts. Gradually, they become increasingly committed, and receive revolutionary tasks of violence." [27]

Finance

The OAS Finance Department had a good deal of success in Algeria proper, and also made determined efforts to collect money in France itself. The network of collectors in Algeria had little trouble, at least at first, in gaining the cooperation of Europeans, who were either in sympathy with the OAS or afraid of it, since the OAS made it clear that it would deal sharply with those *colons* who tried to resist. Bank holdups were common. In metropolitan France the OAS approached businessmen by sending them a note stating that a *percepteur* would soon come calling to collect funds. These people were taxed a specific amount, determined by their income and wealth. Receipts were given everywhere for "taxes" paid.

The money accumulated was used to purchase supplies, pay hired assassins, and give financial relief to the families of the OAS insurgents participating in the movement. Higher officials of the OAS also received specified salaries.

Logistics

Since the activities of the Secret Army were concentrated in metropolitan areas where there were minimal restrictions as far as acquiring the necessities of daily life was concerned, the main effort of logistics was gathering arms for the Secret Army "killer groups." These included plastic bombs, .45 caliber pistols and other small arms, bazookas, and submachineguns. Many of these weapons were brought from French Army depots by the many deserters who joined the movement. Police stations in the metropolitan areas were raided, often with the passive cooperation of the *pied-noirs* ("black feet," or Algerian-born Europeans).

The trademark of the OAS became the *plastique*. These bombs, made of TNT and a putty-like mass which is easily pliable, are safe to carry under varying conditions. The ingredients can be stored easily and special machinery is not needed for manufacture. Bombs and their ingredients were stored in homes, cellars, and cafes secured for the OAS by the Organization of the Masses. Most bombings occurred in Paris; in a city of three million, it was not difficult to find "safe homes" for the terrorists and their equipment. The *plastique* was used mainly for psychological effect because of the noise and damage it produced. It is not a discriminate weapon for assassination, but was quite successful in producing fear among the public.

Security

OAS security measures were less efficient. Double agents working for the French Government were able to infiltrate the OAS on several occasions.

One, acting as OAS courier between Paris and Algiers, provided French police with the information leading to the capture of General Salan.[28]

Originally, the high command of the OAS resided in downtown areas, moving about from one apartment and home to another, protected only by a shield of personal bodyguards. The OAS leaders even went so far as to be interviewed openly by newsmen. The capture of several leaders, however, indicated that the security system was lax. As Salan himself remarked:

> I saw too many people for silly reasons. People that I didn't know . . . that is probably how I was captured. But it was probable now, or later . . . what difference does it make? Everything was collapsing around us.[29]

OPERATIONAL FUNCTIONS

Psychological Operations

The Secret Army at first presented itself not as a group of insurgents but as loyal Frenchmen betrayed by De Gaulle. Subsequently it emphasized the theme that it was merely trying to obtain guarantees for the future for Europeans of Algeria. The Political and Psychological Branch of the OAS was directed by the former director of the French "5 ème Bureau" (Psychological Warfare Branch), and prepared political propaganda.[30]

Posters and painted signs urging the citizens to arm, and suggesting that the OAS "can strike who it wants, when it wants, where it wants," were seen on buildings and streets throughout the downtown areas of Algerian cities.

The OAS succeeded in interfering with broadcasts over the French national radio network, which appealed for support of the central government.

The OAS often used violence to frustrate countermeasures. This was usually done by exploding bombs or machinegunning people. Because the terrorists followed no set pattern, the population was never sure whether the next target would be a government official or a group of school children. This led to mass demonstrations for peace in Algeria, the control of which diverted the government from efforts against the Secret Army. Furthermore, terrorism was violently condemned by many factions of the French political scene, including the Communists; the OAS sought to exploit the many shades of reaction to produce a rift among the political leaders and divide French public opinion. Fully aware that terrorism would provoke the Communists into calling for public demonstrations which could easily turn into riots, the OAS wished to convince other political factions that the Communists were the prime menace. An emotional fear of communism in general might then force President De Gaulle to bring his policies more into line with those of the OAS, thus precluding any agreement between France and the FLN.

Immediately after the cease-fire was agreed on, the OAS began to attack Muslims in hospitals and schools, hoping that the Arab population would retaliate. This would require the French Army to step in and deal with the

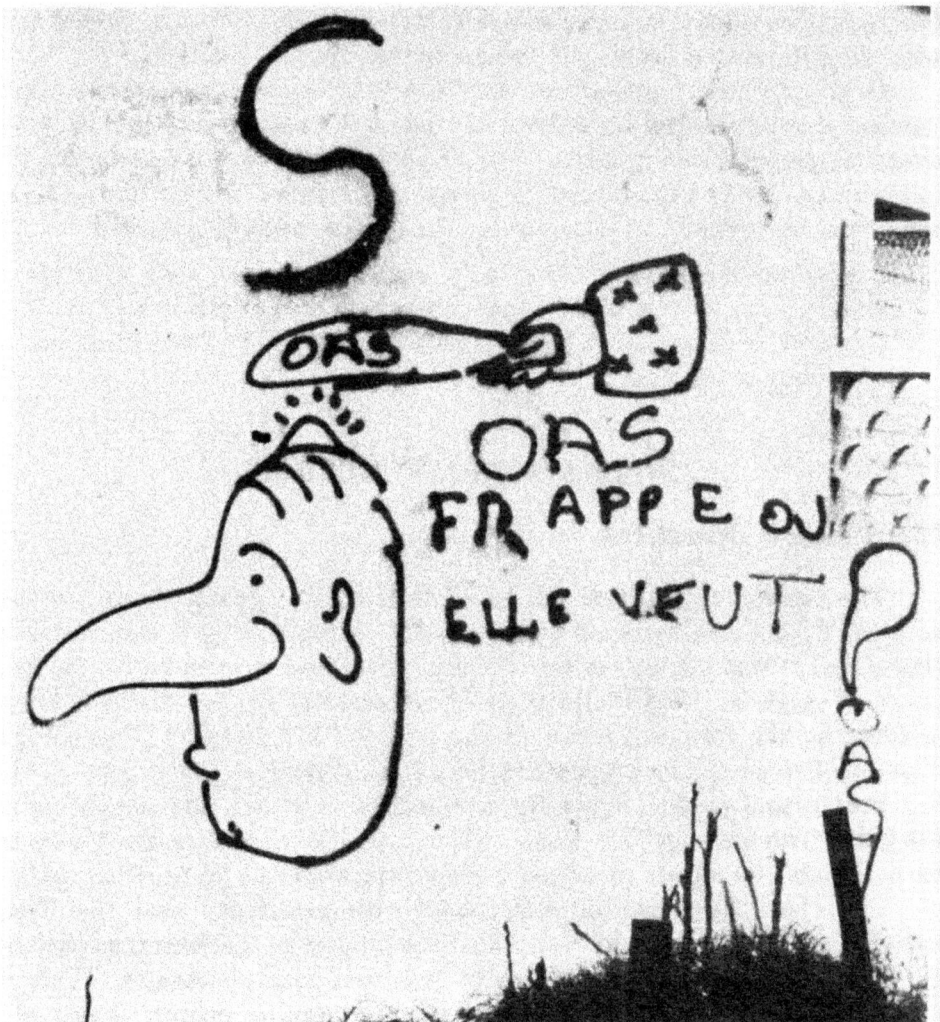

On a wall in Algiers a cartoon depicts French President Charles DeGaulle being struck on the head with a club marked "OAS." The inscription reads: "OAS strikes where it wants to."

Algerians, thus possibly reviving the war. This maneuver had little success, however; both the FLN underground leadership and the French Army saw through it and assigned men to the Muslim sections to ensure that no reprisals were carried out. Attempts were made on the lives of high French Government officials (including De Gaulle) to intimidate the authorities in Paris.

Intelligence

The Organization of Information and Operations section was concerned with assassination as well as intelligence. The combination of these two functions suggests that the main effort of the intelligence service was directed against individuals who were working against the movement. A particular effort was made to track down and murder government counteragents. When discovered, FLN agents were also executed.

The immediate attacks on French undercover agents arriving in Algeria offer some clue to the effectiveness of the OAS intelligence net. The leaders of the Secret Army assumed that the European population would not cooperate with the French authorities trying to stamp out the movement. Most of the *pied-noirs* and local police ignored the presence of OAS members. Therefore, movement was relatively secure in the European sectors of the cities. Native police officers sympathetic to the insurgents turned over records giving important information concerning government and FLN agents. It was reported that the movement had the full cooperation of police, firemen, prison wardens. It was also claimed that Soviet intelligence agents were used to supply information.[31]

Sabotage

Sabotage was used more for psychological reasons than as a weapon for physical destruction. When a plastic bomb was thrown at a cafe or a small store, it was done mainly to disrupt the commercial activity of the area. In addition, many automobiles of government officials were destroyed and police stations attacked in order to destroy security records and other government facilities. The most dramatic sabotage mission in France took place when a truck carrying documents belonging to the Interior Ministry was blown up. Bombs were also placed in the homes of Communists and French officials loyal to De Gaulle.

COUNTERMEASURES AGAINST THE SECRET ARMY ORGANIZATION

The reaction by the army to previous political developments had stimulated the French Government to prepare for violent insurgency on the part of Europeans once the cease-fire was agreed upon.

In Algeria, the French Foreign Legion, composed of foreign professionals except for French commanding officers, was disbanded in 1961. This denied the Secret Army access to many types of military supplies, although many ex-legionnaires wilingly offered their services to the OAS. Regular military units composed of men whose loyalty was questionable were also removed from Algeria and stationed on the continent.

The OAS was declared subversive. This permitted French police authorities to detain and arrest persons suspected of aiding, or sympathizing with, the goals of the clandestine group. Following appeals for mass support of government policies, a popular referendum was held which indicated that President De Gaulle had popular support in the measures he would take to carry out the terms of the cease-fire.

Cases of Undergrounds

French forces cordoned off European sectors of the large urban areas, making building-to-building searches in order to gather arms and make arrests. The city was sectioned off with barbed wire and barricades. Under the terms of the cease-fire, a joint Algerian-French security force was established which had official responsibility for preventing the Algerian population from retaliating against the violence of the OAS.

A counterintelligence force, popularly known as the *Organisation Clandestine du Contingent*, was also established. The government never openly admitted that this organization existed, but members of the OAS (who referred to it as the *Organisation Communiste du Contingent*) and other observers stated that it included many conscripts who did not share the opinions of their professional officers.[32]

The flow of supplies from France to Algeria was restricted to what the French Army itself would use in one month. This measure was taken to prevent the OAS from building up a large inventory of arms, ammunition, and other necessities for a revolt.

FOOTNOTES

1. Michael K. Clark, *Algeria in Turmoil: A History of the Rebellion* (New York: Frederick A. Praeger, 1959), p. 299.
2. Richard and Joan Brace, *Ordeal in Algeria* (Princeton: Van Nostrand, 1960), p. 385.
3. For further details, see Joseph Kraft, *The Struggle For Algeria* (New York: Doubleday, 1961), pp. 65–68.
4. Lorna Hahn, *North Africa: Nationalism to Nationhood* (Washington: Public Affairs Press, 1960), pp. 156–158; see also p. 69, and Joan Gillespie, *Algeria: Rebellion and Revolution* (New York: Frederick A. Praeger, 1960), pp. 94–96. For a thorough account in French of the outbreak of the revolt, see Serge Bromberger, *Les Rebelles Algériens* (Paris: Librairie Plon, 1958), pp. 1–30.
5. Hahn, *North Africa*, p. 158.
6. Kraft, *The Struggle*, pp. 69–71.
7. Ibid., p. 77.
8. Peter Braestrup, "Partisan Tactics—Algerian Style," *Army* (August 1960), 37.
9. Gillespie, *Algeria*, pp. 97–102; Kraft, *The Struggle*, pp. 81–85; see also Brace, *Ordeal*, pp. 106–109.
10. See Bromberger, *Les Rebelles*, pp. 139–209, for a great many pertinent details; see also Charles-Henri Farod, *La Révolution Algérienne* (Paris: Librairie Plon, 1959), pp. 111–118, for biographical data on leading FLN persons.
11. Ibid., pp. 139–209.
12. See Braestrup, "Partisan Tactics," p. 33.
13. Brace, *Ordeal*, p. 113; Hahn, *North Africa*, pp. 158–159.
14. Hahn, *North Africa*, pp. 158–159.
15. Ibid.
16. Braestrup, "Partisan Tactics."
17. Ibid.
18. For a discussion of psychological operations in Algiers, particularly terrorism, see Bromberger, *Les Rebelles*, pp. 161–209.
19. Ibid., p. 152.
20. See Hahn, *North Africa*, pp. 159–165, for a discussion of initial French reaction to the revolt.

21. Raoul Salan, *Directives à la Pacification* (Algiers: Government of Algeria, 1956).

22. See "Synthèse Relative à la Participation de l'Armée aux Tâches Extra-Militaires de Pacification," 1er Semestre 1959, 2/694/EMJ/5/ACT, Algérie, le 18 août, 1959.

23. Kraft, *The Struggle*, pp. 104–105.

24. Ben Welles, *The New York Times*, March 20, 1962, pp. 1, 15.

25. *Time* (January 26, 1962), 22.

26. Ibid.

27. Welles, *The New York Times*.

28. *The Washington Post*, April 22, 1962, p. 18.

29. Ibid.

30. Welles, *The New York Times*.

31. Ibid.

32. Ray Alan, "Brigitte, France and the Secret Army," *The New Leader* (December 25, 1961), 3–5.

CHAPTER 10

GREECE (1945–49)

BACKGROUND

On October 28, 1940, Greece was invaded by Italian troops stationed in Albania. The Greek Army was successful in repelling the invaders and in several months Mussolini's forces were routed. The following spring Hitler sent in German troops to secure his southern flank in preparation for the planned invasion of the Soviet Union; they were successful.

Early in 1942 the British dropped Special Operations Executive (SOE) agents into Greece to destroy the important Gorgopotamos Bridge. After successfully completing this mission, the agents were instructed to remain as liaison officers with indigenous resistance groups. Despite British efforts to bring about some measure of cooperation between the various underground groups, the main efforts of these resistance organizations were directed against one another. Although agreeing not to interfere in each other's resistance activities, the largest resistance groups, the National Republican Greek League (EDES) and the Communist-controlled National Liberation Front (EAM) together with its military arm, the Greek People's Liberation Army (ELAS), fought one another sporadically.

In December 1944 the Communist-dominated National Liberation Front (EAM) made clear its revolutionary intentions [1] by forcibly attempting to take control of the government in Athens. Meeting stiff resistance from British forces stationed in the city, the EAM/ELAS insurgents were ultimately defeated in battle. The insurgents agreed to truce terms in February 1945. By the terms of the Varkiza Agreement, the Communists agreed to disband their armed forces.

After failing to gain control of the government by legal, political, and nonviolent means, the Communists resumed underground and guerrilla warfare. Fighting again broke out in June 1946. Already in *de facto* control of many mountain areas of Greece, the Communists revived their wartime underground organizations and used the newly created Communist states of Albania, Bulgaria, and Yugoslavia as sanctuaries and training bases for many of their 25,000 troops.

The war, which lasted until the fall of 1949, was characterized by Communist use of underground apparatuses in major urban centers for gathering intelligence, instigating strikes, and recruiting popular support in rural areas. Logistical support was received from neighboring Communist countries.

The Communists were aided by the postwar economic chaos which existed throughout Greece. Despite fiscal reforms and extensive United Nations Relief and Rehabilitation Agency aid received in 1946, the county had not rebuilt the harbors or railroads, or replaced the vast quantities of livestock, vineyards, and other resources which had been destroyed. Unemployment was high, and to further complicate matters, radical currency inflation raised prices. Fear of Communist control of labor unions led to repeated attempts by the Greek Government to control or curb labor activities, which only strengthened the appeal of the Communist Party in Greece (KKE).[2] Furthermore, Great Britain, preoccupied with her own problems of recovery, let it be known in February 1947 that she was unable to assist Greece further.

The turning point in Greece's fortunes came when the United States announced the Truman Doctrine in March 1947, and subsequently provided the Greek Government with both financial and military aid with which to fight the Communist Greek Democratic Army (DAS) and the unstable economic situation. This initial grant of 300 million dollars enabled the government to expand its military and security forces to almost 200,000 men.

In 1948 Yugoslavia was expelled from the Cominform. By the summer of the following year, Tito had stopped supplying the rebels in addition to denying them the use of Yugoslav soil for sanctuary and training. Since Bulgarian-Greek animosity had precluded any large-scale participation by the Bulgars in the Greek revolution, Albania now became the only source of outside support. Strategically this forced the guerrillas to maintain themselves in a small mountainous area adjoining the Albanian frontier. Being placed in a position of fighting along conventional military lines, the insurgents were not able to resist the counteroperations of the Greek National Army. The government forces were successful in encircling and cutting off the insurgent escape routes into Albania and finally defeating them when the DAS abandoned their guerrilla tactics and tried to defend their base area in the Grammos-Vitsi Mountain region. This military defeat led to the ultimate collapse of the movement.

In order to isolate the guerrillas from their internal support and food supply, the Greek Government, assisted by U.S. aid funds, cared for the many refugees and displaced inhabitants from areas of insurgent activity. The relocation of about one-seventh of the population into camps prevented the insurgents from maintaining the effective underground support organization which EAM/ELAS had developed during the war. Two other factors were possibly responsible for this collapse. First, the "troika" command system, which had been so successful during World War II, collapsed. This system called for a local "popular leader" to maintain a sympathetic relationship with the rural population, in order to ensure a constant source of supplies. During the postwar period, the *kapitanos* was usually not a local inhabitant, but an outside party member who also served as a political officer. His relationship with the local populace was not one which would stimulate support for the movement, and thus supplies which were more willingly given during the war now had to be confiscated. Second, because of widespread economic chaos and social disorganization, the population lacked the physical, moral, and emotional resources to aid the insurgency even if they had wished to do so. The hardships of 7 years of war and famine had left most Greek peasants apathetic toward armed conflict of any kind. In many cases the peasants refused to cooperate with either the Greek National Army and security forces, or the rebel Greek Democratic Army.

YUGOSLAVIA
BULGARIA
SKOPJE
ALBANIA
VITSI
GRAMMOS
SALONIKA
Konitsa
TURKEY
GREECE
AEGEAN SEA
IONIAN SEA
Athens
PIRAEUS
GREECE
1951 population 7,630,587
sq. miles 51,182

Figure 16. Map of Greece.

Cases of Undergrounds

For the first 2 years of the insurgency, rebel morale was high. They felt that they were acting as patriots seeking legitimate reform in Greece. It was not until the split within the international Communist movement became clear that the image of this "patriotic" movement changed. The Yugoslav-Bulgarian rift over Macedonia [3] was accompanied by internal dissension in the Greek Communist Party.[4] The KKE's final decision to support an autonomous Macedonia was followed by a collapse of rebel enthusiasm. There was no longer a patriotic will to fight.[5]

ORGANIZATION

Although outside interest in the Greek Communist movement had existed before that time, it was not until 1931 that the Comintern intervened in party activities in Greece. Hoping thereby to improve party discipline, it engineered the appointment of the Moscow-trained Nicholas Zakhariadhis as Secretary General of the KKE. The Greek Communists had their first experience in clandestine activity after the party was outlawed by the dictator Ioannis Metaxas in 1936. Using the legal Socialist Party as a front for open activity, the party secretly began to organize cells throughout Greece.

After World War II began, the British Government, seeking to build a resistance movement in Greece, offered to supply arms to any group—including the Communists—which would participate in guerrilla activities. The creation in September 1941 of the National Liberation Front (EAM), which was Communist dominated, gave the Communists a chance to gain control of the activities of various political parties. Also, the KKE was the only party with experience in clandestine operations, a circumstance that proved an enormous advantage. Since EAM's Central Committee was dominated by the Communists it provided the device by which the KKE established control over a number of organizations operating in rural and urban areas. Among the significant organizations controlled by them in this way were the Workers National Liberation Front (EEAM), the National Cooperative (EA), the United Pan-Hellenic Youth Organization (EPON), the Units for Protection of the People's Struggle (OPLA), and a guerrilla commissariat (ETA).[6] All these groups worked to support the largest and strongest guerrilla army in Greece, the ELAS, which at the height of its strength had 50,000 men.[7] The nature of the EAM/ELAS organization varied in different areas. In some villages EAM was represented by only one man, perhaps even a priest or a school teacher. Often it consisted of an expanded unit which included non-Communists. However, the organizer and local EAM secretary was usually a party member, when one was available.

When EAM/ELAS prepared to set up its shadow government in the northwest mountains in 1943, it felt enough confidence in the influence of its subsidiary organizations to try to organize elections secretly behind the enemy lines. This provisional government, the Political Committee of National Liberation (PEEA), was acclaimed by the Soviet press in March 1944, although the first Russian military liaison group (which came to insist that the PEEA cooperate with the Royal Government-in-exile), did not arrive in Greece until July 1944.[8]

'The defeat of the EAM/ELAS forces of Athens by British troops in December 1944 convinced the Communists that revolutionary victory could not be obtained immediately by open force. Failure to obtain Soviet and United States diplomatic support for their claim to rule the country convinced the EAM leaders that they must come to terms with the Greek Government. They therefore signed the Varkiza Agreement of February 1945,[9] which provided for demobilization of the guerrilla forces. The KKE also suffered a major electoral defeat in early 1946. Nevertheless, the elected Greek Government commanded no deep loyalty among the population, and did not solve the difficult economic problems. The weakness of the government and the now tacit support of Russia, together with direct encouragement from other Balkan Communist leaders, persuaded the Greek Communists to resume insurgent action.

The KKE openly assumed responsibility for the rebellion that began in June 1946. Its ruling group was a politburo, composed of seven members, each a leader in his home area. Urban organizations which supported the rebellion were under direct control of the central committee. Branches, or "rays," were organized in each trade and each geographic area. In Athens, for example, one group of agents recruited industrial workers, civil servants, bank clerks, and military reservist groups, while others organized cells in the various quarters of the city.[10]

Originally, military operations at all levels were directed by a tripartite command system. The military commander took charge of combat operations, and the *kapitanos* was responsible for maintaining good relations with the civilian population.[11] Real power, however, lay in the hands of the political representative, who received his orders from the high command of the Communist Party. When this three-man command unit collapsed, the political representative assumed the role of the *kapitanos*.

Some front groups controlled by the party were the Seamen's Partisan Committee (KEN), the Communist Organization of Greek Macedonia (KOEN), secret cells (KOSSA) inside the Greek National Army, a Slav-Macedonian organization (OENA), and the Democratic Women's Organization of Greece (PDEG). In rural areas, the KKE operated quite openly through the facilities and under the cover of the Greek Agrarian Party (AKE), which was responsible for supervising the activities of various "self-defense" groups. This included the Communist youth organization (EPON) and the supply organization of the DAS (ETA). In the "liberated" villages civil administration was usually carried out by officials of these front groups.[12]

In order to acquire an appearance of dignity and officiality, the insurgents had hopes of establishing a "government" in an area near the Albanian border. The town of Konitsa was to be the capital of this "Free Greece." This plan, however, was never realized.

Cases of Undergrounds

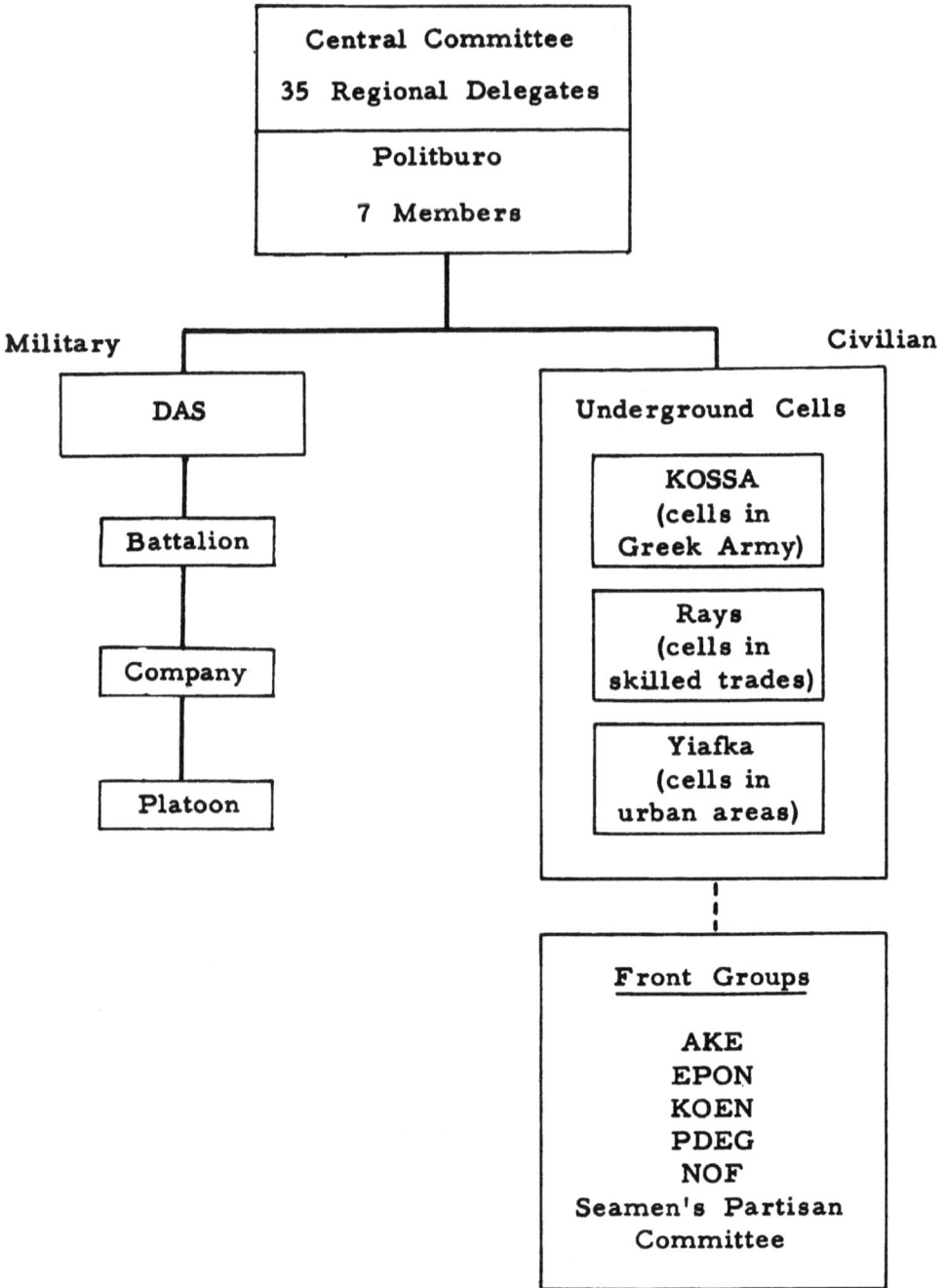

Figure 17. Organization of KKE–DAS.

DAS – Greek Democratic Army
EPON– Youth Organization
PDEG–"Democratic Women's Organization of Greece"
KOEN–Macedonian Communist Organization
NOF –Slav-Macedonian Independence Organization
AKE –Greek Agrarian Party

Markos Vafiades, or "General Markos," commander in chief of the Greek Communist-led guerrillas.

Cases of Undergrounds

Within the revolutionary organization were security units whose identity was unknown to most party members and sympathizers. The most important of these was the *Aftoamyna*, the successor of the OPLA, the security force which had existed during the German occupation. Originally formed to train party members for revolutionary action in towns, the *Aftoamyna* organized its first cadres in 1946. Participants were chosen from among the most fanatical members of the party with preference given to persons with criminal records and those who had distinguished themselves by their daring in conspiratorial work. Members of the *Aftoamyna* were planted in all party organizations from the cell upward to act as observers and informers.[13]

Although the *Aftoamyna* probably penetrated the leadership of the General Confederation of Labor (ERGAS), the KKE's attempt to capture the labor movement failed. ERGAS was reorganized, with the assistance of the British Trade Union Confederation, and became a member of the anti-Communist International Confederation of Free Trade Unions. The *Aftoamyna* scored elsewhere, however. For a time it had access to the files of the Greek General Staff and was able to provide information on the movement of government troops. It also supervised the clandestine traffic in arms, supplies, and recruits. Its mission, as defined in an order signed by General Markos (Commander in Chief of DAS Forces) in December 1948, was to "watch the movement of the enemy of his agents . . . help us call up the andartes [partisans], and send us reliable and suitable men." [14]

UNDERGROUND ACTIVITIES

ADMINISTRATIVE FUNCTIONS

Communications

Although the Communists developed a fairly extensive communications network, it was insufficient to maintain continued and rapid contact between forces scattered in the rugged mountain terrains.

Communication between the underground and the guerrillas, and between guerrilla units in the field, was mainly by courier. Usually couriers were young boys recruited and trained by the village youth organizations.

Direct instructions in code were given through the facilities of the Communist Radio Free Greece, and possibly via the radios of the three neighboring Communist countries. A letter from General Markos to Zakhariadhis (Secretary General of the KKE),[15] while the latter was in Moscow in January 1948, indicates that there was also a system of communication between Communist officials in Moscow and KKE leaders in Greece. It can be assumed that these communications were relayed through Belgrade and Sofia. The guerrillas also made use of a few field radios supplied to ELAS by the Allies during the war and cached after the Varkiza Agreement. One observer noted that, "Throughout Greece, the *Aftoamyna* communicated with one another by Morse, and the whole organization was in direct or indirect communication with the headquarters of DAS." [16]

Recruitment

Beginning with a nucleus estimated as no larger than 3,000 men,[17] the KKE and its affiliates employed four main methods to recruit additional forces and replace losses:

(1) The Communist Party itself supplied "volunteers" from among its ranks. A women's organization, the PDEG, was responsible for recruiting a significant portion of the revolutionary force both by working for the movement and by persuading men to join. The undergrounds (*yiafkas*) of Athens, Salonika, and other cities, channeled new recruits to the fighting units in the hills.[18]

(2) By giving the impression that it supported autonomy for the state of Macedonia, the KKE won the sympathy of some Slavic Macedonians who eventually supplied fighting units from the ranks of the KOEN.

(3) In the mountain villages of northern Greece, the DAS, assisted by local Communists, persuaded or forced able-bodied men to join its ranks. With no government security forces at hand to protect him or his family, a male often dared not refuse.

(4) World War II ELAS veterans were given additional training in Yugoslavia. They returned to Greece with instructions to organize guerrilla bands in the mountain villages of Greece.

The recruiting methods proved to be fairly successful. The rebels were able to maintain an armed force of about 23,000 even after the national government's countermeasures began to be effective. The number of insurgent participants can only be conjectured. One source estimates that at the peak of its strength the Communists mustered no more than 700,000 to 750,000 active members and sympathizers.[19]

A report issued by the United Nations Special Commission on the Balkans investigating charges that Albania, Bulgaria, and Yugoslavia were aiding the rebellion, presented evidence that these nations were giving extensive aid to the guerrilla forces. The training camp at Bulkes, Yugoslavia, was an army camp built for the training of DAS recruits.[20]

Logistics

A major factor in the Communists' defeat in Greece was their inability either to produce sufficient supplies themselves or to obtain continuous logistical support from outside the country.

Although the rebels were required to surrender all arms under the Varkiza Agreement of 1945, much of the material seized from the Italians in 1943 or given them by the Allies during World War II was secreted and used during the period 1946–49. However, the DAS forces evidently had no facilities for producing additional military equipment. Furthermore, they often found it

Cases of Undergrounds

(UPI Photo)

The Greek town of Kalavrita after a guerilla force of 1,000 had attacked it. Forty civilians were killed and the entire police garrison of 20 wiped out.

304

difficult to obtain such basic supplies as food and clothing. Thus, while guerrilla forces operating near the frontier could often obtain supplies from across the border, those in the Pindos Range in western Greece had a more serious problem. They could not live off the country, barren and sparsely populated as it was, and had to be supplied by mule trains. It obviously took a great number of such pack animals to fill even their minimum needs.

The DAS appears to have received the bulk of its support between 1945 and 1949 from other Communist nations who shipped supplies through Yugoslavia and Bulgaria. The U.N. Commission found evidences of arms depots in these countries near the Greek frontier, and witnesses interviewed by the commission stated that they had received Russian small arms while training in Albania. The Joint U.S. Military Advisory and Planning Group noted that this logistical support included not only clothing, rations, arms, and ammunition, but training camps, transit areas, replacement areas, field hospitals, and supply depots— all easily accessible in safe areas across the northern borders.[21]

On August 2, 1947, a convention of Communist military officials* held at Bled, Yugoslavia, drew up the terms by which Communist states would aid the rebellion. In summary, these were:

(1) The Albanian, Yugoslav, and Bulgarian General Staffs undertook to assist the Greek Democratic Army with stores and other supplies, and with technical equipment and instructors.

(2) The same general staffs undertook to organize a rear defense of the DAS and to provide infantry, artillery, and aircraft for the purpose. They also stated their willingness to take part in military action.

(3) The Hungarian and Rumanian Governments also were to be asked to take part in assisting the KKE.

(4) The Albanian Government was to place a naval base at the disposal of the rebels.

(5) Representatives of the Communist governments were to establish contact with the headquarters of the "Greek Democratic Government" as soon as it was formed.[22]

Following Yugoslavia's expulsion from the Cominform in 1948, a series of disputes with the Greek Communists led Tito to alter his policy toward the rebels. Halting supplies in late 1948, he had the borders officially closed between the two countries on July 10, 1949. One reason for this was Yugoslav resentment of KKE statements indicating support of an autonomous, rather than a Yugoslav, Macedonia.[23] With the cessation of Yugoslav assistance and the sealing off of its border, the total amount of outside assistance was sharply reduced, and the internal forces and the underground were unable to fill their logistic requirements. The continuous and insoluble supply problem greatly impeded the Communist war effort.

---

*This convention was attended by members of the KKE, as well as military representatives from Albania, Bulgaria, and Yugoslavia.

Cases of Undergrounds

Security

The *Aftoamyna* (Self-Defense) was charged with internal security. Not only was the identity of its agents within each cell secret, but members were also unknown to each other. When a security measure was to be taken (e.g., eliminating an unreliable individual) a group called a *synergeia* was formed to carry out the deed. The *synergeia* consisted of at least three persons, who were introduced to one another under assumed names by an agent of the central committee. When the mission was accomplished, the members of the group dispersed, changed their addresses, habits, and clothes, and concocted alibis.

It is probable that members of one group, such as the EPON, were urged to spy on suspects within other groups and report any suspects to a higher echelon of the party.

A clandestine judicial system existed to try to judge those accused of a criminal act against the movement. The system was structured around a series of "courts" which judged those accused of noncooperation, treason, and collaboration. These tribunals were primarily interested in restricting the activities of those aiding the Greek counterinsurgent forces and their U.S. advisers.[24]

In many areas of northern Greece, the insurgents lived openly, as there were no government officials or military personnel who could challenge their authority. When they were threatened by government counterinsurgency forces, soldiers fled to the neighboring countries of Yugoslavia, Albania, and Bulgaria. Civilian supporters of the insurgency merely "blended in" with the native population.

OPERATIONAL FUNCTIONS

Psychological Operations

Communist propaganda was addressed to the people of Greece, to Communist Party members and sympathizers, and to governments and peoples of other countries. The overall propaganda line was that the monarchy was alien (the King was a Dane) and Fascist, and that its officials were puppets of the British Foreign Office and the U.S. State Department. Propaganda addressed to KKE members and sympathizers and to prospective recruits emphasized the true "patriotism" of the insurgent movement.

For the first 2 years of the postwar insurgency (1946–48), the call to patriotism kept the morale high among the *andartes*. The Communists lost the initiative when they openly supported a Bulgarian plan for the establishment of an autonomous Macedonia. Few Greeks, with the exception of the Slavic Macedonians, could support such a plan.

Although word of mouth was the major instrument of Communist agitational and propaganda activity, the KKE and its subsidiary organizations also used radio and published materials. After the party was outlawed in December 1947, its operations were conducted clandestinely and, to an increasing degree, from foreign soil.

Until they were suppressed in December 1947, two daily newspapers published in Athens, *Rizospastis* and *Eleftheria*, were in effect organs of the KKE. Afterward they continued to appear in various forms at various places in "the free hills of Greece." [25]   Actually, much of the literature appears to have been printed in Belgrade, Sofia, Bucharest, and Moscow.   One publication printed in Yugoslavia was called *The Voice of Bulkes* and was directed primarily to members of the DAS and rural inhabitants of northern Greece.   The Yugoslav League of Anti-Fascist Women undertook drives to furnish printing supplies to the illegal newspapers.

Little information is available concerning the distribution of these clandestine publications.   Probably the party's youth and women's organizations took charge of circulation in Greece, and the soldiers themselves brought in pamphlets printed in Yugoslavia and Albania.   The couriers who maintained communications between guerrilla units may have distributed printed material while on their rounds.

The KKE operated a radio station, *Eleftheria Ellada* (Free Greece)— also called Radio Markos after the DAS commander—which purported to be on Greek soil but probably was located in Albania and later in Rumania. Radio Belgrade, Radio Tirana, and Radio Sofia also reported the activities of the "Free Greek Government."

Besides disseminating propaganda, the Communist youth movement, EPON, served as a school for Communist indoctrination and prepared its members to become agitators, saboteurs, and *andartes*.[26]   It gave the young people responsible tasks, such as distributing leaflets and watching persons suspected of being traitors to the Communist cause.

The public reaction to all this propaganda failed to meet Communist expectations; there were constant complaints that the masses "were not responding."

Intelligence

The Communists maintained an elaborate network for intelligence in the mountains of Northern Greece.   In rural areas, "self-defense" personnel operated clandestinely through local *yiafkas* (cells which kept the guerrillas informed of all government troop movements and locations).   When government forces arrived in the vicinity, the armed partisans would flee and the local resistance leaders would blend in with the population.   Usually fear of reprisal prevented local inhabitants from exposing the Communists to the loyal troops.

The KOSSA was responsible for gathering and transmitting counterintelligence information.   The members of this organization were expected to relay to the DAS high command military information which might affect the plans of the revolutionaries.

COUNTERMEASURES

The Communists' first attempt to seize control of Greece, in December 1944, was defeated by British forces, who, after the German withdrawal, had moved in elements of two divisions.

Thus they were able to meet open force when it materialized. They followed up their military victory with an extensive reconstruction and rehabilitation program.

On the heels of their defeat in the general elections in early 1946, the Communists suffered a further blow by the reorganization, under British auspices, of the Communist-dominated labor confederation in Athens. This move was arranged by delegates of the British Trade Union Confederation, who came to Athens and urged the Greek Government to oust the Communist labor leaders. The net effect of this reorganization, plus the apathy with which it was regarded by the urban workers, was a serious setback for Communist hopes of an early victory.

The KKE complained constantly that the masses were not responding. It felt that without political strikes in the towns, without active unrest among the urban workers, no revolution was possible. If the DAS had received half as much support from labor as the French and Italian Communist parties, the revolution might have succeeded.[27]

Notwithstanding the success of the British thrusts against the insurgents, the Greek Government itself was unable to institute effective countermeasures during 1946–47. A basic problem was the low morale of the army. Many Greek soldiers were not convinced the government would win, many disliked shooting at fellow Greeks, or preferred simply to take no action at all. Army units frequently allowed guerrillas to escape.

Another factor was that individual Greek politicians demanded military security for their home districts without concern for the total problem. This prevented any coordinated effort of government forces against the rebels. Political interference of this nature destroyed much of the initiative the Greek National Army might have had. At the same time, British military advisers had great difficulty in establishing the necessary cooperation between the army and the police forces. They were even less successful in convincing Greek military strategists that small, mobile units were required to suppress the insurgents.[28]

By 1947 the British were unable to sustain the financial burden any longer, and the United States, under the provisions of the Truman Doctrine, assumed the responsibility of maintaining the Greek economy and financing and equipping the Greek Armed Forces. The initiation of an expansive U.S. aid program, which was to total $500 million by 1952, coupled with the establishment of a U.S. military advisory group, headed by General James Van Fleet, gave the Greek Government the facilities it needed to begin an effective program against the DAS. Between 1947 and 1948 the Greek Army and gendarmery reached a total of 182,000 men.[29]

With the help of U.S. aid, the Greek Government was able to care for about 700,000 refugees and others who had evacuated territories of guerrilla activity. In this way the insurgents were denied a prime source of supplies and recruits.[30]

At the suggestion of Van Fleet, a reindoctrination center was established on the island of Makronesos to rehabilitate captured insurgent soldiers. Its success was attested to by the violent reaction it provoked from the Communists.[31]

In 1949, in order to isolate the DAS forces, the government carried out a series of mass arrests designed to destroy the Communist intelligence network. "As a result the armed guerrillas, operating without their eyes and ears, could no longer avoid surprise attacks by government (GNA) forces. Guerrilla leaders and their forces were killed or captured in a number of quickly executed operations beginning in the Peloponnesus and working toward the satellite borders."[32] These counterinsurgency operations were successful largely because of the quasi-military units set up under military command.[33]

Certain happenings within the European Communist bloc also worked to the advantage of the Greek Government. Friction developed between the leaders of the Greek Communists and Stalin because the latter did not provide large-scale material support. Furthermore, the border nations of Albania, Bulgaria, and Yugoslavia failed to support the DAS adequately under the provisions of the joint agreements made by their general staffs. When Yugoslavia ceased its direct support in 1948, the guerrillas were denied their major supply routes and sanctuary. And without a reliable system of supply, the DAS was unable to maintain large-scale operations.

Defeat in conventional military battle in the Grammos-Vitsi Mountain area led the insurgents to announce in the fall of 1949 that they were halting military operations. The rebellion is considered to have been terminated at that time.

FOOTNOTES

1. Armed conflict between British forces and those of ELAS was an unplanned development. When armed ELASites approached government buildings, they found British soldiers accompanying Greek police guarding these establishments. Exchange of fire between the British and some of the insurgents took place at this time. See William McNeill, *The Greek Dilemma* (New York: J. B. Lippincott, 1947), chapter VII, esp. p. 175; see also C. M. Woodhouse, *Apple of Discord* (London: Hutchinson and Co., 1948), pp. 217–218.
2. Hugh Seton Watson, *The East European Revolution* (New York: Frederick A. Praeger, 1951), p. 333.
3. This problem of postwar Macedonia began long before Tito's conflict with the Kremlin. Traditionally, it has been an area of dispute between Bulgaria, Greece, and Yugoslavia. As early as 1943, Tito announced Macedonia to be one of the six "federal republics" of the new Yugoslavia. In the summer of 1945, partisan troops were sent to parts of Macedonia to insure Yugoslav control. See "The Conference of Berlin, 1945," Foreign Relations of the United States, Diplomatic Papers, Vol. 1 (Washington: Government Printing Office, 1960), pp. 666–668. In August of 1947 President Tito and Georgi Dimitrov of Bulgaria drew up a plan for the division of Macedonia which did not include

Cases of Undergrounds

Greece. In November 1948 a Bulgarian Communist organization announced their support for an "independent Macedonia." This plan was supported by the Greek Communist Party in March 1949. On July 10 of that year Tito sealed the borders between Yugoslavia and Greece. (See *Encyclopaedia Britannica*, Vol. XIV, p. 563.) See also McNeill, *Greek Dilemma*, pp. 261–269. For a detailed discussion of the Macedonian question, see Elizabeth Barker, *Macedonia* (Oxford: Oxford University Press, 1950); Christopher Christides, *The Macedonian Camouflage* (Athens: The Hellenic Publishing Company, 1949); and H. F. Armstrong, Tito and Goliath (New York: Macmillan Co., 1955).

4. Dimitrov Kousoulas, *The Price of Freedom* (Syracuse: Syracuse University Press, 1953), p. 178.

5. See William McNeill, *American Aid in Action* (New York: Twentieth Century, 1957), p. 43.

6. Woodhouse, *Apple of Discord*, p. 63.

7. E. R. Wainhouse, "Guerrilla War in Greece, 1946–1949: A Case Study," *Military Review* (June 1957), 22.

8. See Woodhouse, *Apple of Discord*, p. 112.

9. For further discussion on this matter see D. M. Condit, *Case Study in Guerrilla War: Greece During World War II* (Washington: Special Operations Research Office, 1961), pp. 98–100; C. M. Woodhouse, "The Greek Resistance 1942–1944," in *European Resistance Movements 1939–1945* (New York: Pergamon Press, 1960), p. 389; D. G. Kousoulas, *The Price of Freedom*, p. 120.

10. F. N. Voight, *The Greek Sedition* (London: Hollis and Carter, 1949), p. 213.

11. See Woodhouse, *Apple of Discord*, pp. 67 and 144; see also Condit, *Case Study in Guerrilla War*, pp. 149–150.

12. Wainhouse, "Guerrilla War," p. 19.

13. Voight, *The Greek Sedition*, p. 215.

14. Ibid., p. 219.

15. Ibid., p. 254–258; see also *Documents of International Affairs*, ed. Margaret Carlyle (London: Oxford University Press, 1952), pp. 318–320.

16. Voight, *The Greek Sedition*, p. 219.

17. Capt. Labignette Ximenès, Capt. A. Souyris, and H. Carrère d'Encausse, "The Communist Insurrection in Greece (1946–1949)," Intelligence Translation No. H–2060 (tr. from *Revue Militaire d'Information* No. 281) (Washington: DA, OACSI, November 3, 1958), p. 40.

18. Wainhouse, "Guerrilla War," p. 22.

19. Wainhouse, in "Guerrilla War," suggests 750,000 (p. 22).

20. Harry Howard, *U.N. and Problems of Greece* (Department of State publication 2909, Washington, 1947), p. 16.

21. Wainhouse, "Guerrilla War," p. 19.

22. Voight, *The Greek Sedition*, pp. 207–209.

23. See footnote 3.

24. Voight, *The Greek Sedition*, pp. 180–186.

25. Floyd Spencer, *War and Postwar Greece* (Washington: Library of Congress, European Affairs Division, 1952), p. 110.

26. Voight, *The Greek Sedition*, p. 144.

27. Ibid., p. 205.

28. Spencer, *War*, pp. 101–102.

29. Wainhouse, "Guerrilla War," p. 22.

30. See McNeill, *American Aid*, p. 40.

31. Ibid., p. 111.

32. Wainhouse, "Guerrilla War," p. 24.

33. See Alexander Papagos, "Guerrilla Warfare" in *Modern Guerrilla Warfare*, ed. Franklin Mark Osanka (New York: The Free Press of Glencoe, 1962), p. 238.

CHAPTER 11

THE PHILIPPINES (1946–54)

BACKGROUND

After the Japanese invaded the Philippines, in 1942, resistance was carried on by many Filipino guerrilla bands. One of these was the Communist-directed *Hukbong Bayan Laban Sa Hapon* (Hukbalahap), or People's Anti-Japanese Resistance Army. During the height of their World War II activities, in the year 1943, the "Huks" numbered about 5,000 full-time officers and men.[1] Immediately after the war, Communists were refused seats in the newly-formed Parliament to which they had been elected, and the Huks resorted to rebellion. Against the rebels, the government marshalled 24,000 poorly-equipped paramilitary police of the Department of the Interior's Constabulary. It was this force that had to carry the brunt of the campaign against the Huks during the years 1946–50. They opposed insurgents who, at their apex, consisted of about 12,000 armed guerrillas and 100,000 active members.[2] The regular Philippine Army was ill-equipped, unorganized for sustained combat, and staffed by many inept and corrupt officers, and it played only a minor role until 1950.

Hampering government operations were certain social and economic conditions which alienated the large rural peasantry. In the rural areas, government posts were monopolized and political leadership was dominated by small groups of wealthy landowners. The tax system favored the wealthy. Personal influence frequently superseded statute law; hence nonpayment of taxes by landowners and failure to enforce the minimum wage law was usually the rule. Government officials were generally related and nepotism was common. Some officials served without pay and were susceptible to bribery. In many cases, promotions within the military were obtained through influence rather than merit.

Supplies for military units were inadequate. On arriving in a barrio* the officer in charge would often confiscate food and material from the local populace. Troop morale was consequently very low, and civilian-military relations were poor. Furthermore, the elections of 1949 were characterized by intimidation of voters, assassination of candidates, and general corruption.

The Huks were in almost virtual control of central Luzon (called "Huklandia") in late 1950, when Ramón Magsaysay was appointed Minister of Defense by President Quirino and took charge of the counterinsurgency program. To aid the new Defense Minister, the United States sent to the Philippines a Joint U.S. Military Assistance Group (JUSMAG) and substantial loans. One of the first acts by Magsaysay was a reorganization of the entire armed forces of the country. The Constabulary was placed under Department

---

*A barrio is a village or subdivision of a town and usually includes large areas of adjoining arable land.

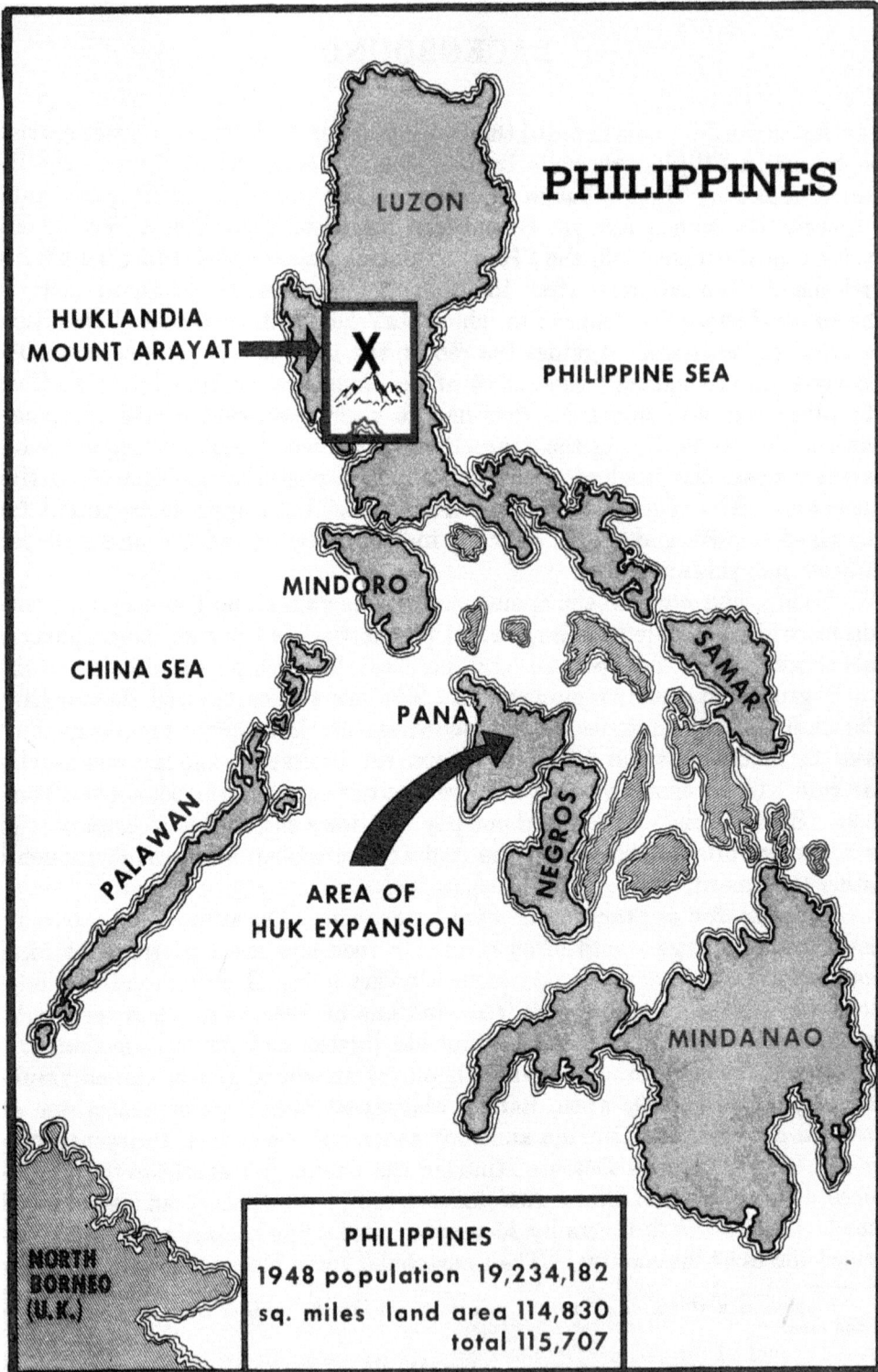

Figure 18. Map of the Philippines.

of Defense command; many policemen were transferred into the regular army; and out of the army, 26 battalion combat teams (BCT's), totalling about 26,000 men, were formed to serve as the core of the new counterforce. Total government fighting strength (regular army and Constabulary) now figured at about 30,000 men.[3] Incompetent and corrupt army officers were dismissed. Provisions were made to ensure honest elections. Resettlement areas for surrendered and captured Huks were established by the Economic Development Corps (EDCOR). Habeas corpus was suspended in order to detain government officials suspected of corruption as well as suspected Communists and Communist sympathizers.[4] Magsaysay's counterinsurgency program was expanded after his election to the Presidency in 1953. In 1954 Luis Taruc, commander of the Huk forces, surrendered, and by the following year the rebellion was no longer a serious threat to the constitutional government of the Philippine Republic.

ORGANIZATION

The first Communist organizers in the Philippines were William Janequette (alias Harrison George), a representative of the U.S. Communist Party,* and Tan Malaka (alias Elias Fuentes), an Indonesian who had been expelled from his homeland by the Dutch and subsequently became the Comintern's agent for Southeast Asia and Australia.[5] In 1928 two future Philippine Communist leaders, Crisanto Evangelista and Cirila Bognor, went to Moscow for a brief period of training. After returning home, Evangelista in 1930 founded the Communist Party of the Philippines (CPP), the stated aim of which was "to overthrow American imperialism in the Philippines."[6] In 1932 it was admitted to the Comintern.[7]

Until the war, party membership was confined to small groups of individuals, many of whom were schooled in Moscow. The party was supported by a few intellectuals, and at times had a fairly substantial following among workers and peasants. Having maintained contact since the 1920's with the Pan Pacific Secretariat of Labor Unions, a branch of the Red International of Labor Unions (Profintern), many Communist Party officials spent much of their time before the war engaging in labor activity,[8] both in Manila and in the interior of Luzon Island. The Printers Union of the Philippines, influenced by Mariano Balgos, and the League of the Sons of Labor, headed by Evangelista, achieved some significance in Manila before the war. The League of Poor Laborers (AMT), predecessor of the Confederation of Peasants (PKM), was to be one of the principal mass-support bases of the insurgent movement from 1946 to 1954. It organized rural elements of the Philippine population for use in "front" activities.

---

*The Comintern charged the Communist Party in the United States with responsibility for the activities in the Philippines.

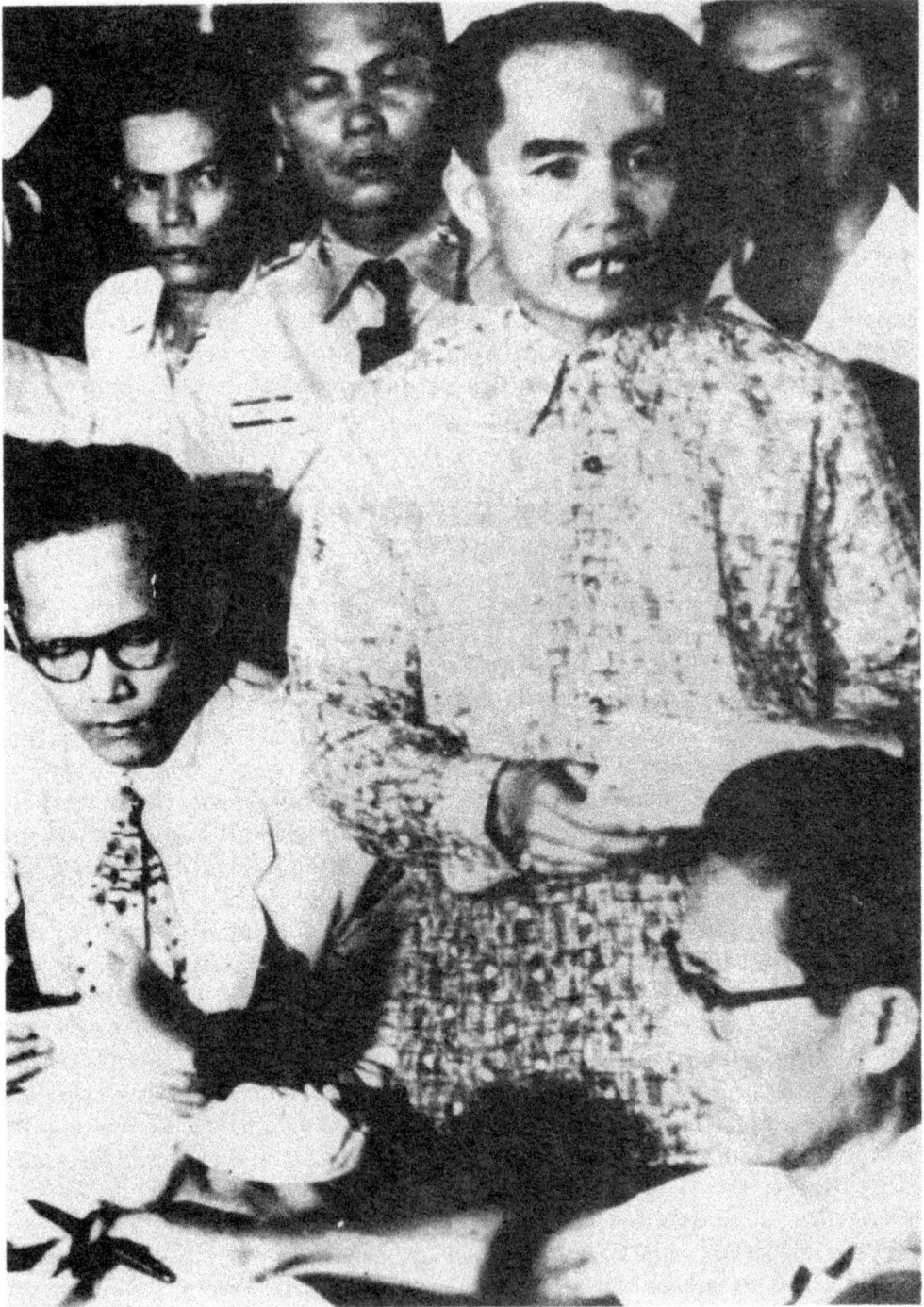

(UPI Photo)

Luis Taruc, after pleading guilty to a charge of rebellion, delivers a speech on Marxist doctrine and the history of the dissident movement in the Philippines.

The first attempts to gain mass support led to several violent strikes in 1931, after which party leaders were arrested. The following year the party was outlawed, but its activities continued underground, and through various legal peasant and labor organizations.[9]

In 1938 the imprisoned leaders were released and immediately began to infiltrate the legal Socialist Party, which had been growing steadily. One observer notes, however, that "Regardless of the official legal interrelations, the Communist Party and its various labor organizations by 1939 operated openly, and by 1941 had achieved legal standing and enjoyed the rights of a minority party under the law."[10]

At the beginning of the war, the CPP pledged its loyalty to the Philippines and the United States. Its cooperation as a unit was refused by the U.S. military authorities,[11] however, and thus its plans for possible united front activities were curbed. When Japanese invasion came, the Communists made no effort to support the Fi-American group,* but laid plans for forming their own resistance group. The result was the official formation on March 29, 1942, of the People's Anti-Japanese Resistance Army, or "Hukbalahap" (the abbreviation of the name in Tagalog language), under the command of Luis Taruc.[12] Although the Huk army, which numbered approximately 5,000 by 1943, eventually swelled to about 10,000, the number of party members within it remained small, and most members, including Taruc, concealed their party affiliation.[13]

The Communists organized the Barrio Units Defense Corps (BUDC) and although these units had some responsibility for maintaining public order (similar to that of other "home guard" units), their prime purpose was to be the civilian counterpart of the Huks and to supply an intelligence organization for the Huk guerrillas. Since the BUDC units were first organized in villages where the AMT peasant unions had operated before the war, many of the BUDC leaders were probably former union officials. Both the Huk and Barrio groups facilitated the development of various united-front activities, such as popular agitation for land reform. During the war, the Hukbalahap claimed to have killed 25,000 men, only 5,000 of whom were Japanese. The other victims were evidently Filipino "obstructionists."[14]

During the brief and confused period immediately following the liberation of the Philippines, the central government attempted to obtain the cooperation of the Communists. The latter ostensibly cooperated by participating in such open activities as campaigning for elections. They did not, however, scrap their plans for ultimately trying to overthrow the government. Thus, although the government tried to disarm the Huks, most of the Huk veterans kept their arms, and maintained their control in most areas of Huklandia.[15] Other Huks, as well as other Communists and sympathizers, became active in the Communist-dominated labor organizations, particularly the PKM and the Federation of Philippine Workers, and in youth groups. Most of the members of the Communist Politburo in Manila, in fact, were officers in the

*The joint Filipino-American Guerrilla Command under General MacArthur.

Cases of Undergrounds

unions affiliated with the Congress of Labor Organizations (CLO). Thus when the Communist leaders decided to open "phase one" of their plan for takeover, which included commencing open warfare, in 1946,[16] it was not very difficult for them to reconstitute the Huks as a fighting organization and build an underground support network on the order of the BUDC. Plans for warfare were intensified after the international Communist meetings at Calcutta in 1948, which directed South Asian Communists to resort to open warfare for "national liberation."

The Huks, whose name was changed to People's Liberation Army (HMB) in 1950, eventually had three types of forces: mobile striking units under Luis Taruc, operating as a regular military force; seven regional commands, each under the leadership of a top Communist official, and local self-defense corps.[17] Their overall structure was a combined political and military command with a general headquarters located in the Sierra Madre Mountains in central Luzon. This was considered to be the "Politburo-out," and the national headquarters of the HMB. The regional commands, under the control of the Communist Party Regional Committees (RECOs) were autonomous, but were obligated to report to the general headquarters in the Sierra Madre or to the underground sub-headquarters (Politburo-in) in Manila. Local militia functions, as well as support activities for the military effort, were performed by local groups organized according to the system used by the BUDC during the war. District

THE NATIONAL ADMINISTRATIVE COMMITTEES

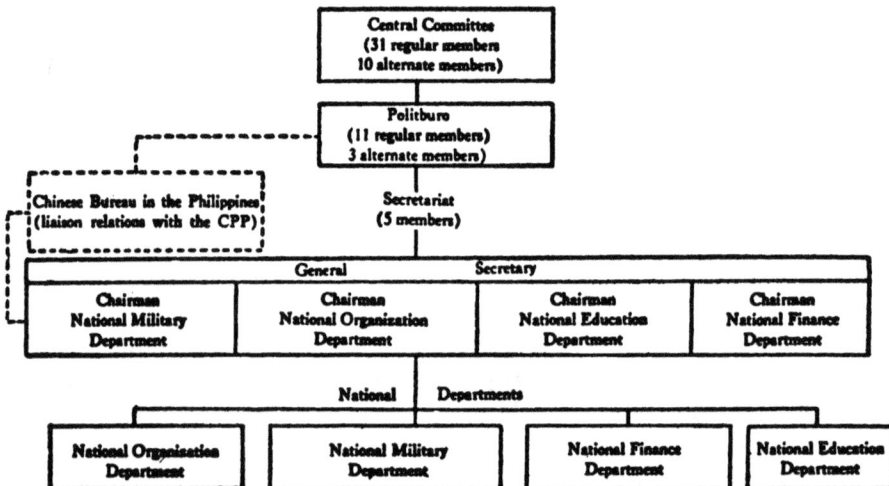

Source: Document captured by Task Force "GG" at Lagrimas, Twin Falls, Mount Dorst, August 12, 1951, based on "Political Resolution 12," March 1951 Central Committee Conference.

Reprinted from *The Philippine Answer to Communism* by Alvin H. Scaff with the permission of the publishers, Stanford University Press. © 1955 by the Board of Trustees of the Leland Stanford Junior University.

Figure 19. National Organization of the Communist Party of the Philippines.

318

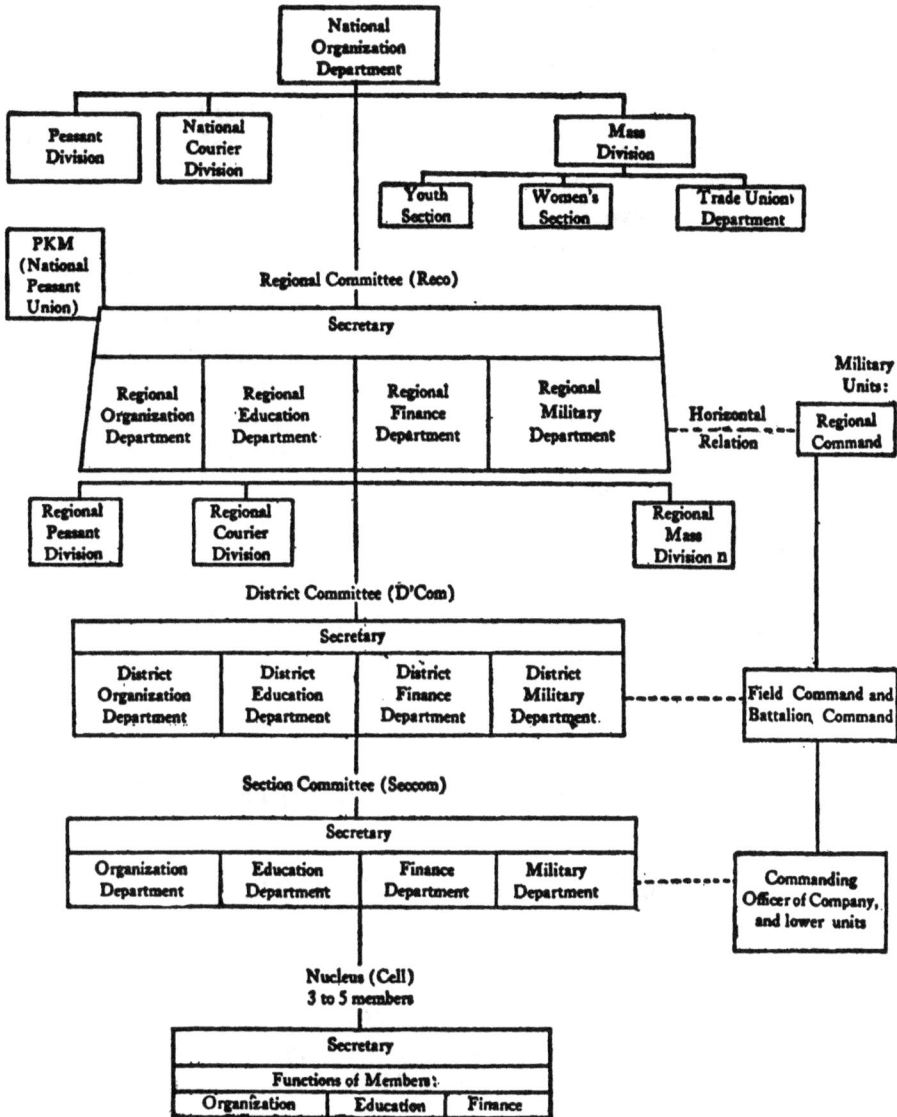

Source: Documents captured by the Philippine Armed Forces, 1951.

Reprinted from *The Philippine Answer to Communism* by Alvin H. Scaff with the permission of the publishers, Stanford University Press. © 1955 by the Board of Trustees of the Leland Stanford Junior University.

Figure 20. The Organization Department.

committees, under direct orders from the RECOs, generally organized these support units so that they included at least one party member.[18]   In a Huk-controlled barrio, the police and municipal administrators and sometimes members of the Philippine Constabulary were in the service of the rebels.

In central Luzon, which they controlled almost completely as a "liberated area," the Huks put to use the administrative experience they had gained during the war.   They appointed civil officials, collected taxes, established courts and administered justice, set up schools, and organized the people into groups whereby they could receive indoctrination and also help the revolutionary movement.   There were also many fringe areas which were controlled by government troops during the day, but where the Huks took control at night.[19]

When the Huks first entered a new town, a meeting would be called at a place such as a schoolhouse, where a Huk leader would explain the movement, and steps would be taken to establish Huk administrative control.   Usually a Huk military commander announced himself as governor of such a region. Often the Huks did not replace existing officials; they simply bribed them to carry out Huk-directed policies.

A special "expansion force" was formed to initiate Huk movements on islands other than Luzon.   The only area where it made any headway was Panay, and even there it never developed into a substantial threat.   The main reason for Huk failure in this venture was probably the lack of support the movement received among local officials, who had been of prime importance to the Huks in gaining their grip on Luzon.

UNDERGROUND ACTIVITIES

ADMINISTRATIVE FUNCTIONS

Communications

The Huks employed the usual type of shortwave radio connections between Huk headquarters and guerrilla units, between these units and barrio-supported groups, and between party headquarters in Manila and Huk headquarters in Mount Arayat.   Within each RECO, communications were the responsibility of a "director of communications," who remained in contact with Huk headquarters.

Additional communication was by standard courier system under the overall direction of a courier division established within the organizational bureau of the party in Manila.   Couriers operating under orders from the national headquarters or the RECO director carried out their missions alone or through a system of relays.

They used flags and flashlights for messages.   Sometimes, in rural areas, they signaled by imitating animal calls and banging two bamboo sticks together.

Recruitment

According to Philippine Army estimates, the Communists were eventually able to enroll approximately 100,000 followers, including 12,000 armed, active soldiers. The nucleus of the revived Huk organization were veterans of the World War II Huk fighting forces and Barrio Unit Defense Corps, many of whom resumed fighting because they felt the government had neither given them due recognition for their services during the war nor fulfilled their hopes of postwar reforms, including land reforms. Individuals recruited by the Huks included—

(1) those who were embittered at the fact that the 70/30 crop sharing law which was supposed to give the peasant 70 percent of the crop he grew,[20] was never put into effect;

(2) those who were dissatisfied with the inequities of the land-reform program, and with a judicial system which favored the landholders in disputes;

(3) veterans who had not received backpay;

(4) those who were forced to bribe government officials in order to receive this pay;

(5) those who joined accidentally (e.g., a man would join a barrio youth organization and later find out that his job was to supply the Huk band in the rear with food);

(6) those who joined because of verbal persuasion;

(7) a "natural leader" in a barrio (who) would have pressures put upon him by the local population to became commander of a local Huk unit;

(8) those who were recruited through coercion, many having to join at gunpoint, or because of threats to the family;

(9) those who were blackmailed into serving the Huks (an individual would be asked to perform what he thought was an innocent task for "a friend," only to discover that he would have to continue his services or be exposed to the authorities for aiding the rebels);

(10) those who were recruited after a Huk unit entered a barrio and demanded that a certain number of people accompany them back to the jungle for training and indoctrination.[21]

Finance

The channeling of funds for the Huk movement was the responsibility of the National Finance Committee of the Politburo-in located in Manila. Taxes were levied by the Huk high command in rebel-controlled areas of central Luzon, while in other areas Communists would illegally present themselves as government tax collectors. There were cases of outright confiscation, which included raids, holdups, and train robberies. Although there is no direct evidence that union funds were used to finance the movement, it is highly probable that the members of CLO-controlled unions were tapped for "donations." The PKM (Confederation of Peasants) placed its financial resources at the disposal of the Huks.

Cases of Undergrounds

Many gifts were received from the 20,000 Chinese in Manila. One wealthy Chinese, Co Pat, is alleged to have supported by himself the entire Wah Chih Chinese guerrilla unit.[22] The Huks apparently received direct financial support from the 3,000-member Chinese Communist Party of the Philippines, which is said to have received some $200,000 in funds from China.[23]

Logistics

In Huklandia, where Huk control was complete, food was supplied by the local population, and it was common practice to take a certain percentage of the annual crop. In areas where the Huks were not in complete control, the members of the BUDC were responsible for supplying the guerrilla units. As the countermeasures introduced by Magsaysay began to be effective, the Huks began to raid barrios simply to obtain supplies, especially food.

Most of the weapons used by the Huks were obtained during the war. Additional arms and ammunition were supplied by or captured from the Municipal Security Forces, who were either apathetic toward the central government, or susceptible to bribery.

Security

Each guerrilla band maintained a special "terror force," which took violent security measures against the local population. An "enforcing squadron" punished offenders, usually by performing an act which would be visible to observers. Among the guerrillas themselves, "treason" or desertion could lead to execution.

In Manila the labor unions and youth organizations were infiltrated wherever possible, and the offices of union officials were used as meeting places and storage areas for Huk documents. The guerrillas were instructed not to operate in zones where underground headquarters were located and nonmilitary functions were being carried out. The fact that they kept written documents, however, was disastrous to the party apparatus in Manila. The operations of the Hukbalahap virtually ceased in Manila when the leaders and many of their documents were captured in June 1950.

OPERATIONAL FUNCTIONS

Psychological Operations

One authority has pointed out that the usual "agitprop" organization in the Philippines took the form of PEIRA (the Political, Economic, Intelligence, and Research Association), which was the body primarily responsible for the dissemination of Huk propaganda. In addition to a continuing flow of posters, pamphlets, and leaflets designed for the general public, four major publications by PEIRA were *Ang Kommunista, Titus, Mapappalaya,* and *Klayarn.* Almost everything was mimeographed, and the pamphlets were usually illustrated with crudely drawn cartoons. The contents were a mixture of political analysis, comments on the news, stories of peasant exploitation "at the hands of the landlords," and appeals to support the "national liberation movement." This literature was widely circulated throughout central Luzon by young courier agents of PEIRA.

The propaganda line varied according to current objectives and targets of attack. Certain themes, however, reappeared constantly. These were: corruption, bribery, and injustice of the federal government; land reform; charges that U.S. interests always represented imperialism and colonialism. Local and temporary issues were also seized upon, such of the irregularities and intimidation of voters and candidates in the federal elections of 1949, and the brutality of EDCOR "concentration camps." It should be noted, however, that no charges were leveled against municipal employees, as this was a group potentially useful to the Huks. Except for the anti-United States themes, the Huk propaganda lines were fairly effective. An investigator who interviewed ex-Huks to determine the most important aspects of Huk appeal found a variety of contributing factors. Some were forced to join at gunpoint; others joined in order to avoid government prosecution on other criminal charges; others indicated primary concern for agrarian reform; others were influenced by verbal persuasion.[24] Personal friendship with an insurgent, Communist promises of land reform, the hopes for a more democratic government, and mere curiosity drove many to the ranks of the insurgents.

A great deal of emphasis was placed on the indoctrination of new Huk recruits, which was the responsibility of the RECO Educational Department. These recruits, as well as populations of Huk-controlled areas were put through a well-developed Communist indoctrination program.[25]

In addition to these mass schools, special military schools were operated. One of these was "Stalin University," located in the vicinity of Mount Arayat, which was designed to train military unit leaders in guerrilla tactics.

In the immediate postwar period the use of terror to further the revolutionary movement was sporadic; it evidently occasioned great debate within the ranks of the Central Committee.[26] A series of terrorist raids on villages took place immediately after the resumption of hostilities in the spring of 1946. The major purpose of these attacks seems to have been to show the local inhabitants that the insurgent forces could not be defeated by government army and security forces, and thus to forestall popular cooperation with government officials. The terrorist attacks against individual vehicles and individuals traveling in areas occupied by insurgent forces were intended to impress upon the public the strength of the clandestine organization. This practice was largely abandoned after the attack on the widow of ex-President Quezon, an incident which aroused great public indignation. In 1949, after the election of Quirino and the escape of several Communist leaders to the hills, the Huks decided to intensify their efforts, and stepped up their terrorist attacks on villages. These attacks were accompanied by others on farmland and forest areas, to extend the area of Huk control and to demonstrate concretely their aversion to landlords.

Cases of Undergrounds

Intelligence

Until Magsaysay became Defense Minister in 1950, intelligence efforts were directed at gaining the cooperation of government officials who had access to pertinent government information. Until Magsaysay was able to install and enforce some measure of security within the government bureaucracy, obtaining valuable information usually required merely a bribe.

In the "liberated areas," municipal officials, especially the town mayor and police chief, were usually in the service of the insurgents. They were often able to inform on government agents and military officials assigned to the area. It was not until the government was able to deny the Huks the aid of local officials that any counterinsurgency measures showed signs of success.

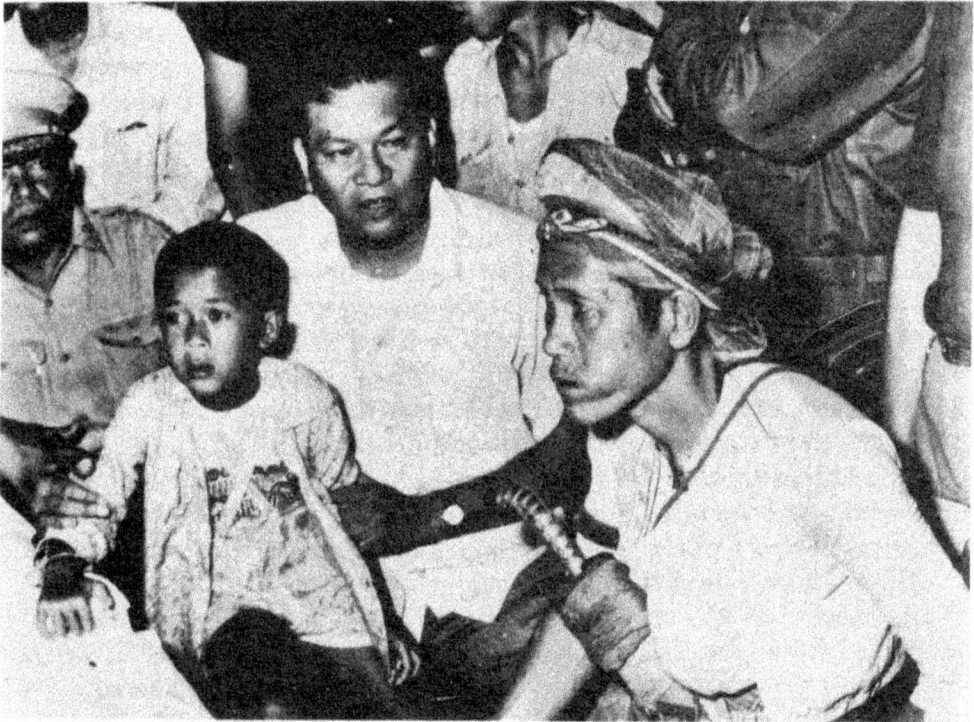

(UPI Photo)

Datu Kamlon (seated in foreground), notorious brigand, surrenders to Defense Minister Ramon Magsaysay (seated center) in ceremonies at Lahing Beach. The boy on Magsaysay's lap is the bandit's 6-year-old son, who was sent to a government patrol ship as hostage when the surrender talks began. Terms of the surrender included a $20,000 payment to Kamlon by the government for damages caused in his territory by army operations, and the release of a number of his men serving time for sedition.

COUNTERMEASURES

Upon his election in 1949, President Quirino initiated an extensive program of economic, political, social, and military reform. Most of the reform measures were conceived and administered by Ramón Magsaysay when he became Defense Minister in 1950, and completed by him when he became President in 1953. Magsaysay made frequent, unscheduled trips into the hinterland to observe the progress of the reform measures. Utimately, through a combination of personal integrity, determination to end the revolt, and demonstrated interest in public welfare, he restored popular confidence in the government and its leaders.

POLITICAL REFORM

The presidential election of 1949 was effectively exploited by the Philippine Communists, who convinced the population of Luzon that it was natural to assume that the existing government would always be corrupt. In contrast, during the elections of 1951, government authorities stationed teachers as poll clerks, used ROTC cadets to guard polling places, and directed soldiers to prevent intimidation of voters and guard ballot boxes. Magsaysay had announced these measures previously, as he wished to assure the voters that their voting rights would be guaranteed. He announced that government officials who had abused their responsibilities would be tried, and urged all citizens to report directly to him and his personal staff any complaints they might have concerning the behavior of government troops and officials. Reforms initiated during these tense years of Magsaysay's leadership led to the turning of the tide of public opinion toward the government.

MILITARY REFORM

The Philippine Constabulary had not been able to stem the growth of the Huk movement. Its company-size units were inadequately equipped for the task and not able to conduct sustained antiguerrilla operations. It also frequently failed to obtain the cooperation of the populace. It was resented by many because some of its officers and men had served in the Bureau of Constabulary under the Japanese. As it stood, the regular army was hardly better prepared to combat the insurgents. Promotions were not awarded on the basis of merit but on the strength of political influence, thereby creating a morale problem among the officers and men. Corruption was prevalent throughout the army. Then too, the army was not organized for extensive counterguerrilla operations. It consisted mainly of administrative, service, and training units, and had only two infantry battalions ready for combat.

Magsaysay's solution was to bring a reorganized Armed Forces into the counterinsurgency campaign, replacing the Constabulary as the prime counterforce. The first step was to remove the Constabulary from the Department of the Interior and place it under the command of the Department of Defense. Constabulary forces were greatly reduced and many were transferred into regular army units. The army was reorganized for counterguerrilla operations. Twenty-six battalion combat teams of about 1,047 officers and men each were formed. Each BCT contained—besides infantry—intelligence, psychological warfare, and medical units that could be assigned to particular sectors. Squad-sized ranger teams that could stay on patrol for as much as 7 days were formed in order to achieve the mobility and jungle penetration capacity necessary to better pursue the guerrillas.[27]

Officers who were not performing their duties were removed, promotions were made on the basis of merit, and soldiers who were caught stealing were punished in the presence of the villagers. The determined effort to rid the Philippine Armed Forces of inefficient and unreliable officers led to the dismissal of personnel of all ranks, from the Army Chief of Staff, who was an influential personal friend of the President, to local battalion commanders who were slow in carrying out counteroperations against guerrilla bands. Magsaysay's obvious determination to rid the Armed Forces of undesirable elements, and his success in doing so, increased public confidence in national government.

Patrols had been in the habit of falsifying their reports to cover the fact that they did not seriously attempt to search out the enemy. To end this, Magsaysay required the patrols to take photos of the enemy dead. Enlisted men were given a promotion for every guerrilla killed and received a personal letter from Magsaysay.

He had the military perform civic and social welfare missions in addition to military operations against the Huks. Each military unit was assigned a civil affairs officer who maintained liaison with local barrio police officials and civilian home guard units. In this manner integration and cohesion of all government and quasi-government agencies was achieved. To free soldiers for patrol and other combat duty, local defense units of civilian commandos were formed with regular army men as leaders. Civil advisory committees were also formed to resolve disputes between regular forces and the civil defense units.

Authorities developed a program to gain the necessary intelligence. Intensive interrogation of the family and friends of known Huks was undertaken. The interrogators urged these people to encourage the guerrillas to abandon the Communist cause, and sought to persuade them to act as undercover agents and infiltrate Huk units and their civilian front organizations. Intelligence was transmitted to the proper authorities in a variety of ways. One method was to fly a reconnaissance plane over a farm where the farmer had arranged farm equipment or other commonplace possessions in such a way as to transmit information about guerrilla strength and direction of movement. This enabled the government to maintain a network of agents, widely distributed over a large rural area, who could quickly provide information about guerrilla actions without compromising their roles as informers.[28]

As a psychological device, booby-trapped weapons were deposited at points where it was known that rebels would find them. When they exploded, it made many rebels hesitant about using stolen firearms or captured weapons.

Although the Army Intelligence Corps was responsible for the overall collection of information, each troop unit gathered and utilized such intelligence data as was necessary for its own operations. Government troops dressed as Huk guerrillas visited towns in "liberated" areas, and noted underground participants and officials friendly to the guerrillas. In this way they acquired information about activities of the insurgents and identified collaborators. Also, once other villages were aware of such operations, they hesitated to assist the Huks. A graduated system of rewards was instituted for information leading to the capture, dead or alive, of ranking leaders. To stimulate the flow of information for the populace, immediate cash payment or partial payment was made to informers; the responsibility for immediate payment was extended to lower echelon field commanders.

A fee was paid for any unregistered firearm turned in to the Defense Ministry; 60,000 arms were turned in at 75 pesos each. Rewards were also offered to civilians for killing or capturing Huk guerrillas. To lessen the danger of Huk reprisals against those cooperating with the government, civilians were urged to bring complaints of any kind directly to Defense officials or to Magsaysay himself.

An agent who infiltrated the Communist National Finance Committee was able to collect records which led to the arrest of 1,175 party members. Rewards were offered to individuals who furnished information leading to the arrest of active Communists. The high-ranking Communist official, José Lava, for example, was worth $50,000 to anyone who turned him in. The climax to these efforts came when, acting on information supplied by a Huk informer, the government captured the Communist Politburo and the underground party headquarters in Manila. Magsaysay was contacted by the informer and asked to meet him alone. From the informer, Magsaysay learned the identity of an underground courier in Manila. The members of the central secretariat had taken the precaution never to hold meetings in the same place twice; they met at the homes of the various Politburo members. However, government agents followed the woman courier, who posed as a peddler delivering vegetables to these homes. In the raid, the government captured 12 members of the Politburo along with other high-ranking officials and many party records. They found 42,000 pesos, electric mimeograph machines, typewriters, subversive publications, submachineguns, pistols, and grenades. The records captured included a complete roster of all members of the party in the Philippines, party sympathizers, and active Huk supporters, who included many Chinese and Filipino businessmen. Additional documents described the management of the organization. The information gained from the raid was significant in the subsequent breaking of the insurgent movement.[29]

After 1950 the security officials also concentrated on improving counterintelligence activities, since many government agents who had infiltrated the ranks of the Huks had been discovered.[30]

JUDICIAL REFORM

Before 1952 the Communists exploited the fact that small landowners rarely had a chance to gain justice when they were abused by large landowners. Magsaysay, however, made provisions for the peasant to have the right of legal counsel at government expense if he so desired. Also, Magsaysay held open trials for captured Hukbalahap leaders and government officials accused of dishonesty.[31]

ECONOMIC REFORM

The army also launched an economic betterment program as part of its counterinsurgency campaign. As the Huks received most of their support from tenant farmers, the first civil actions of Magsaysay and the army were agrarian reforms. Initially, rural civil betterment activities were conducted under the heading of Psychological Warfare, but later they were administered as Civil Affairs. At the battalion level, battalion commanders and their civil affairs officers would meet with barrio heads and other civilian leaders to plan for the defense of farmers during their work in the fields and the implementation of barrio self-defense procedure. This initial contact between military and the civilians led to further discussions of numerous barrio needs and to the initiation of more measures by the army. Department of Agriculture agents were escorted by army personnel into rural areas so that they could acquaint farmers with newer agricultural techniques. Eventually, troops were used to construct barrio grammar schools, to drill pure water wells, and to carry out other public work projects. Also, civilians wounded in the crossfire between government troops and the guerrillas were treated in army hospitals.

Probably the major civic effort of the Armed Forces was the work of its Economic Development Corps (EDCOR), whose aim was to rehabilitate and resettle Huk prisoners and their families and, by this example, to induce defections from the guerrilla ranks. The program was initiated near the end of 1950. Four years later, EDCOR could point to the successful completion of several major projects: four farm communities for former Huks were established, a vocational training center was begun, and one complete barrio was moved to a more favorable site.

The first two farm communities were constructed in 1951 in Mindanao. Along with the selected Huk prisoners, volunteer retired military personnel and civilian applicants were settled in these communities. The first step in the construction of each of these settlements began with the arrival on site of small army units, usually numbering 12 officers and 91 enlisted men. These units scraped dirt roads, set up security procedures, and constructed initial housing to receive the settlers. The troops then worked with the settlers upon their arrival to clear the land for farming, to build family houses, village centers, school buildings, chapels, and dispensaries, to set up sawmills, to drill wells, to build markets, and to construct sanitary facilities. The army also helped the settlers earn title to the land by handling the legal matters involved.

Similar procedures were later followed in 1953 and 1954 in establishing two more communities. As of December 1958, there were 5,175 settlers in these communities. By 1959 EDCOR reported that all four communities were approaching independence; the settlers were able to begin payment for their land; political stability was established in the areas of the communities; and there was no indication of organized resistance among the former dissidents.

Another project of EDCOR was the establishment of a Huk Rehabilitation Center. This organization was founded to train surrendered Huks in the skills necessary in order to obtain regular employment. The center was begun in part of an army warehouse, where the men were trained in woodworking. The center eventually produced furniture for army barracks and officers' quarters, and the profits from these sales were retained by the workers.

The EDCOR villages in Mindanao were quite distant from the major area of insurrection in central Luzon, and it was felt that the success of a similar project in Luzon would be much more damaging to the guerrilla cause. The army, therefore, selected an economically poor Luzon town, located in the middle of Huk territory, as the target for a civil betterment program. The chosen town was San Luis, Pampanga, the birthplace of Luis Taruc and the place of residence for many families of Huks. The plan called for the relocation of some townspeople into a settlement to be established nearby. The army drained some swampland across the river to prepare for the settlement; it built a bridge to link the town with the new settlement; and it moved townspeople's houses across if possible, and, if not, it built new housing to replace the houses which could not be transported. Army personnel also drilled fresh water wells, built schools for children, and helped the farmers by providing seed and agricultural advice. Word of this project was widely disseminated by word-of-mouth, and was followed by the surrender of many Huks who stated their reluctance to fight against troops who were aiding the families of Huks. Magsaysay felt that this project was more effective in combatting the Huks than the application of several fighting battalions.[32]

Although the program was not by any means large, the psychological effects of the EDCOR operations upon the Huk movement and the populace were great. It provided the people with new respect for the government and it offered the rebels an alternative other than resistance. Already assured of amnesty and protection from fellow insurgents, EDCOR now provided the rebels with a hope of future economic security. Huk defections eventually provided the government forces with the intelligence necessary to break the movement and by 1954, the Huk insurgents had lost most of their effectiveness as a revolutionary force.

FOOTNOTES

1. Maj. Kenneth M. Hammer, "Huks in the Philippines," *Modern Guerrilla Warfare*, ed. Franklin Mark Osanka (New York: The Free Press of Glencoe, 1962), p. 179.
2. Alvin Scaff, *The Philippine Answer to Communism* (Stanford: Stanford University Press, 1955), pp. 27–28.

3. Maj. Boyd T. Bashore, "Dual Strategy for Limited War," *Modern Guerrilla Warfare*, ed. Franklin Mark Osanka (New York: The Free Press of Glencoe, 1962), pp. 196–197.

4. Charles T. R. Bohannan, "Anti-Guerrilla Operations," *The Annals of the American Academy of Political and Social Science* (May 1962), 27.

5. See Jesus Vargas, *Communism in Decline—The Huk Campaign* (Bangkok: SEATO, 1957); see also Col. Ismael Lapus, "The Communist Huk Enemy" in Seminar, *Counter-Guerrilla Operations in the Philippines, 1946–1953* (Fort Bragg, North Carolina: 15 June 1961), p. 11; and J. H. Brimmell, *Communism in South East Asia* (Oxford: Oxford University Press, 1959), p. 101.

6. Lapus, "The Communist Huk," p. 12.

7. A. Doak Barnett, *Communist China and Asia, A Challenge to American Policy* (New York: Random House, 1961), p. 491.

8. U.S. Department of Labor, *Summary of the Labor Situation in the Philippines* (Washington: August 1956), pp. 5–6.

9. Scaff, *The Philippine Answer*, pp. 12–13.

10. Ibid., pp. 151–152.

11. It is noted that by 1943 the Huks were not fighting the Japanese, but the guerrilla forces sponsored by the U.S.A. Malcolm Kennedy, *A History of Communism in Southeast Asia* (New York: Frederick A. Praeger, 1957), p. 318.

12. Russell Fifield, "The Hukbalahap Today," *Far Eastern Survey* (January 24, 1951), 14; and Lapus, "The Communist Huk," p. 14. Fifield states that although Taruc was a Communist since 1939, most of the Huk leaders were Socialists or peasant leaders, and that the movement could not be characterized as "Communist" at this time.

13. Fifield, "The Hukbalahap," p. 15; Lapus, "The Communist Huk," p. 16.

14. Luis Taruc, *Born of the People* (New York: International Publishers, 1953), p. 209

15. Taruc, however, claims that the Huks officially disbanded in early 1945 and maintained contact through veterans organizations. Ibid., p. 217. For an objective account of events of the period, see Scaff, *The Philippine Answer*, pp. 25–30 ff; Fifield, "The Hukbalahap," pp. 14–15.

16. According to captured Communist documents, the period 1946–51 was to be the "period of preparation," to be followed by a military offensive aimed at "the seizure of national power." See Uldarico S. Baclagon, *Lessons from the Huk Campaign in the Philippines* (Manila: M. Calcal & Co., Inc., 1960), p. 12.

17. Lapus, "The Communist Huk," p. 22.

18. Scaff, *The Philippine Answer*, p. 34.

19. Ibid., p. 34.

20. Fifield, "The Hukbalahap," pp. 13–14.

21. Scaff, *The Philippine Answer*, Chapter 10.

22. A separate Chinese Communist Party in the Philippines worked closely with the Huks during the insurrection period. Communism was never, however, very strong among the Chinese community. See Sheldon Appleton, "Communism of the Chinese in the Philippines," *Pacific Affairs* (December 1959), 367–391.

23. Ibid., p. 379.

24. Scaff, *The Philippine Answer*, pp. 116–120.

25. Ibid., pp. 31–32.

26. Lapus, "The Communist Huk," p. 17.

27. Ibid., p. 61; see also Bashore, "Dual Strategy," p. 197.

28. Ibid., p. 45.

29. C. Rómulo, *The Magsaysay Story* (New York: John Day, 1956), pp. 114–117.

30. Lapus, "The Communist Huk," pp. 44–45.

31. Ibid., pp. 56–57.

32. Col. E. G. Lansdale, USAF, "Civic Activities of the Military, Southeast Asia," (Anderson: Southeast Asia Subcommittee of the Draper Committee), 13 March 1959; see also Operations Research Office Staff Paper ORO–SP–151, "Civil Affairs in the Cold War," (Bethesda, Maryland: Johns Hopkins University Press, 1961), pp. 185–189.

CHAPTER 12

PALESTINE (1945–48)

BACKGROUND

When the Zionist campaign for the return of the Jews to their Biblical homeland began in 1897, there were only 47,000 Jews living in Palestine. In 1917 the British issued the Balfour Declaration requesting that "a national home for the Jewish people" be established in Palestine. Five years later the declaration was incorporated into the League of Nations Mandate for Palestine. Under this mandate, the United Kingdom had full responsibility for the civil administration of Palestine; a Jewish agency was to be established to advise the administration on economic, social, and other matters affecting the establishment of a Jewish national home.[1] The Palestinian Jews and Zionists in other countries interpreted the Balfour Declaration as promising them a homeland and rejected British explanations that it really did not mean quite that. At the same time, the Arabs of Palestine were pressing more and more insistently for a constitution and eventual independence.

In August 1929 the World Zionist Congress, meeting in Zurich, created the Jewish Agency with the participation of non-Zionist Jews. This agency was to become the organ through which world Jewry would deal officially with the *Yishub** community governments and national council, the mandatory authorities, and the League of Nations. Many non-Zionists, while opposed to the creation of a Jewish state, "were nevertheless anxious to assist in the further growth of a national home as a Jewish cultural and religious center."[2] In the early years of World War II, however, non-Zionist participation in the agency virtually ceased to exist.

The British tried repeatedly, but without success, to find a formula that would satisfy everybody. Three attempts (in 1923, 1929, and 1935)[3] to set up a legislative council foundered, and several commissions were sent to investigate the reasons for the failures. In response to the recommendations of one of them, the British issued, in 1930, a white paper providing for stricter control of immigration and containing measures to protect Arab peasants and tenants. It evoked Zionist protests, and was, in effect, rescinded. As Nazi persecution swelled the influx of immigrants, the situation worsened. In 1937 a commission reported that partition was the only solution. Another white paper in 1939 dealt what appeared to be a death blow to Zionist dreams of a Jewish state. It provided for an independent Palestine after 10 years, with a permanent Arab majority, but with protection for Jewish rights. To preserve that Arab majority, it decreed that no more than 1,250 immigrants per month or 15,000 per year could enter the mandate area.

The immediate Zionist reaction to the white paper of 1939 was to make preparations to increase clandestine immigration, and to expand the illegal

---

*The *Yishub* was the Jewish community in Palestine.

Cases of Undergrounds

Jewish-community defense force, the Haganah.* The official attitude of the Jewish quasi-government in mandatory Palestine after September 1939 was essentially that expressed early in the war by the then chairman of the Jewish Agency Executive, David Ben-Gurion, who said that the Palestine Jews would fight the war as if there were no white paper, and would fight the white paper as if there were no war. The Haganah did in fact cooperate closely with the Palestine Government and Britain in the Allied war effort. At the same time it continued its clandestine operations connected with enlarging its organization and increasing its supplies of military materiel. That relations were far from harmonious during the war could be seen in the second half of 1943 in the anger of the Palestine Jews over a series of arms trials and searchers implicating the Haganah.*

At the end of the war, Palestine had a population of two million, of which some 1,200,000 were Arabs, 600,000 Jews. Because of the plight of the European refugees, Zionist organizations around the world urged the creation of an official Jewish state, with or without the cooperation of mandate authorities. The Haganah, with a membership close to 10,000, began to step up its underground activities against some 90,000 mandate troops, and to expand the illegal immigration of Jews to Palestine.

In addition to the Haganah, there was another underground group, the *Irgun Zvai Leumi* (National Military Organization). The *Irgun* had been organized in 1937 after a disagreement within the Haganah over the question of armed reprisals against the Arabs. This group conducted an extensive campaign of sabotage and terror, a policy not favored by the sponsors of the Haganah, the Jewish Agency, or the World Zionist Organization. At the beginning of World War II, the *Irgun* began to cooperate with the government, which led to another internal disagreement. A group of about 50 broke off from the *Irgun* and regrouped themselves into what was to be called the "Stern Gang," which carried on a campaign of terror. In 1944 the *Irgun* again undertook a campaign of violence and sabotage against the government.

By 1947 the increased tension and clashes between Palestinian Arabs and Jews, and increasing diplomatic pressure from both Zionists and independent Arab states, made it obvious that some sort of settlement was necessary. Great Britain referred the whole Palestine matter to the General Assembly of the United Nations, which appointed a special committee of inquiry, the United Nations Special Committee on Palestine. In August of that year this commission recommended termination of the British mandate at the earliest workable date, and a tripartite partition of Palestine into an Arab and Jewish state with Jerusalem as an international zone.

On May 14, 1948, Great Britain officially terminated her responsibilities for the security of Palestine and announced the withdrawal of her military forces. The same day, the State of Israel was proclaimed by the Jewish Provisional Council of Government as the state called for by United Nations General Assembly Resolution 181(II).

---

*The Haganah had been organized in 1920 to protect the Jewish community in Palestine from Arab raids.

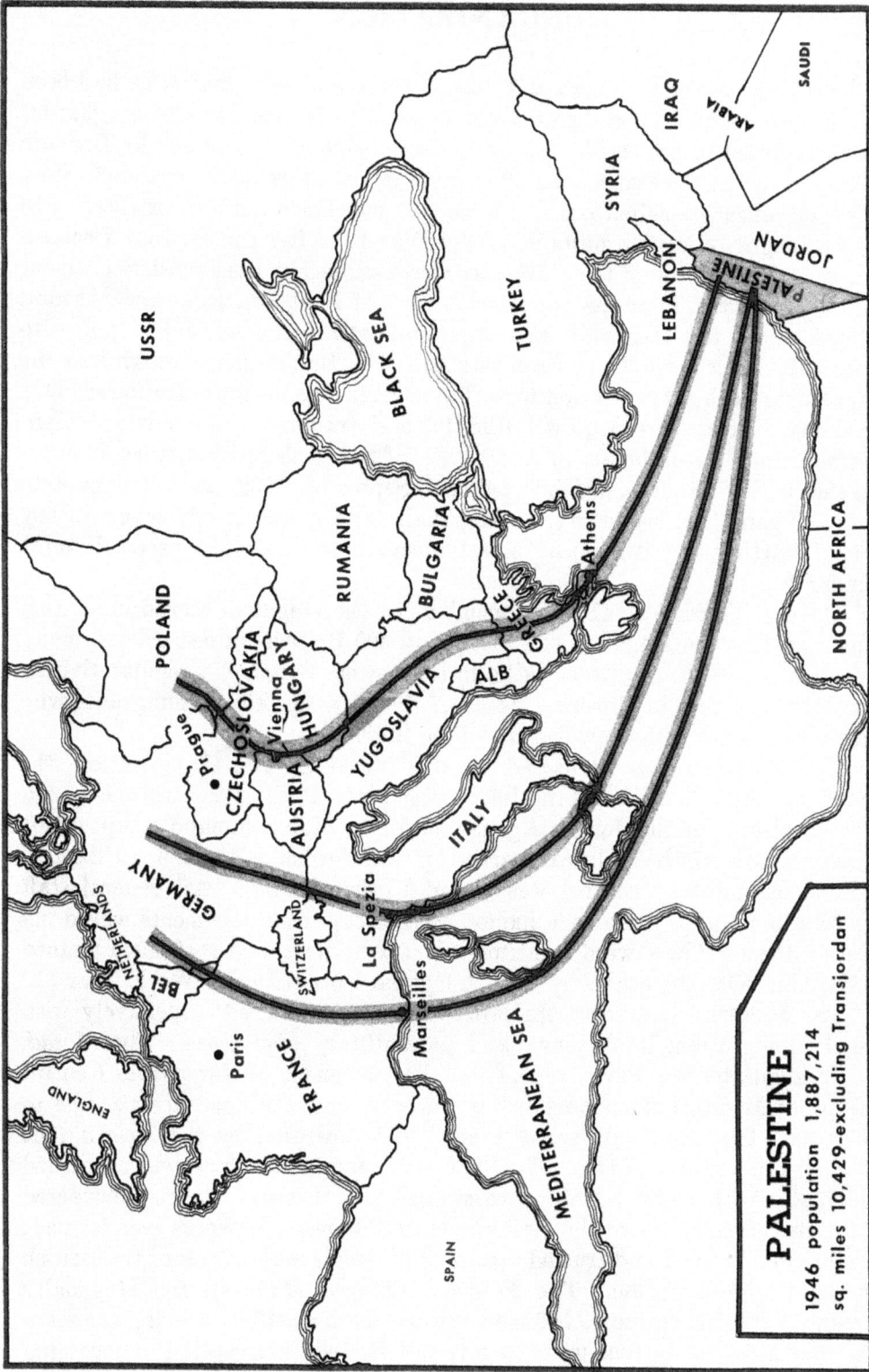

Figure 21. Map of Palestine.

PALESTINE
1946 population 1,887,214
sq. miles 10,429-excluding Transjordan

ORGANIZATION

Friction betwen the Arab and Jewish communities in Palestine had been endemic for years. When the area was under Turkish rule, *shomrim* (guards) were appointed to protect Jewish settlements vulnerable to attack by Bedouin (nomadic Arab) tribesmen. In 1907 these guard units were expanded into a larger organization called the *Hashomer*, or watchmen. Many of these militiamen fought in British units in World War I. After the Balfour Declaration in 1917, the duties of the *Hashomer* were assumed by the Jewish Settlement Police. However, after the anti-Jewish riots of 1920, an illegal underground defense corps, the Haganah, was organized within the settlement police to protect the settlements from Arab attacks. The British never recognized the Haganah as the legal protection force, but nevertheless the organization secretly expanded. Arms were acquired illicitly, and training was conducted clandestinely until the outbreak, in April 1936, of the Arab revolt against Zionism and the British mandate, which lasted until September 1939. A British captain Orde Wingate, then began to train Haganah units openly in offensive military tactics, starting in 1937 with 75 volunteers organized into "special night squads." [5]

During World War II many members of the Haganah served in British units. In 1945 the Haganah had more than 300 British-trained officers, many of whom had learned military skills in the Jewish Brigade, a regular British Army unit fighting in Europe. Some of the 20,000 postwar immigrants who had served in other Allied armies joined the Haganah.

The Haganah was sponsored by the World Zionist Organization, the Jewish National Council, and the Jewish Agency.[6] Political control was vested in the chairman of the Jewish Agency, and in a high command composed of representatives of the political parties of the Jewish settlement (*Yishub*).[7] Professional military control was under a general staff.[8] All general staff members had one or more code names, so that captured documents would not identify them. These were full-time salaried soldiers, many of them trained by Captain Wingate, others by British Special Forces during World War II.[9]

The Haganah had three specialized groups, concerned respectively with illegal immigration, intelligence, and paramilitary operations. The *Mosad*, with headquarters in Paris, supervised the movement of Jews from Central Europe to Mediterranean ports after the war, and arranged for passage to Palestine. The principal "secret roads" were operated by the *Bricha* (the "escape" unit of the *Mosad*). The *Shai*, operating in Palestine, was concerned principally with counterintelligence against the British. It was considered by the British to be one of the most efficient intelligence networks ever formed; its 2,000 well-trained and trusted agents infiltrated every branch of the British Mandate Administration. The *Palmach* (*Plugot Mahatz*), the Haganah's commando branch, trained 2,500 men and women by 1948.[10] During the mandate they acted as defense units to forestall British interception of incoming refugee ships.

Brig. Gen. Orde Wingate (1944).

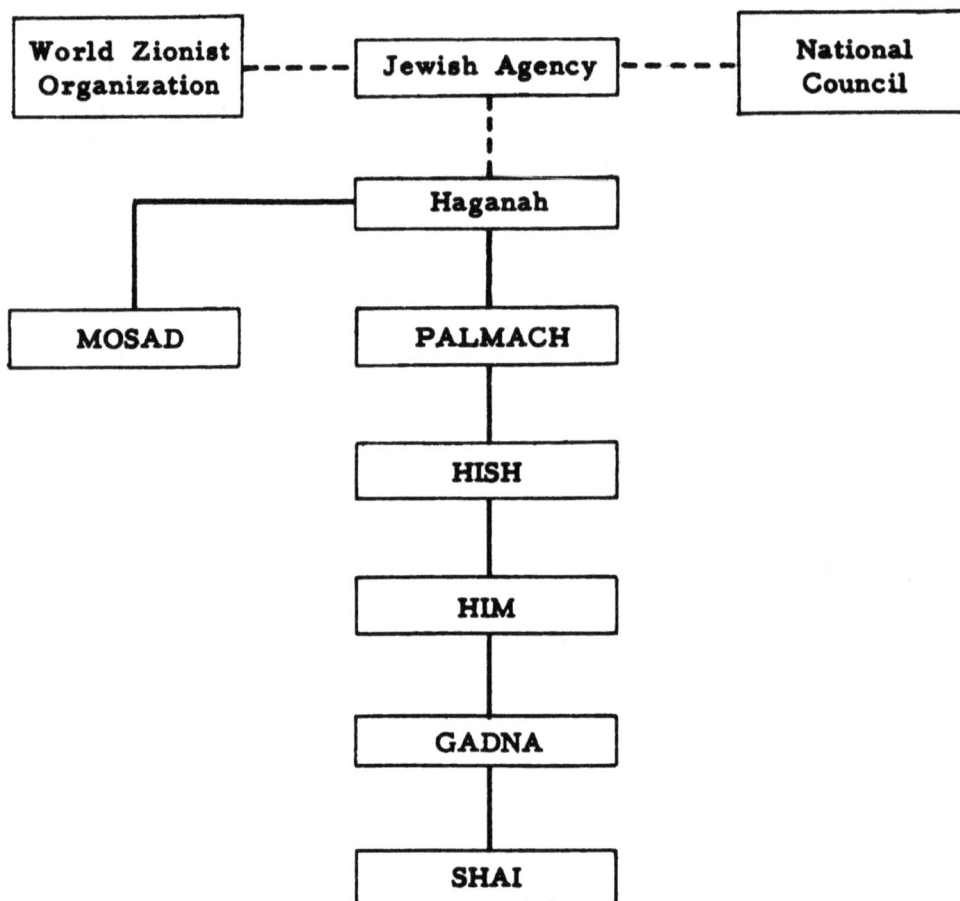

Figure 22.   Haganah.

Auxiliary military units were the *Hish* (specially trained field units total-
ing 9,500 men) and the *Him* (a local defense organization of 32,500 men and
women).   The Haganah also supervised the training of the *Gadna* youth
group.   At the height of its strength, it is estimated that the Haganah could
call on about 45,000 persons.[11]

UNDERGROUND ACTIVITIES

ADMINISTRATIVE FUNCTIONS

Communications

A system of communication between the settlements employed Aldis lamps,
heliographs, flags, and light signals.   A wireless telegraphy system was used

for operational purposes, supplemented by carrier pigeons for longer range communication. An elaborate hand-courier system was developed, including a special corps of children trained for message carrying. Women were also used for traveling longer distances, using taxis or public buses.[12]

Clandestine communications of the Haganah were concerned, for the most part, with coordination of *Shai* and *Mosad* operations. For a successful voyage of an illegal immigrant ship from a Mediterranean port to the shores of Palestine precise information was required as to: (1) the location and condition of available vessels, (2) number of passengers and sailing date (this was often fixed by British security measures in European ports), (3) estimated time of arrival of the Haganah ship, and (4) precautions to be taken in order to ensure a safe landing and successful transport of passengers to a safe area. Ship-to-shore contact was maintained between the ship and agents in Palestine once the vessel was at sea. Many of the radio operators had been trained by the British during the war, and had defected with their equipment once the emigration of Jews from Europe had begun.[13]

One participant states that a coordinating center for immigration operations was set up at Istanbul.[14] It is also that members of the Jewish Brigade established a competent wireless service throughout Europe to communicate with the Haganah High Command in Tel Aviv.[15] The same source states that a French officer in Cairo acted as a courier for the *Mosad* between that city and Tel Aviv.[16] It is likely, however, that *Mosad* and *Shai* agents traveled or sent their messages by normal channels of communication.

Recruitment

After World War II the Haganah needed large numbers of specialists and technicians. It attempted to recruit Jews from other countries, acting on the assumption that the great majority of Jews throughout the world would support any undertakings of the Jewish Agency. Threats or violence were never employed, although some of the promises made to prospective recruits may have been misleading. The Jewish Agency based its appeals on two general premises:

(1) The Haganah was the official organ of the quasi-government in Palestine, which was sponsoring a movement entitled to worldwide moral, political, and financial support.[17]

(2) This was to be a nonviolent movement; condemnation of the terrorist activities of the rival *Irgun* and Stern Gang was implicit.

In order to recruit qualified persons to carry out specific duties concerned with illegal immigration, the Jewish Agency acted, in effect, as an open employment agency. Many radio operators, seamen, and military personnel had been trained by British Forces during the war. These men offered their services when full-scale resistance began.

Those Jews who would not associate themselves with Zionism were urged to support programs aiding the health and welfare of world Jewry.

Cases of Undergrounds

Finance

After 1930 the Haganah received funds through the Jewish Agency, which had offices or representatives throughout the Western world. In 1942 a large-scale effort was undertaken to enlist the support of the 5,000,000 Jews living in the United States. Such Zionist leaders as David Ben-Gurion, chairman of the Jewish Agency, traveled throughout the country seeking financial aid. Appeals for money were made in newspaper advertisements, at charity balls and other social events.

Logistics

Arming the *Palmach* and other groups, in defiance of British regulations which limited the arms of each village defense group to a few shotguns, was a large-scale operation. By 1946, however, almost every combatant had a personal weapon (rifle, submachinegun, or revolver). Other weapons frequently used included: (1) rifle grenades and other types of grenades; (2) large numbers of medium machine guns; (3) two- and three-inch mortars; (4) antitank mines and Molotov cocktails; (5) homemade flamethrowers; and (6) heavier weapons such as bazookas, light antitank guns, and light field pieces. Ammunition for these heavier weapons was in short supply, but stocks of light ammunition were usually adequate. Certain weapons and ammunition, such as mortars and mortar-bombs, grenades and submachineguns, were manufactured in secret factories, and well-concealed caches existed in almost every settlement and in the large towns. Armored cars were improvised by nailing armor plating to commercial vehicles.[18] Although the British made systematic searches, they were unsuccessful in the control of illegal arms.

Arms came from four sources:

(1) *Jewish immigrants.* Those coming illegally brought in whatever they could. Those entering under the quota would hide firearms in suitcases, baskets, and jars. Often they bribed Arab customs officials, to whom they posed as smugglers of trade goods. Members of the *Mosad* in Europe purchased arms in large quantities and shipped them to Palestine with immigrants.

(2) *Illegal purchases from Arab tribesmen.* These were weapons abandoned or sold as surplus after the North African campaign or left behind by the British, who had used the region as a training area until the end of 1943.

(3) *Planned raids on British arms depots.* These were often arranged by members of the *Shai* or the Jewish Settlement Police, who could ascertain the quantity and types of supplies at specific depots.

(4) *Clandestine production.* During the Arab rebellion (1936–39) the Haganah began to establish workshops for the clandestine production of arms. These were usually located in rural areas in locations where small workshops would not attract attention. Materials were purchased most frequently through regular trade channels, and established firms were used as "covers." After World War II larger shops were established. The workshops produced legitimate as well as illegal items. An intelligence network was organized, and

340

lookouts were provided for each plant; such a warning system allowed sufficient time to switch from illegal to legal production in case of a raid. This was possible in parts-producing plants, where arms components did not constitute the entire output. Arms assembly plants, on the other hand, had to be carefully concealed, for in these the identity of the product was unmistakable.

Workers for the arms plants were carefully chosen after extensive security checks on their background, behavior, etc. Since their occupations had to be kept secret, they were encouraged to limit contact with outsiders, thus lessening the chances of security leaks. To minimize the possibility of compromise, the top management of arms production was limited to a few men who knew the locations and operational features of the various plants, and controlled the movement of work among plants; and of these men, only a very few of the most trusted knew more than one link in the system. The head office of the clandestine arms production unit was located within the office of a large enterprise and had a cover title within the legal organization. Records of arms production and financial transactions were carefully maintained in code, with each military item having a cover name in the company's files. Officers of the arms system were also employees of legitimate firms or institutions and were paid as such.

The Haganah maintained a transportation system as well as an assembly plant for weapons, parts of which were produced all over the country. One assembly plant was located less than 100 meters from the headquarters of British Forces in Tel Aviv, near a power station and central bus station, in the vicinity of garages and auto workshops where day and night truck traffic was normal and did not arouse suspicion.[19]

Individual arms were stored in parts of household furniture, while large caches of arms were usually put in reinforced concrete underground storage chambers. This involved problems of temperature and humidity control and camouflaged means of ventilation.

Because the British were taking measures to halt the flow of homeless Jews from inland Europe to the ports of the Mediterranean, supplies for the voyage to Palestine, as well as the vessel itself, were acquired clandestinely by *Mosad* agents. Most of the ships were obtained from shipowners in Greece and Italy, who often charged exorbitant fees.[20] Soon after the war a Jewish transport unit of the British Army in Italy was infiltrated by Haganah agents. These men gave their time and money in order to obtain supplies for *Mosad* operations.[21]

Illegal Immigration

After the war the United Nations requested European countries to cooperate in providing free transportation for displaced persons. Only Poland refused to make facilities available to the Jews. This created one of the most difficult problems of the *Mosad*, the agency charged with supervising illegal emigration. Emigration of Jews from Poland usually involved long marches to the Czech border, where the border guards had to be bribed or eluded. However, the Czech Government recognized the *Mosad* as a legal organization, and did not interfere with its agents.

Cases of Undergrounds

The British tried unsuccessfully to persuade other countries to help stem the illegal immigration. French civil authorities made no secret of their refusal to heed official British requests that they discourage Jews from traveling to Mediterranean ports, where they could seek passage to Palestine. Every British attempt to arrest or slow the flood of emigrants brought charges, inspired by Jewish Agency propaganda, that they were "behaving like Nazis."

The facilities of the International Red Cross, United Nations Relief and Rehabilitation Agency, and Roman Catholic charities [22] were put at the disposal of *Mosad* agents. U.S. occupation officers often provided transportation for Jewish emigrants. The SHAEF Command offered the service of U.S. naval vessels. It is even alleged that the British Ambassador in Turkey issued visas for Palestine without permission,[23] and some individuals within the British Army saw that *Mosad* agents were provided with supplies ranging from trucks to telephone wire.

In accordance with international maritime laws, the British never interfered with Palestine-bound ships outside of territorial waters. Haganah headquarters maintained constant radio contact with refugee ships, and instructed vessels where to land when they were one day out. The site chosen would depend on the tides, the disposition of British military units and naval craft, and the proximity to a Jewish settlement. Steps would then be taken to defend the beachhead. Haganah men would go to the site by individual routes to avoid detection. Once at their destination, they would assemble their weapons, which had been carried in parts for better concealment. All roads approaching the landing area would then be sealed off by improvised roadblocks. Should word be received that the ship had been intercepted or the landing abandoned for some reason, the Haganah men would disperse and avoid contact with any British patrols. If the enemy arrived during an actual disembarkation, however, they would be engaged by the Haganah.[24] British communications were also jammed to cripple efforts to intercept shiploads of illegal immigrants.

OPERATIONAL FUNCTIONS

Psychological Operations

The *Shai* operated several radio transmitters known as "The Voice of Israel," or "*Kol Israel*." *Kol Israel* began broadcasting in March 1940. After several months of operation, it went off the air in compliance with a wartime truce with the British. It began transmitting again in the spring of 1945, broadcasting daily for about a half hour in English, Hebrew, and Arabic.[25] Members of *Gadna*, a Haganah youth group, would circulate handbills announcing the time of coming broadcasts. The transmitter was moved often, and the British never caught up with it.

The Haganah published at least one clandestine newspaper, *Ashnab*.

In Palestine, the Haganah had to compete with the *Irgun* and the Stern Gang for the sympathy and support of the population. It sought to convince the Jewish population that the Zionist goals could best be achieved by non-violent methods. The Stern Gang's terroristic attacks on British officials and the sabotage instigated by the *Irgun* were condemned as unnecessary provocations which would only postpone the establishment of a free Jewish state. The Haganah insisted that it condoned military operations only for the purpose of self-defense.

Intelligence

The intelligence unit of the Haganah was the *Shai*. Many members of the *Shai* had gained practical experience during 4 years of underground and intelligence work in the various European resistance movements. The main task of this intelligence unit was to infiltrate every possible channel of British security, and to gather information on—

(1) Location of British supplies and munitions,
(2) Name, location, and rank of British Criminal Investigation Department (CID) agents,
(3) Proposed British countermeasures,
(4) British communication and codes,
(5) Collection of British circulars and handbooks.[26]

Much of this information was contributed by Jewish policemen and government employees who were secret members of the Haganah. The majority of these were recruited by underground agents after they had secured positions with the Palestine colonial administration. Some who were not members of Haganah also supplied information at their convenience. Such information had to be checked by a *Shai* agent to ensure that it had not been planted by the British to impede resistance operations. The Haganah tried to convince all Jews that it was their patriotic duty to supply the underground with information, without payment. Experience with contracting shipowners had shown that those who participated in clandestine activities for financial gain were likely to play a double game.

When a person showed interest in joining the *Shai*, attempts were made to determine his sincerity. One method was to have a *Shai* agent pose as a British agent and arrest the prospective recruit. His reaction to intensive questioning was taken as an indication of his loyalty, or lack of it.

The *Shai* was also responsible for maintaining the security of Haganah activities, including:

(1) Guarding the operations of *Ta'as*, the secret arms factory. Only those immediately concerned with production knew its location. *Shai* agents were expected to see to it that no routine patrols approached the site. The *Shai* was also responsible for the importation of machinery and raw materials and for safe delivery of the weapons.

(2) Guarding the *Makals*, or training schools for new members, while classes were in session.

(3) Warning Jewish communities of forthcoming British searches. This information was usually gathered by agents working in the CID, or by

monitoring and decoding internal British communications. *Shai* agents monitored 74 radio stations operated by the CID.[27]

The British authorities admitted that often members of the Haganah high command knew of decisions made by the highest level of colonial administrators long before orders implementing these decisions had reached their destination.

Sabotage

Unlike its rivals, the *Irgun* and the Stern Gang, both of which sabotaged railroads, bombed hotels, kidnapped mandate officials, and assassinated "obstructionists," the Haganah maintained a general policy of abstaining from indiscriminate sabotage and terror. The Haganah did, however, engage the British forces interfering with illegal immigration. Sabotage was carried out against coastal radar stations and refugee detention camps. In 1945 members of the Haganah's *Palmach* cooperated with the *Irgun* in sabotaging a railway line.

COUNTERMEASURES

During 1945–47 the British Armed Forces were increasingly active in carrying out government policy in Palestine and curbing underground activities. By 1949 their forces totaled 90,000 troops, including almost three infantry divisions, several RAF squadrons, approximately 7,400 soldiers of the Arab Legion, and 3,000 members of the Transjordanian Frontier Force. Also available were the services of 4,000 British members of the Palestine Police Force, plus the Mediterranean Fleet.[28]

The British in Palestine were organized under a dual civil and military administration. The civil authority was headed by a High Commissioner, who in turn supervised the activities of the military commanders in Palestine and Transjordan. The Commissioner enacted regulations which gave military officials powers equivalent to martial law. These powers were in turn delegated to the military commanders in the various sectors. Although the military commanders had the power to deal with problems without recourse to higher authority, political expedience sometimes restricted the actions taken.[29]

Counteroperations against the Haganah had two main purposes: to find and confiscate arms caches, and to halt illegal immigration. To accomplish the first objective curfews were imposed, road traffic was restricted, and the power of search and arrest was exercised freely. Patrols operated day and night to keep the illegal forces on the run. There were frequent spot checks of hotels, cafes, and vehicles.

Frequent searches were made of large sections of cities and entire villages. Generally, a cordon and search resulted after an attack by underground members. The commander would cordon and search an area in which the suspects were believed to be hiding. Planned searches were also made when the police or military intelligence received information on the possible location of wanted suspects. General searches were made for illegal arms and ammunition.

Initially searches were conducted by the police. However, strong resistance rendered these ineffective. The police were refused entrance to homes and hand-to-hand fighting with bricks and sticks broke out in the streets. When troops arrived on the scene, a signal was sounded by alarm gongs or sirens, and the villagers from a nearby settlement came in a body to break into the area being searched. During these encounters the Jewish community even organized first aid stations to care for Jewish casualties. In most cases they stopped short of armed resistance. However, through this technique, they succeeded in preventing the capture of wanted individuals and supplies which could be moved during the confusion.[30]

Planned searches by military units were also difficult, since Jewish intelligence and the Jewish community would provide neighboring villages with a minute-by-minute account of British movements. To counter these activities, the British employed cordon and search procedures tailored to fit this unique situation. Since information invariably leaked out to the underground through civilian employees working on the military base, it was necessary to seal the base before an operation. Civilians were kept on the base under guard until the operation was completed. Since much of the preparation went on in view of the civilian populace, elaborate cover plans were devised to conceal the purpose of the military activity. Reconnaissance was usually not possible without giving away the mission; therefore planning was accomplished through the use of maps and photos. Written orders were kept to an absolute minimum, and were usually handed out just before the mission. Neither radio nor telephones were trusted for communication, since they might be monitored by *Shai* agents. The troops participating in the operation were not alerted until they were awakened at midnight or later. Troops were assembled under cover of darkness, and plans were made for the troops to arrive at the search area before dawn.[31]

The village was surrounded while it was still dark and the search began at dawn. Success depended on the search being well coordinated and carried out with the utmost speed.

Roadblocks were set up. An inner cordon of troops surrounded the area to be searched in order to seal it off and prevent escape. An outer cordon was placed at strategic points some distance from the village to prevent interference from neighboring villages, and to act as reserves. Cages or enclosed areas were erected to which the inhabitants were brought for interrogation. Search parties collected the inhabitants of the village and moved them to the cages for detention. Other search parties looked for hidden arms. Screening teams checked identification cards against photos and lists of wanted men and suspects. After these search operations, suspects were taken to permanent detention camps.[32]

British warships and airplanes patrolled the coastline constantly, ready to intercept immigrant ships as soon as they entered territorial waters. Between February 1947 and January 1948, 22 ships carrying 40,000 immigrants were stopped,[33] and their passengers were sent to detention camps in Cyprus.

FOOTNOTES

1. Frank Sakran, *Palestine Dilemma* (Washington: Public Affairs Press, 1948), p. 108.
2. J. C. Hurewitz, *The Struggle for Palestine* (New York: Norton, 1950), p. 40.
3. Ibid., p. 23.
4. Ibid.
5. See Richard Crossman, *Palestine Mission* (New York: Harper, 1948), p. 178; for a discussion of the conditions under which this training was carried out, see Christopher Sykes, *Orde Wingate* (New York World Publishing Co., 1959), Chapter VII.
6. Ibid., pp. 129, 170; see also John Marlowe, *Rebellion in Palestine* (London: Cresset Press, 1946), p. 228.
7. One author states that the Haganah was sponsored by the *Histradruth*, a Palestine labor federation. See A. M. Hyamson, *Palestine Under the Mandate* (London: Metheun & Co., 1950), p. 188.
8. For a detailed account, see Netanel Lorch, *The Edge of the Sword: Israel's War of Independence, 1947–1949* (New York: G. P. Putnam's Sons, Putnam & Co., Ltd., 1961), pp. 43–50.
9. See M. P. Waters, *Haganah* (London: n.d., Newman Wolsey Ltd.), Chaper III.
10. Sgan Aluf Ramati, *The Israel Defense Force* (Jerusalem: Published by "Israeli Digest," 1958), p. 6.
11. Ibid., p. 6; for a thorough account of organizational changes after 1947, see Hurewitz, *The Struggle*, pp. 75–84.
12. Efriam Dekel, *Shai-Exploits of Haganah Intelligence* (New York: Thomas Yoseloff, Inc., 1959), p. 140.
13. John Kimche, *Secret Roads* (New York: Farrar, Straus, 1955), p. 66.
14. Ibid., p. 109.
15. Ibid., p. 66.
16. Ibid., p. 114.
17. Hurewitz, *The Struggle*, pp. 38–42.
18. Dekel, *Shai-Exploits*, p. 140.
19. Gershon Rivlin, "Some Aspects of Clandestine Arms Production and Arms Smuggling," in *Inspection for Disarmament*, ed. S. Melman (New York: Columbia University Press, 1958), pp. 191–202.
20. Dekel, *Shai-Exploits*, p. 248.
21. Kimche, *Secret Roads*, p. 109.
22. Ibid., p. 135.
23. Ibid., p. 66.
24. Moshe Pearlman, *The Army of Israel* (New York: Philosophical Library, 1950), p. 72.
25. Dekel, *Shai-Exploits*, p. 97.
26. Ibid., p. 344.
27. Ibid.
28. Lorch, *The Edge*, pp. 50–51.
29. R. D. Wilson, Cordon and Search (Aldershot, England: Gale & Polden, Ltd., 1949), pp. 17–18.
30. Ibid., p. 33.
31. Ibid., pp. 36–37.
32. See R. N. Anderson, "Search Operations in Palestine," *The Army Quarterly* (January 1948), pp. 201–208.
33. Wilson, *Cordon*, p. 249.

INDEX

INDEX

Index

Index

Index

Index

Index

* U.S. GOVERNMENT PRINTING OFFICE : 1964 O—710-146

www.ingramcontent.com/pod-product-compliance
Lightning Source LLC
Chambersburg PA
CBHW052108020426
42335CB00021B/2679